The Science and Business of Drug Discovery

Edward D. Zanders

The Science and Business of Drug Discovery

Demystifying the Jargon

Second Edition

 Springer

Edward D. Zanders
CEO
PharmaGuide Ltd.
Cambridge, UK

ISBN 978-3-030-57816-9   ISBN 978-3-030-57814-5   (eBook)
https://doi.org/10.1007/978-3-030-57814-5

This Springer imprint is published by the registered company Springer Nature Switzerland AG
The registered company address is: Gewerbestrasse 11, 6330 Cham, Switzerland

To Rosie

Preface

Jargon, according to the *Concise Oxford Dictionary*, can either mean unintelligible words, or gibberish, barbarous or debased language, or else a mode of speech only familiar to a group or profession. Anyone trying to approach the drug discovery industry from the outside might have some sympathy with all these definitions, particularly if required to deal with industry insiders on a professional basis. The language may indeed seem barbarous or like gibberish, but mostly of course it is the mode of speech familiar to the scientists, clinicians, and businesspeople who are responsible for discovering and developing new medicines. All the different professional groups that deal with the pharmaceutical industry will be exposed to the jargon at some point, because the business is highly technical. It is true that a non-scientist in, for example, a technology transfer office, will not be expected to have a detailed knowledge of a product or service being offered to a pharmaceutical company, because that is normally left to technical colleagues. On the other hand, he or she should at least be able to recognize where these offerings fit into the bigger picture of drug discovery and why their clients might be interested in taking discussions to another level. In some ways, listening to scientists talking in a business meeting is the same as listening to conversations in a foreign language; just having a sense of the meaning rather than the full detail is enough to avoid feeling excluded. These general principles apply to other professions as well, such as recruiters and translators who of course have their own specific issues with jargon. So, the need for a guide to the drug discovery industry for non-specialists is clear enough, but what form should it take?

One possibility is a training program like the *How the Drug Discovery Industry Works* course that I have been running in the United Kingdom since 2004. Although it is quite possible to cover the main points about the biopharmaceutical industry within a single day, only a limited amount of information about such a vast subject can be conveyed to delegates without all concerned feeling that they had just finished the New York marathon. My thoughts turned to producing something that could be hosted online. This has obvious attractions in terms of distribution and reach, but also runs the risk of being submerged in the vast oceans of information available in cyberspace. Since there is something quite comforting about reading the printed word on paper (or e-Reader), I resisted the temptations of the new and decided to write a book instead. The aim is to provide a thorough review of the technical and business aspects of drug discovery in a way that can be understood by a reader with little scientific knowledge, while still retaining the jargon and terminology that is actually used in the pharmaceutical industry. This jargon and terminology can be daunting even to a trained scientist, so in keeping with the second part of the title *Demystifying the Jargon*, the meanings behind the key terms and phrases are explained in simple terms and placed in the relevant context.

There is no single source of information about all the activities occurring within the pharmaceutical industry, as the sheer number and variety of different processes make this impossible. These activities include such disparate topics as the biology of an infectious microbe or the leakage of contaminating chemicals from bottle stoppers. Reference material about drug discovery and development is of course readily available on the Internet and elsewhere, but this is both a curse and a blessing. When approaching the subject for the first time, it is difficult to put the information in context, to find authoritative sources, and to discriminate between what is important and what is not. On the other hand, once the path through the maze of information has been mapped out, the available resources are incredibly powerful and can provide detail on almost any topic. This book focuses on the most important elements of drug development by laying out a smorgasbord of the topics that underpin discovery, clinical trials, marketing, and the pharmaceuticals business, without going into excessive detail about specific points. The vast subject of biochemistry, for example, is covered in about two pages, but the information given is sufficient to give the reader a sense of the essence of the subject so they are in a position to make an informed (rather than random) search of outside sources.

In writing this book, I have drawn upon experiences gained while working in the pharmaceutical and biotechnology industries for over 20 years since leaving academia. I discuss the technical aspects of chemical and biological research from the perspective of a lab scientist and cover more commercial and strategic issues from a research management background. The great challenge is to convey this knowledge in a way that is intelligible to non-scientists and PhD-level scientists alike. I hope that I have been able to achieve this by offering a choice of material that can be used or bypassed according to the reader's experience. Chapter 3, for example, covers the chemistry of small and large molecules in a very basic way and will probably be glossed over by anyone with a science background. However, even in a chapter like this, there will be material that is tailored specifically for some aspect of drug discovery and its jargon, so it will still be useful to those with a more advanced knowledge of chemistry.

Science and business move at such a rapid pace that it is sometimes difficult to keep up with events. Despite this, every effort has been made to keep this book (in its second edition) as up to date as possible on both the technical and commercial aspects of drug development. New technical areas (or rebranded old ones) are covered in various chapters as well as the full range of molecular entities that have pharmaceutical potential, including nucleic acids and stem cells. Attention is also given to the major structural upheavals underway in research-based pharmaceutical companies and how these create both opportunities and barriers to those who deal with the industry.

Finally, to make the demystifying process less arduous, this book intersperses factual information with lighter comments and asides gained from personal observations of the pharmaceutical industry and the behavior of the participants in this fascinating and important world.

A Brief Note About Terminology

The names used to describe the drug discovery industry and the companies that form it are used interchangeably according to context:

> Drug discovery industry/company
> Pharmaceutical industry/company
> Pharma industry/company
> Biotechnology industry/company
> Biotech industry/company
> Biopharmaceutical industry/company
> Big pharma
> Research and development organization
> R&D organization

The context should be obvious in most cases. For example, a big pharma company like Pfizer is clearly not the same as a small biotechnology company, although it may use the same technologies. The term "biopharmaceutical company" is a useful term for companies of all sizes that research and develop new medicines, so this term will be used from now on as a generic name for a drug discovery organization. Clearly a Research and Development organization (or R&D organization) is not restricted to pharmaceuticals, but the term is still used in practice.

Cambridge, UK   Edward D. Zanders

Acknowledgments

I should firstly like to acknowledge my former employers and colleagues in the biopharmaceutical industry who gave me the opportunity to learn about drug discovery both as a lab scientist and as a manager. I would like to thank Dr Alan Williamson for opening the doors of large pharma to me and Drs David Bailey and Philip Dean for doing the same with the biotech world.

This book grew out of my drug discovery training courses and it would not have been possible to write it in its present form without helpful discussions and feedback from my delegates, including Dr Graham Wagner and Adrian Bradley. I am grateful to John Harris and Sarah Walker from the Oxford Biotechnology Network (OBN) who have organized and hosted my drug discovery courses over the last 5 years.

I thank Carolyn Spence from Springer Science + Business Media LLC for agreeing to publish this book as a second edition and Jeffrey Newton who coordinated its production.

I have also had particularly useful discussions with pharmaceutical translators, who keep me on my toes by picking me up on my use of English during the drug discovery courses tailored to their interests. I very much appreciate the assistance of several pharmaceutical translators who have helped me with these courses in general and specifically Chap. 20 of the book. They are Christine Kirkham, Maria Wyborn, Rebekah Fowler, also Shelley Nix and some members of her ITI Pharmaceuticals Special Interest Group.

I have been greatly helped by Drs Wendy Snowden and Eddie Blair, who have provided helpful comments and additions to the chapters on clinical trials and diagnostics.

Some individuals and publishers have kindly supplied figures and data and I have tried to ensure that these have been properly acknowledged in the text. These include Drs Francesco Falciani, Andrew Filer and Dagmar Scheel-Toellner from the University of Birmingham, Dr Philip Dean from Cambridge, and the following organizations and publishers: The FDA, EMA and ICH, Our World in Data, the USA's CDC, Oxford University Press, EFPIA/PwC, KEGG, Springer Nature,

Wiley, ICS, Taylor and Francis, AAAS, Alacrita, EspaceNet, Informa, IQVIA, Pharmsource, and AUTM.

Lastly and by no means least, I am grateful as always to my wife Rosie for her support and encouragement, particularly as she has been experiencing the same ups and downs while writing a second edition of her own book as I have with mine.

Contents

Chapter 1
Introduction

Most people reading this book will be doing so because they want to know how medicines are discovered and developed by the biopharmaceutical industry. They will already know that success and failure in drug development costs money and that there is currently no political or economic will among governments for all the burden of medicine development to be funded by the taxpayer[1]. This means that private drug companies are here to stay for the foreseeable future, despite the less than flattering image that some of them may have acquired over recent years. Whatever the rights and wrongs of the many viewpoints expressed about the pharmaceutical industry, the fact remains that millions of people have firsthand evidence of the power of modern medicines to improve and even save their lives. It is beyond the scope of this book to discuss the different viewpoints in any detail. Having worked in both a major pharmaceutical company and smaller biotechnology start-ups, I can only offer the perspective of a scientist with firsthand experience of what actually goes on inside these organizations and the motivations of the people who work for them. These employees display the range of human personalities found in all walks of life, from the well-adjusted to the perhaps not quite so well-adjusted. All these people have one thing in common; they are enthusiastic about their work and the fact that they might be able to make a positive contribution to human welfare. Sometimes this last feeling is reinforced when patients write to the company to express their appreciation for a medicine used to treat their illness. Despite these fundamentally positive aspects, the challenges facing the biopharmaceutical industry in image and substance are very real. These challenges, and the industry's responses, are discussed further in Chap. 17.

---

[1] This situation has changed somewhat with the COVID-19 outbreak in 2020.

E. D. Zanders, *The Science and Business of Drug Discovery*,
https://doi.org/10.1007/978-3-030-57814-5_1

1.1 The Benefits of Medicines

Is it possible to measure how much use the biopharmaceutical industry has been to society? One way is to look at the increase in life expectancy at birth that has occurred from 1900 to 2019. Table 1.1 shows life expectancy for a small group of countries representing different levels of economic development (from Roser et al. 2013 (that includes 2019 data)).

These data show a dramatic increase in life expectancy, albeit with a worrying disparity between developed world countries and Nigeria (as an example from Africa). Infant mortality decreased over this time, so in the UK in 1900, 0.8 children per woman died before the age of 5, 0.08 in 1950, and < 0.01 in 2016 (Roser et al. 2013).

How much of this is attributable to better medical intervention? Despite the view held by prominent medical scientists and others that medicines have made a major contribution to increased life span and decreased mortality, there is considerable debate among historians as to exactly how much of this is due to new medicines and how much is a result of improved nutrition and hygiene. There can be no doubt that the introduction of new medicines in the form of vaccines and antibiotics has contributed to a decline in mortality by controlling infectious diseases. There is however a clear distinction between longevity alone and quality of life. There is not much point in extending the life span in old age if that means having to put up with chronic disability and suffering. Without wishing to go too much further into this complex subject, it is interesting to note the work of epidemiologists who have studied the contribution of medical intervention to health outcomes. For example, J. Bunker attempted to quantify these issues, although he recognized that these estimates are based on incomplete data (Bunker 2001, OUP with permission). To quote this author:

> The gains in life expectancy and quality life that I credit to medical care should be seen in the context of other determinants of health. Compared to the very large gains in life expectancy in the first half century that resulted from improvements in public health, the contribution of medical care is relatively small. With improvements in public health largely complete, medical care is now the major determinant of life expectancy, its impact substantially greater than that of the social environment or lifestyle.

As much as half of this increased medical benefit (post 1950) has been due to the reduction of deaths from heart disease or stroke (cardiovascular diseases); this has

Table 1.1 Life expectancy at birth for selected countries for years where data available. (From Roser et al. 2013, under CC-BY license)

Country	1900/2001	1950	2019
USA	49.3	68.2	78.9
UK	46.9	68.7	81.3
Japan	38.6	61.1	84.6
China	–	42.9	76.9
Nigeria	–	33.1	54.7

been achieved by both antihypertensive (anti-high blood pressure) drugs and cardiac surgery. The remaining 50% increase is due to improved treatments for many other conditions, none of which has individually made such an impact upon life expectancy:

> Everything in life that's any fun, as somebody wisely observed, is either immoral, illegal or fattening.

These words from the humorist P.G. Wodehouse (Wodehouse 1970) have a certain ring of truth to them; leaving out the immoral and illegal bit, this summarizes the dilemma of those with an affluent Western lifestyle who pay for it with a high incidence of chronic disease, such as obesity and diabetes. Major causes of death have changed markedly between 1880 and 1997, most noteworthy being the increase in cancer and cardiovascular disease and the significant reduction of infectious diseases through vaccinations and improvements in hygiene. This is illustrated in Table 1.2 which shows the top causes of deaths in the USA in 2017.

From a biopharmaceutical industry perspective, the chronic illnesses shown in the table offer the greatest commercial potential for treatments designed to make a positive impact upon human health and well-being (see Chap. 6. for a more detailed discussion). The only infectious disease in this list is influenza and the associated pneumonia, but of course the SARS-CoV-2 virus has changed the picture, such that at the time of writing (2020), the COVID-19 disease has resulted in over 200,000 deaths in the USA alone.

These statistical data, although informative, are also rather impersonal. Another way of assessing the benefit of medicines is simply to look at one's own life and ask whether it would be significantly different if the treatments were not available. I have been fortunate enough to have enjoyed reasonable health from childhood to "maturity" without (so far) any serious chronic illness, so the different medical treatments I have required over the years are not very remarkable (Table 1.3).

It is hard to avoid the conclusion that my chances of reaching my present age would have been slim without the vaccines and antibiotics. Furthermore, the control of blood pressure by ACE inhibitors has made it more likely that I can postpone a

Table 1.2 Top causes of death in the USA in 2017. (Data from National Center for Health Statistics 2020)

Cause	Number of deaths
Heart disease	647,457
Cancer	599,108
Accidents	169,936
Lower respiratory	160,201
Stroke	146,383
Alzheimer's disease	121,404
Diabetes	83,564
Influenza and pneumonia	55,672
Renal diseases	50,633
Intentional self-harm	47,173

Table 1.3 A list of conditions and drug types used to treat them. This is a personalized illustration of the health benefits of modern medicines

Medicinal product	Benefit
Anesthetic	General anesthesia for operations (tonsillectomy, dental abscess) local anesthesia-dentistry
Antibiotics	Control of numerous infections, including bronchitis and pleurisy
Vaccines	Freedom from polio, smallpox, diphtheria, tetanus, etc.
Antipyretics	Aspirin, paracetamol for fever and acute pain relief
NSAIDs	Anti-inflammatories for muscle strains and gout
Allopurinol	Freedom from gout
ACE inhibitors	Normalized high blood pressure
Opiates	Pain relief for slipped disc
Inhaled steroids	Control of seasonal rhinitis

heart attack or stroke for a few more years at least. The other medicines have enhanced the quality of my life rather than saved it. The anti-inflammatory and analgesic medicines have made it more bearable, as anyone who has suffered an acute attack of gout will testify, and the allopurinol has effectively eliminated this disease, and the accompanying risk of kidney stones, for as long as I take the tablets.

This of course is one person's luck of the draw; all of us have lost friends or relatives to cancer, and as we get older, we become more aware of the scourge of dementia. This should focus the mind on what the biopharmaceutical industry is ultimately in business for. The technical and commercial challenges are enormous, but ultimately surmountable, if past experience is anything to go by.

1.2   Economic Health

What about the contribution of the biopharmaceutical industry to economic wellbeing? The industry is mainly comprised of individual businesses that must trade at a profit to support their existence through innovation and by attracting investment from the financial markets. Although its primary role should be to improve human (and animal) health[2], the economic contribution by pharmaceutical and biotech companies to countries, organizations, and individuals can be substantial. A survey of the industry for the EU in 2016 showed €206Bn added value to the economy and 2.5 million jobs, of which 45% were held by women. The gross value per employee of €156,000 is higher than other key R&D sectors. In a group of over 650,000 people with breast cancer or HIV, pharmaceutical interventions between 2007 and 2017 added two million healthy life years €27Bn to the economy and saved €13Bn in

---

[2] This book does not cover veterinary medicine and drugs, but the scientific principles are the same for humans and animals.

healthcare costs accrued due to complications (EFPIA PwC report 2019 with permission).

The global economy has undergone some major changes since the turn of the twenty-first century which clearly influences all sectors, including biopharmaceuticals. There is a feeling in scientific circles that the twenty-first century is the century of biology, just as the twentieth century was dominated by physics. This has caught the attention of governments worldwide, who consider investment in the life sciences to be critical for the future economic well-being of their countries.

1.3   The Developing World

The Westernized "developed" economies are the largest markets for prescription medicines by a significant margin. It is therefore inevitable that any coverage of the biopharmaceutical industry will assume that its research and development activities are directed almost exclusively at these affluent nations. The problem for millions of people in the developing world is that treatments for tropical diseases like malaria are not economical to develop and that medicines for "Western" diseases are too expensive. This situation is now changing because of economic, political, and social factors, including the rise of "venture philanthropy" and new pricing models. Perhaps most significantly, rapidly growing economies (e.g., China and India) are sustaining many people with Western lifestyles and the diseases to match. Multinational pharmaceutical companies invest in these countries and in association with non-profit organizations are providing medicines to treat infectious diseases such as malaria. The area of pharmaceutical markets and commercial trends will be covered in Chaps. 16 and 17.

1.4   Why Can We Put a Man on the Moon But Still Not Cure Cancer?

The answer to this question is fundamental to understanding the technical challenges that are particular to drug discovery and the life sciences. Put simply, we do not have enough understanding of how living things operate at the molecular level to make precise predictions of what will happen if we perturb them with a drug. This is despite the extraordinary progress that has been made in analyzing the components of living systems in fine detail as well as being able to manipulate processes using, for example, genetic engineering. The problem lies with the fundamental basis of biology, namely, Darwinian evolution. Complex life results from what might be called "tinkering" with molecules available from the earliest stages of life (Jacob 1977). This is a random "see what fits" process compared with an engineering approach where things are designed from scratch. For example, adenosine triphosphate (ATP) is used as a key energy-producing molecule, a component of nucleic acids (RNA) and a signaling molecule acting on purinergic receptors, three completely different biological functions that "worked." The properties of a system can be defined by the components that make it up, but not the other way round, so predictions of function based on a "parts list" is not straightforward in biology. In other words, it is not possible to "reverse engineer" the system.

Evolution throws up several obstacles, for example, drug resistance by microbes and tumor cells caused by mutations that confer a growth advantage. In contrast to biology, physics is underpinned by well-established laws backed up by precise measurements that are often accurate to many decimal places. I have heard a famous physicist say that the subject is quite simple. I would not personally go that far (think of quantum theory), but she has a point if a comparison is made with biology. If we look at how physical laws are applied, in electronics, for example, the basic idea of digital information being represented by the presence or absence of electrical charge is easy enough to grasp. Combining this with materials science, we get solid-state electronics combined with miniaturized power supplies to design computers, mobile phones, and the like. The laws of physics create an upper limit to how far these devices can be improved, but it is still possible to say that the limits of present technology have not yet been reached and that better devices will come onto the market. We can almost guarantee that a new electronic device will operate as specified, but we simply cannot do the same for a medicine designed to treat a complex disease. A useful analogy comes in the form of two US initiatives from the second half of the twentieth century, the Apollo moon landings and the war on cancer. President Kennedy delivered an address to Congress in 1961 that included the sentence "I believe that this nation should commit itself to achieving the goal, before this decade is out, of landing a man on the moon and returning him safely to the earth." As we know, this was achieved in 1969, through an impressive display of technical skill, project management, and bravery on the part of the astronauts. The point here is that the technology was in place to be able to turn an ambitious proposal into an achievable objective within a relatively short period of time. Again, the laws of physics were understood and properly exploited. In signing the National Cancer Act in 1971, President Nixon expanded the remit of the National Cancer Institute and enshrined cancer research and prevention into federal law, effectively declaring a "war on cancer." Although it was then understood that the elimination of the disease would take longer to achieve than landing astronauts on the moon, it is now obvious to anyone that nearly 50 years later, despite huge advances in cancer medicine, we are still a long way off the original goal. The blame does not lie with the skillful scientists who have made huge strides in understanding the cellular and molecular biology of cancer and the clinicians who deliver treatments based on drugs, surgery, and radiation. Instead, it must lie with the sheer complexity of the disease, with its widespread genetic abnormalities and myriad interactions between cancer cells and the cellular environment in the rest of the body.

So, despite the drug development process being technically complex and very costly, it is still successful in producing effective medicines for a wide range of diseases. Just how this is carried out forms the basis for the rest of this book.

References

Bunker JP (2001) The role of medical care in contributing to health improvements within societies. Int J Epidemiol 30:1260–1263

EFPIA PwC report (2019). https://www.efpia.eu/publications/downloads/ Accessed 2 June 2020

Jacob F (1977) Evolution and tinkering. Science 196:1161–1166

National Center for Health Statistics (2020). https://www.cdc.gov/nchs/ Accessed 2 June 2020

Roser M et al (2013) Life Expectancy. https://ourworldindata.org/life-expectancy Accessed 2 Jun 2020

Wodehouse PG (1970) The woman in blue. Hutchinson, London

Part I
How Drugs Are Discovered and Developed by the Pharmaceutical Industry

Chapter 2
Introduction to Drugs and Drug Targets

Abstract This chapter lays out some formal definitions of a drug or medicine and introduces the concept of a drug target. It then describes the wide range of drug types that have been produced by the biopharmaceutical industry. These include orally available drugs, proteins, nucleic acids, cells of various types, and vaccines. Some background on all these different types of medicine is provided to create a foundation for the remainder of the book.

2.1 Introduction

The focus of this book is the discovery and development of prescription only medicines (POMS)[1] with some description of the diagnostics being developed to support their use in the clinic. Medical devices, such as metered dose inhalers and osmotic pumps, which are important for delivering drugs to the right places in the body, are only briefly mentioned.

The terms drug and medicine are used interchangeably, although the word "drug" has the connotation of an illegal substance, such as cocaine or heroin (controlled drugs in the UK). The American Food and Drug Administration (FDA) (Drugs@ FDA (2020)) defines a drug as follows:

- A substance recognized by an official pharmacopoeia or formulary.
- A substance intended for use in the diagnosis, cure, mitigation, treatment, or prevention of disease.
- A substance (other than food) intended to affect the structure or any function of the body.
- A substance intended for use as a component of a medicine but not a device or a component, part, or accessory of a device.
- Biological products are included within this definition and are generally covered by the same laws and regulations, but differences exist regarding their manufacturing processes (chemical versus biological).

---

[1] Once drugs have been approved for use without prescription, they become over-the-counter medicines (OTCs).

E. D. Zanders, *The Science and Business of Drug Discovery*,
https://doi.org/10.1007/978-3-030-57814-5_2

Fig. 2.1 The dart board analogy of drugs binding to their target. Those drugs that bind strongly and selectively to their biological targets are analogous to a dart that sticks firmly to a dart board in a high-scoring position. Many drugs bind weakly to secondary targets, giving rise to both desirable and undesirable side effects

A more scientific definition might be as follows:

A drug is an agent which modifies a drug target in order to bring about a change in the functionality of that target. Drugs may reduce or accelerate target activity.

A drug target can be thought of as a dart board, where the drug molecules are the darts (Fig. 2.1). Strong, accurate binding of a drug to its target is important for successful activity; by analogy, hitting a high-scoring section of the dart board (like the bull's eye in the middle) helps to win the game. The real nature of drug targets and how they are discovered will be covered in the following chapters.

2.1.1 Different Types of Medicines

Many people think of drugs as medicines that are swallowed in the form of tablets or capsules[2]. I generally get this answer when I ask my course delegates what comes into their minds when they hear the word drug (leaving aside illegal products). The biopharmaceutical industry was built upon the discovery of orally active medicines, and this is still the preferred outcome for any drug development program. The medicines can be self-administered in a regular way (perhaps once or twice daily) with consistent dosing and high patient compliance. Other routes of administration, such as injection, inhalation, or topical application, are used to ensure that certain drugs have a chance to enter the circulation without being broken down in the stomach or liver, but these are simply not as straightforward as oral delivery. While working on drug discovery programs for asthma, I was told that the ideal drug for a worldwide market would be delivered orally, partly as the result of cultural issues in some countries regarding the use of inhalers. Inhaled drugs are actually quite effective in

---

[2]The word "pill" is used colloquially to describe tablets and capsules.

treating asthma, but the point was made that we should always try to develop a pill for this disease if possible; indeed, this was the desired objective for all our research programs.

Although orally active small molecules are preferred for new medicines, they are far from the only products being developed by the biopharmaceutical industry. Proteins now take up a significant part of drug development pipelines, and a more recent category of "advanced therapy medicinal products"[3] covers medicines made up of genetic material or even intact cells. Finally, novel vaccines are being developed to treat serious infectious diseases and cancer. The remainder of this chapter provides a summary of each of the drug categories mentioned above.

2.1.1.1   Small Molecules

These drugs are usually taken by mouth, although other routes of administration may be required. The chemical definition of a small molecule will be covered in Chap. 3, but drugs of this type are small enough to cross the alimentary canal (stomach and duodenum) after being swallowed. They can then enter the bloodstream and pass into the liver. They are then distributed through the body via the circulatory system (Fig. 2.2). The target for the drug is associated with the cells that make up the organs and tissues of the body.

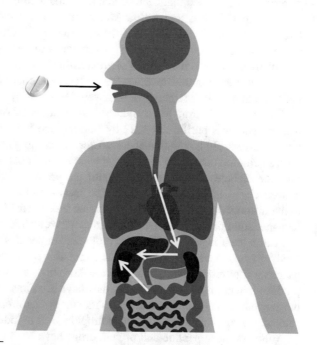

Fig. 2.2 Passage of orally available drugs into the body. After swallowing a tablet or capsule, the drug is dissolved in the stomach fluids and is passed through the intestines where it is absorbed by the blood vessels entering the liver (hepatic portal system). After the drug has been exposed to metabolic systems in the liver, it passes into the general circulation. Once it reaches the areas where the drug target is expressed, it binds to it and exerts its biological effect

---

[3] ATMPs – term adopted by European Medicines Agency (EMA); see Chap. 13.

The oral (bio) availability of small molecules that cross the stomach into the liver can be reduced dramatically by metabolism which can cause their rapid breakdown and excretion from the body; in addition, drugs can strongly bind to proteins in the blood, thereby reducing the amount available to interact with the drug target. Both metabolism and protein binding contribute to the pharmacokinetic properties of a drug, an important area that is covered in detail in Chap. 11.

If small molecule drugs are adversely affected by this first pass metabolism, alternative routes of administration ensure that the drug passes directly into the general circulation. Apart from injection, drugs can be delivered transdermally or subcutaneously (i.e., through or under the skin, respectively). Sometimes drugs are administered rectally in the form of suppositories, particularly if the intended target is associated with gastrointestinal disease. Another route of administration is under the tongue (sublingually), where the drug can reach its target without first passing through the liver. An example of this is the sublingual delivery of nitroglycerine, an important heart medicine when not being used in high explosives.

2.1.1.2 Proteins

The word protein was first used in 1838 by the Dutch chemist Gerhard Mulder because of his studies on biological products such as silk, blood, egg white, and gelatin (Vickery 1950). Although not aware of the exact chemical nature of these materials, he reasoned that each source harbored a common "radical" in combination with phosphorus and sulfur. Mulder named this radical "protein" after the Greek word *proteios* meaning "of the first rank or position." This seems entirely appropriate, as it reflects the central importance of these large molecules in the function of living organisms. From a pharmaceutical perspective, these molecules are both targets for drugs and are drugs themselves. The chemical nature of proteins and their function as drug targets will be covered extensively later in this book.

The first protein drug to be injected into a patient (if we discount vaccines) was insulin. This was purified in 1922 by Banting and Best in Canada who used it to treat a 12-year-old boy with diabetes. After overcoming some initial problems with severe irritation caused by impure samples, the scientists managed to successfully treat the diabetes for several years until the premature death of the patient in a motorcycle accident (Sneader 2005). The success of this, and subsequent trials, led to the introduction of pure forms of porcine insulin (from pigs) and, subsequently, human insulin produced using recombinant DNA technology.

Insulin itself is part of a group of biological molecules called peptide hormones, small proteins that are secreted into the circulation by specialized organs such as the pancreas or pituitary gland. Because these hormones are small and relatively easy to produce in natural or synthetic form, they have been investigated extensively by the biopharmaceutical industry. Examples include somatotropin (growth hormone) for stunted growth in children and gonadotrophin used to induce ovulation.

Although these peptide and protein drugs have been marketed for many years, there was, until recently, little incentive to develop a protein if a small molecule could be found to do the same thing. However, the explosion of information about drug targets brought about by advances in cell and molecular biology in the 1990s led to the realization that not all of them could be influenced by small molecules. Some important drug targets in major diseases like cancer and arthritis can only be affected with large protein molecules, so the biopharmaceutical industry has been forced to take them seriously as drugs. Modern protein drugs fall into the following categories:

- Hormone-like molecules, including those that stimulate the growth of blood cells after cancer therapy.
- Restoring proteins that are absent in deficiency diseases, e.g., coagulation factors in hemophilia.
- Protein decoys that mimic the drug target to prevent the natural protein from binding to the target or that neutralize it by removing it from the circulation.
- Antibodies, normally produced by the immune system to fight infection, but which are instead directed against specific drug targets.

The size range of protein drugs is quite wide: insulin, for example, is 25 times smaller than a full-size antibody. What they have in common, however, is a lack of oral availability, since they are both too large to pass through the stomach and are broken down by the digestive system. This means that they must be delivered into the circulation by injection or other means. Currently, over 12 billion injections are made annually, a figure that is likely to increase substantially as new drugs based on proteins and other large molecules are introduced to the marketplace. Much effort is being expended by the medical devices industry to find effective means of delivery that can be performed without either physical or psychological discomfort to the patient. Devices such as autoinjectors have made subcutaneous injection (*sub cut*) a straightforward procedure for self-injection, but attention is currently being focused on needle-free devices. These use high pressures to drive the protein through the skin; although this can cause more bruising than with using needles, the technology to improve this situation is being advanced all the time (Aroroa et al. 2007). Other delivery methods for proteins are being investigated, including transdermal patches, implants, intraocular administration, inhalation, and even oral delivery, if the protein is small enough, using encapsulation devices.

2.1.1.3 Nucleic Acids

Genes lie at the heart of biology in both health and disease. Through a digital code based on four "letters" used in groups of three, each gene specifies a protein with a distinct function in the cell or in biological fluids such as blood. The path to identifying the chemical nature of the gene has been a long one, starting in 1869 with Miescher's isolation of nuclein from the pus in the bandages of soldiers fighting in the Crimean War. The name nuclein was later changed to nucleic acid, and less than 100 years later, the double helix structure of deoxyribonucleic acid, or DNA, was announced by Watson and Crick in Cambridge. I regularly pass through the unassuming site where they did this work and look at the blue commemorative plaque on the Eagle pub (now selling "DNA beer") where Francis Crick announced to the (apparently underwhelmed) drinkers that they had "discovered the secret of life." The structure led to an immediate realization of how genetic information could be passed from cell to cell through the generations, an essential prerequisite for a living organism.

DNA is one member of the family of nucleic acids, large molecules with structures that can pair with each other in a highly specific manner. This phenomenon, called hybridization, is essential for the natural function of nucleic acids, but it can also be exploited in a wide variety of laboratory investigations. Furthermore, hybridization can be exploited to target the activity of specific genes and therefore has potential use in drug development. For example, if a gene carried by a virus is silenced by a drug, the virus may be unable to survive in the cell that it has infected and will therefore die. Alternatively, a cancer cell that is growing uncontrollably because a gene is permanently stuck in the "on" position could be stopped by arresting its expression in a similar way. Drugs of this type are at an early stage of development and are based on ribonucleic acid (RNA). This versatile molecule is essential for transferring the genetic code from DNA and translating it into a specific protein. From more recent work, it appears that RNA is also actively involved in the regulation of genes, i.e., the process of switching them on or off at defined times and locations within the living cell. This has implications for diseases like cancer, where many genes are deregulated, leading to uncontrolled cell growth.

Figure 2.3 gives a simple illustration of the relationship between DNA, RNA, and the proteins that provide most targets for both small molecule and protein drugs.

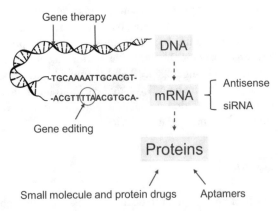

Fig. 2.3 Targets of different drug types. One strand of the DNA double helix is transcribed into a messenger RNA (mRNA) copy that specifies the amino acids that will be incorporated into proteins during translation in the cell. The proteins themselves are targets for small molecule and protein drugs (including antibodies) as well as aptamers that consist of nucleic acid sequences. Antisense and small interfering RNAs (siRNAs) interfere directly with the mRNA or its transcription, thus suppressing the production of a protein target. Gene therapy and gene editing act directly on the DNA itself to correct mutations in genes, remove genes altogether, or introduce new functions

RNA-based drugs (except for aptamers) provide therapeutic benefit by influencing the expression of the protein target itself rather than by directly interacting with it. Gene therapy and editing involves altering DNA to correct mutations or to manipulate the functions of cells for therapy.

> **Some Terminology**
> The four letters ACGT that make up DNA (in RNA, T is replaced by U) are small molecules from the nucleotide family. These can be added together in the laboratory to form chains of different lengths. Short chains are oligonucleotides (*oligo*, few), and longer ones are polynucleotides (*poly*, many). The length of the chain of any nucleic acid is measured in bases or base pairs (bps) depending on whether the chains are single or paired (single stranded or double stranded). The order of letters in any of these chains is called the sequence.

What follows is a summary of the main types of nucleic acid drugs with pharmaceutical potential.

Antisense DNA

This technology was developed in the 1980s as a tool to silence genes in cells isolated in the laboratory. Antisense molecules are modified oligonucleotides (about 25 bases) that bind to a specific gene sequence copied in the form of an RNA molecule. Once bound, this sequence signals the cell to break down the RNA at that point and thus stop the production of the protein specified by that gene. Alternatively, the cellular process of creating protein from the mRNA is physically blocked by the antisense/mRNA hybrid structure. Antisense molecules are illustrated in Fig. 2.4 below.

Antisense DNA has had a checkered history because of issues concerning stability of the drug in the patient and delivery into living cells via the parenteral route.

The pioneering antisense company Isis Pharmaceuticals (later Ionis) developed mipomersen to lower cholesterol in patients by reducing the levels of a key protein involved in the production of the so-called "bad" cholesterol (Akdim et al. 2010). Although working in patients who do not respond to standard cholesterol lowering drugs and demonstrating a proof-of-principle for this type of medicine, mipomersen has poor bioavailability, making it technically and commercially unviable. However, success has been achieved with inotersen (Tegsedi®) an ASO that blocks the production of transthyretin in the liver thereby reducing the neurological defects caused by a mutant form of this protein in hereditary transthyretin amyloidosis. The marketing approval for Tegsedi® granted by the FDA and EMA in 2018 marks the coming of age for this type of nucleic acid drug, several of which are now in clinical development.

Peptide nucleic acids (PNA) are antisense molecules containing both oligonucleotides and unnatural amino acids (i.e., those not found in normal proteins). This

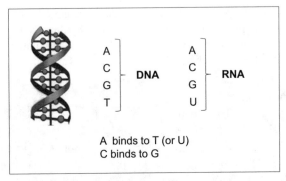

A binds to T (or U)
C binds to G

Natural double-stranded DNA

Sense	-ACGTTTTAACGTGCA-
Antisense	-TGCAAAATTGCACGT-

RNA with synthetic antisense oligonucleotide

Sense	-ACGUUUUAACGUGCA-
Antisense	-TGCAAAATTGCACGT-

Fig. 2.4 Sense and antisense in DNA sequences. Nucleic acids (DNA and RNA) are large molecules built up from four chemical building blocks (nucleotides) written as A, C, G, and T (or U in RNA). Each strand of a DNA double helix is a linear string of nucleotides written out as in the figure. The sequence of the "sense" strand is retained in the mRNA used to code for protein, while the "antisense" strand is complementary to the sense strand. This is because A only hybridizes to T (or U) and C only to G. An antisense drug molecule is based on a specific antisense sequence as it will select its complementary sense sequence out of millions of others

combination gives them the same gene targeting effects as antisense nucleic acids but makes them much more resistant to breakdown in the blood. They also have the unique ability to directly interact with DNA, unlike the other nucleic acid drugs which only interact with RNA. Despite these attractive properties, interest in PNAs appears to be because of their diagnostic potential rather than their use as therapeutic drugs since there are significant problems with water solubility and delivery into cells (Montazersaheb et al. 2018).

Locked nucleic acids (LNAs) are oligonucleotides modified with a chemical group to "lock" the ribose component into a configuration that enhances stability. The biotech company Santaris (now part of Roche) is evaluating the LNA miravirsen in clinical trials for hepatitis C virus (HCV) infection. As well as being chemically distinct from other ASOs, this compound also targets a non-protein-coding RNA microRNA in the human liver called miR-122 (Janssen et al. 2013). These microRNAs were only discovered in 2001 but have caught the attention of scientists because of their ability to regulate the expression of a large variety of genes in the body and through being implicated in several diseases, including cancer and infection. It is sobering to reflect upon all the years that scientists, including myself, have concentrated on the larger RNA involved in producing proteins and in the process have literally thrown these smaller molecules down the sink without realizing their importance.

Small Interfering RNA

siRNA is used in the same way as antisense, and there are similarities between the two. For example, both types of nucleic acid force the breakdown of RNA specific for the gene of interest, must be chemically modified to improve resistance to metabolism, and must be introduced parenterally. The difference lies in the fact that siRNA is introduced as a double stranded, rather than single-stranded oligonucleotide, and is processed by the cell in a different way to antisense molecules. This phenomenon of RNA interference was demonstrated by Fire and Mello in animal cells in the late 1990s and led to its rapid adoption by the biomedical research community. In fact, the two scientists received the Nobel Prize in Physiology or Medicine in 2006, which is a remarkably short time after their initial discoveries.

siRNA targets specific genes very efficiently and is used as a tool in basic research (Petrova et al. 2013). However, clinical development of siRNA has been more problematic as early interest from investors and big pharma was not maintained because problems emerged with delivery and toxicity. Alnylam, a pioneering company in this area, kept faith with the technology, and now its siRNA drugs are entering the clinic. One of these is patisiran (Onpattro®) that targets transthyretin production in hereditary transthyretin amyloidosis (see also ASOs above) and was approved in 2018. Inherited diseases of this type are comparatively rare, so the application of ASOs and siRNA drugs to common indications is awaited with interest. For example, late-stage clinical trials of inclisiran in patients with hypercholesterolemia have shown great promise. Inclisiran targets PCSK9 which is elevated in the condition; if approved in the future, it could be marketed to a substantial patient population depending on dosing and cost.

Aptamers

So far, we have considered nucleic acid-based drugs that bind to other nucleic acids like RNA. Aptamers (Latin, *aptus*, to fit) are different in that they are designed to bind to protein targets just like small molecule and protein drugs. The chemical nature of nucleic acids makes them ideally suited for this purpose. Aptamers are produced in the laboratory as a mixture of trillions of different molecules, only a few of which will strongly bind to the drug target and prevent it from working. There are several aptamer drugs in clinical development, including one on the market for treating age-related macular degeneration. This drug pegaptanib, marketed as Macugen® by Pfizer, works by binding to VEGF, a protein responsible for the abnormal growth of blood vessels in the eye; it is the first of possibly many aptamer drugs that may find clinical use in the future (Nimjee et al. 2017).

2.1.1.4 Gene Therapy and Editing

The nucleic acid-based drugs in the previous section are designed to inhibit the function of genes involved in diseases like cancer or AIDS by interacting with messenger RNA or directly with a protein target. Gene therapy on the other hand works at the level of the gene itself by directly modifying the DNA sequence. For example, many distressing genetic diseases resulting from mutations (Latin *mutare*, to change) in a single gene, such as hemophilia or cystic fibrosis, are prime candidates for this approach. Gene therapy covers the following procedures: replacing entire stretches of DNA with working copies, editing out individual nucleotide bases, and genetically modifying cells taken from the patient before reintroducing them by infusion. These areas are covered in the following sections on gene editing and cell therapy.

The DNA used for gene therapy is a synthetic molecule that can be designed to encode any desired protein; this is relatively straightforward, but problems arise when attempting to efficiently deliver DNA into the body. If not enough gets into the cells, there will not be enough protein expressed to compensate for the faulty version produced by the patient. Even if the DNA is effective upon first injection, it may not be possible to maintain effective levels of the new gene to keep the disease at bay. This problem has proven to be the bane of several gene therapy trials.

To make gene therapy work, the DNA must be transported in a "vector" (Latin, one who conveys or carries). This is itself a DNA molecule, often based on the DNA found in viruses. This is because viruses must infect human (or other) cells to make proteins from their own genes because they do not have the machinery to do this independently. The components of the virus that are used to infect cells can be purified in the laboratory and reassembled into a new virus containing the gene therapy DNA. This virus is then introduced into patients. In the process of creating the modified virus, any elements that might cause disease are removed. A commonly used vector is adeno-associated virus (AAV) which efficiently delivers DNA to a variety of cell types and which provokes a modest immune response in patients. This is important, because in a small number of cases, earlier generations of vectors had caused fatal inflammatory reactions thereby casting a shadow over the whole field of gene therapy. Another major problem arises when the viral nucleic acid literally integrates with the DNA in the patient's own cells and causes cancer. This happens because the new DNA switches on genes that are normally silenced to avoid inappropriate cell division.

Despite these caveats, gene therapy drugs are being evaluated in over 3000 clinical trials (The Journal of Gene Medicine Clinical Trial site 2020), and some have been given marketing approved by the regulators. For example, in 2017, voretigene neparvovec (Luxturna®) from Spark Therapeutics was approved for use in an inherited form of blindness caused by deficiency in the RPE65 gene. In this case, the correct copy is delivered to the retina via the AAV vector without being incorporated into the genome.

Zinc Finger Nucleases

The following four sections cover agents used to edit the DNA sequence of a gene in situ.

Zinc finger nucleases recognize specific DNA sequences in double-stranded DNA and then change the DNA to either enhance the activity of a gene, inhibit it, or change it through mutation. The zinc finger nuclease is part protein and part polynucleotide. The "zinc finger" is a fingerlike structure adopted by the protein component when modified by the addition of zinc atoms. ZFNs are being developed by the Californian company, Sangamo Biosciences Inc., for several diseases including SB-FIX, a treatment for hemophilia B that replaces a defective blood clotting gene (F9) with a functional copy that is expressed in the liver (Sangamo 2019).

Transcription Activator-Like Effector Nucleases (TALENS)

TALEN®s work on the same principle as ZFNs in that they comprise protein elements (TAL effectors from *Xanthomonas* bacteria) engineered to recognize any nucleotide sequence fused with an enzyme (FokI). This directs cuts in DNA to a region of choice. The same issues arising in gene therapy in general also apply to TALEN® technology, namely, delivery, immunogenicity, and non-specific effects on the genome at large. There are no approved drugs based on this technology at the time of writing, but it has been used in cell therapy for a child suffering from acute lymphoblastic leukemia (Qasim et al. 2017).

CRISPR-Cas9

Gene editing technologies, like the ZFNs and TALENs above, exploit the intrinsic repair mechanisms for correcting the DNA damage in living cells that arises during the normal process of cell division. As is so often the case, biological phenomena are observed in the lab or outside world and then the key elements of these processes isolated and used for a specific purpose. For example, with CRISPR-Cas9 gene editing, it was noted that bacteria infected with viruses (bacteriophages) incorporate viral DNA sequences into their genomes in the form of CRISPRs, i.e., short randomly interspaced short palindromic repeats. These turned out to be a sort of adaptive immunity in which the bacteria "remember" the phage infection and can eliminate a subsequent infection by using the enzyme Cas9 to cut the DNA. This CRISPR-Cas9 system provided a toolbox that rapidly became the standard method for DNA editing due to its simplicity and reliability (Broad Institute CRISPR timeline 2019). There is inevitable interest in using CRISPR editing for treating diseases, and recently CRISPR Therapeutics, in conjunction with Vertex Inc., have trialed CTX-001 as treatment for genetic disorders of hemoglobin, namely, sickle cell disease and β-thalassemia (CRISPR Therapeutics 2019). We can expect many more applications of this technology, although it has not been without controversy as it can be used to alter the germline DNA such that any edited changes are passed down subsequent generations. This area is the subject of scrutiny by governments and organizations involved in the legislation of medical applications.

Emerging Base Editing Technologies

The nature of the nucleic acid molecule (DNA and RNA), and the way that it binds (hybridizes) to its target sequence, means that sometimes the binding may occur to other regions, thereby giving rise to so-called off-target effects. Furthermore, the DNA repair process in which the sequence is filled in around the edited gene is itself imperfect. Ideally, it would be desirable to edit individual DNA bases in a highly precise way. This is where emerging technologies such as prime editing are attracting interest as future therapeutics. Prime editing employs a genetically engineered Cas9 that cuts one strand of the DNA double helix (duplex) and has a reverse transcriptase activity to produce another strand of DNA from an RNA template. The technology is described in (Prime editing 2019).

RNA Editing

The above technologies use DNA as the target, but there are attractions to instead editing the transcribed messenger RNA (mRNA) for the defective gene of interest. There are currently no marketed therapies based on RNA editing, but a promising method called LEAPER (leveraging endogenous ADAR for programmable editing of RNA) was published in 2019 (Qu et al. 2019). Based on the Cas13 enzyme that targets double-stranded RNA, LEAPER also incorporates part of the ADAR enzyme in a combination that selectively converts adenosine (A) at a specified position to inosine (I) that is recognized as guanosine (G) by the cell. In effect this is a base editor for converting A to G in any RNA derived for a gene in which A is present as a deleterious mutation. Like the prime editing above, this is in its early stages but shows great promise in preclinical models and may therefore become another medicinal product in the future.

2.1.1.5   Cell Therapy

Until around the 2010s, it would have been unthinkable that whole cells would be considered as pharmaceutical products to be sold by large pharma companies with a current and historical interest in small molecule drugs. This situation is beginning to change through the approval of cell-based therapies based on modified cells of the immune system or stem cells for the hematopoietic system. These areas are covered in the next two sections.

Stem Cell Transplantation

The human body contains roughly 200 different cell types and yet originates from only one cell, i.e., the fertilized egg (ovum). This means that there must be some process operating during development of the embryo (and later the adult) that generates these different cell types. This process relies on pluripotent stem cells in the embryo which turn into different cell types, such as nerve, muscle, and blood during development. Adult stem cells replenish mature cells that have a limited life span, such as blood and skin cells, and are present throughout life. Since many diseases can result in permanent tissue damage, any therapeutic approach that reverses or repairs the damage is going to be of interest to the biopharmaceutical industry, hence their involvement in stem cell research. In fact, stem cell therapy is not new; bone marrow transplantation to restore normal blood has been performed for decades. In the case of leukemia, for example, blood stem cells are purified from the blood or bone marrow of the patient (autologous), or a donor (allogeneic), and can be stored outside the body during which time radio and chemotherapy are used to remove all blood cells, including the leukemia. The non-cancerous stem cells are then reintroduced into the body, where they divide and repopulate the blood. Although these procedures are hazardous and often used as a last resort, they do create an opportunity to introduce specific genes into the blood of patients by using gene therapy to modify the stem cells used in the transplant. Clinical trials are underway using this approach to treat several blood diseases (see CRISPR Therapeutics 2019), for example. AIDS can be treated by removing the CCR5 receptor for the HIV virus by editing hematopoietic stem cells. As a result, the mature T cells (and monocytes) that develop from these stem cells will no longer express the receptor targeted by the virus.

The transplantation described above uses adult stem cells, but the dream of restoring damaged tissue like the heart, muscle, or brain will require stem cells derived from embryos or adult cells produced by complex manipulation in the laboratory (see Chap. 8).

Chimeric Antigen Receptor T cells

T (thymus-derived) lymphocytes, generally known as T cells, are central players in fighting infection as well as identifying and removing tumors. This latter process is generally successful, but of course tumor cells do escape and lead to cancer. New approaches to cancer therapy have become apparent as more becomes known about the relationship between tumors and the immune system. For example, antibodies directed against "checkpoint" molecules are proving so successful in controlling some cancers that huge efforts are being made by pharmaceutical companies to understand how they work and how they can be applied to a broad range of tumor types. Another therapeutic approach is to genetically engineer T cells from the individual cancer patient and target these cells to specific tumor antigens (generally proteins) on the cell surface. For example, these chimeric antigen receptor-modified (CAR) T cells have been generated against the CD19 antigen present in acute lymphoblastic leukemia (ALL) and kill the target B cells causing the disease; this has led to disease remission in patients who did not respond to standard therapies. The success of CAR T cell therapy in treating refractory ALL has been such that the engineered cells are being marketed as tisagenlecleucel (Kymriah®) by Novartis and have been approved for clinical use by the FDA (Novartis 2019). This is a good illustration of how diverse the product range of major pharmaceutical companies has become over a relatively short time, ranging from small molecules to genetically engineered cells.

Microbial Cell Transplantation

Drug development for the microbial world is mostly concerned with anti-infective agents directed against pathogenic bacteria, fungi, protozoa, and viruses. Most of these agents are small molecules that inhibit cellular (or viral) processes without much effect on the host (human) cells. Alternatively, vaccination harnesses the immune system to engage and kill infectious organisms (see later). In recent years, however, genetic tools have allowed investigators to study the microbial world (microbiome) in unprecedented detail. The human microbiome comprises (commensal) bacteria that reside in the gut, mucosal tissues, and the skin. While most bacteria live in basic harmony with the human tissues they inhabit, any imbalance in this can lead to conditions such as diarrhea and inflammatory bowel diseases. Furthermore, the resident microbes in the gut can also affect the metabolism of therapeutic drugs, so microbiome research has taken on increasing importance in the biopharmaceutical industry and is leading to new types of drug product based on intact bacteria. The idea is to deliver individual species of bacteria into the gut via fecal transplantation (FMT). Although this area is in its infancy, the possible applications include scavenging of toxic molecules, in situ production of therapeutic molecules, intracellular delivery of genetic payloads, displacing infectious pathogens, immune system modulation, and reestablishing a standardized microbiome (Jimenez et al. 2019). Some clinical success has been achieved when using fecal transplants of healthy stools into patients suffering from the results of infection with *Clostridium difficile* (*C. difficile*) bacteria. However, the procedure is not without risks, particularly the transfer of antibiotic-resistant bacteria as reported in 2019. In this case, two patients were seriously ill, one of whom died. This has inevitably led to scrutiny by the FDA and a review of FMT procedures in general (DNA Science blog 2019). Despite this setback, the various issues associated with FMT, and microbial therapeutics in general, will be addressed in future clinical trials leading to another form of biotherapeutic to stand alongside proteins, nucleic acids, and human cells.

Therapeutic Viruses

Viruses bind to the surface of target cells and subvert their machinery to replicate their own genetic material, build new virus particles, and ultimately break out to spread the infection. These properties may be exploited for gene therapy (see above) or targeted with vaccines and small molecule antiviral therapies. Another possibility is to use the destructive power of viruses to destroy specific cell types such as those present in tumors. Such oncolytic viruses are under clinical development, for example, talimogene laherparepvec (TVEC), trade name Imlygic®, for melanoma. This is an attenuated herpesvirus engineered to produce GM-CSF that allows stimulation of the patients' immune system in addition to the killing of melanoma cells (Conry et al. 2018).

2.1.1.6 Vaccines

Although vaccination has been performed for hundreds of years, it only came to the world's attention in 1796 after Jenner's pioneering work on smallpox. He coined the word "vaccine" (Latin *vacca* for cow), because of the cowpox virus he used to immunize his subjects. Since then, vaccination has, along with better hygiene and nutrition, become arguably the single most effective public health measure in human history. The influenza, HIV, SARS, MERS, Ebola, Zika, and now SARS-CoV-2 viruses have brought home the fact that infectious diseases are still capable of catching us off guard[4]. An HIV vaccine has been particularly challenging to develop as the virus mutates rapidly to evade the immune system. However, some broadly neutralizing antibodies can be generated that recognize a wide range of HIV types, so there is much work being done to find a vaccine that will stimulate the production of these unusual antibodies in the general population. Of course, it must not be forgotten that the developing world still must live with the scourge of tuberculosis, leprosy, malaria, and other tropical diseases, so vaccine research is vital here as well.

Like other medicinal products, vaccines are produced by the biopharmaceutical industry and must meet the same standards of safety and efficacy as any other drug. All medicines carry some risk, which must be balanced against the benefits provided. There are issues with vaccines, however, since most are designed to provide protection against future infections (prophylactic vaccination) and are therefore administered to healthy people who must take the risk (admittedly small) of unwanted side effects due to immune system activation. This, along with the relatively low financial returns of vaccines, used to discourage the biopharmaceutical industry from working in this area of drug development. When I joined the industry in the 1980s, working in an immunology department, there was absolutely no interest in developing vaccines. Ironically, I had previously worked in a tumor immunology unit whose long-term aim was to discover how the body uses the immune

---

[4]An understatement in the case of the COVID-19 pandemic

system to fight cancer. Current therapies based on checkpoint inhibitor antibodies and CAR T cells point to the success of tumor immunology research, but it might be possible to use therapeutic vaccines as well. Needless to say, the emergence of the COVID-19 pandemic has now put vaccine research front and center of global research efforts.

How Vaccines Work

The immune system has evolved to provide protection against invading organisms and is divided into two main areas: innate and adaptive. The innate system is the first line of defense against attack by bacteria and viruses, but it has no memory of the encounter with these agents. The adaptive system is brought into play after the initial infection and reinforces the attack that, if successful, will clear the infection from the body. This is where antibodies and white blood cells called lymphocytes appear on the scene. The adaptive system retains a memory of the encounter so that a further infection will be cleared rapidly and efficiently, possibly some decades later. Until quite recently, the two arms of the immune system were seen to be separate entities; from an immunology researcher's point of view, the adaptive system was the most challenging and interesting, and the innate system was frankly considered a bit boring. Times have changed, and the field has been energized with new discoveries about how the innate system recognizes patterns of molecules on invading bacteria and viruses and how closely it is integrated with the adaptive system. This is highly relevant to vaccine research since the adaptive arm is responsible for creating an "immunological memory" to be activated upon later encounter with the infectious agent. Vaccines are made up of two components, an antigen combined with an adjuvant. Antigen is a general term for the agent that provokes a specific immune response through recognition by T and B lymphocytes. The adjuvant literally acts as a helper to enhance that response. This is particularly useful when then antigen alone may not be very immunogenic (i.e., does not provoke a strong immune response), and it also means that amount of antigen per dose of vaccine can be kept to a minimum. Adjuvants work in part through stimulating the innate immune system, which in turn enhances the adaptive arm. This means that new findings about this aspect of immunology are being translated into a new generation of adjuvants with superior performance to existing molecules.

Types of Vaccine

When microorganisms, such as a bacteria or viruses, infect the body, antibodies and lymphocyte responses are produced naturally to allow clearance of the invader. Vaccines fool the body into thinking that it is being invaded because they mimic the ability of the microorganism to stimulate an immune response but without causing disease at the same time. In practice, the vaccine may be anything from a live attenuated virus to fragments of viral or bacterial DNA.

- Live Attenuated Virus

This vaccine uses the organism that it designed to protect against to create immunity without triggering disease. This means that the organ in the vaccine is diluted or attenuated. Recipients of these vaccines are inoculated, rather than immunized. Examples include measles and chickenpox.

- Inactivated Vaccines

These are whole organisms that have been rendered uninfectious by treatment with heat or chemicals. The term "inactivated" applies to vaccines derived from viruses, while those from bacteria are known as "killed." Because these products are less potent than live vaccines, a higher dose must be administered. Polio and hepatitis A vaccines are commonly used examples of this type, where the virus has been inactivated by formalin (formaldehyde) treatment.

- Toxoid Vaccines

Certain bacteria, such as tetanus, diphtheria, and cholera, produce proteins called toxins that are responsible for the characteristic symptoms of these diseases. These proteins are purified from bacteria grown in culture and converted by formalin treatment into "toxoids" that are devoid of harmful activity. These toxoids are used as vaccines in combination with adjuvants.

- Subunit Vaccines

Bacteria and viruses are quite different in appearance and life cycle, yet both are covered with proteins that assist in gaining entry into human cells. Influenza viruses are a good example of this as they express two proteins hemagglutinin (H) and neuraminidase (N) which can be purified and used in combination as a vaccine. Antibodies generated against each of these proteins prevent the virus from entering the cells of the respiratory tract. Subunit vaccines can be purified from whole viruses or bacteria or else produced in cell cultures using recombinant DNA technology. Some bacterial vaccines are based on surface carbohydrates (sugar-like molecules) instead of proteins.

- DNA Vaccines

Whether the vaccine is based on whole organisms or subunits, the cost of manufacture and safety testing can be considerable. Stability of the product where there is no refrigeration can also be a problem. DNA vaccines offer a possible solution to this because they are relatively simple to produce and are quite stable. Furthermore, their ease of design and manufacture makes them ideal for rapidly responding to new infectious agents such as the SARS-CoV-2 virus. The idea for this approach to immunization came from gene therapy, where it was noticed that DNA could provoke an immune response, which could be exploited in vaccination, rather than just dismissed as a side effect. DNA vaccination requires the same kind of vector as used for gene therapy and is introduced into the body by injection or the lungs by aerosol. If the DNA is designed to code for a protein normally found in a subunit vaccine, it will produce it directly in the human tissues and provoke an immune response. Only one DNA vaccine has been approved by the FDA (for West Nile virus in horses), but clinical development for human diseases continues apace (Hobernik and Bros 2018).

- mRNA Vaccines

Interest on messenger RNA (mRNA) vaccines has centered on the fact that unlike with their DNA equivalent, the encoded protein is transiently expressed and can be designed to cover a wide variety of peptide sequences. As with all nucleic acid-based drugs, delivery is a problem, particularly with highly unstable RNA molecules; nevertheless, companies are actively developing mRNA vaccines for infectious diseases and cancer (Pardi et al. 2018).

- Cancer Vaccines

A cancer vaccine for the human papilloma virus (HPV) that causes cervical cancer was introduced in 2006, and its success is one of the reasons why industry enthusiasm for vaccines is at an all-time high. However, this vaccine is still based on the conventional principle of immunizing healthy individuals prophylactically to prevent possible infection by the virus in the future. The point of a tumor immunology approach is that it should be possible to develop a therapeutic vaccine to treat the disease itself once it has become established. Attempts at producing true cancer vaccines along these lines have been made for many years now, and the work is slowly producing encouraging results. For example, apart from the conventional vaccination approach (e.g., with HPV), it is possible to isolate whole cells from the cancer patient, process them, and reinfuse to stimulate an immune response against the tumor. The cells may themselves be part of the tumor or part of the immune system in the form of dendritic cells. The latter cells are in effect the sentinels of the immune system that survey the body for infectious agents (or tumors) containing foreign antigens and then activate a T-cell response to these antigens. Sipuleucel-T (Provenge®) is a dendritic cell vaccine from Dendreon Inc. in the USA that was approved by the FDA in 2010 for a personalized treatment of prostate cancer. White blood cells are isolated from the patient and incubated with a protein that is highly expressed in prostate cancer combined with a cytokine (GM-CSF) that stimulates the growth of dendritic cells. These activated cells are then reintroduced into the patient to attack the prostate tumor. Although clinical trials of sipuleucel-T gave sufficiently compelling results to support marketing approval, uptake of the vaccine has not been as high as initially expected. However, as with many of these situations, a closer look at the scientific background may provide clues for maximizing the benefit. For example, there are different dendritic cell types (cDC1, 2A, 2B) which may explain variable responses to the vaccine. Furthermore, as in many similar situations with other drug trials, a close examination of the clinical trial data may reveal clues for future improvements (see also Chap.12).

Vaccines are also being developed for other chronic diseases which may have no obvious association with infection. Alzheimer's disease is one example, where much effort has been expended into producing vaccines against the pathogenic amyloid and tau proteins, but with limited success so far, due in part to the dangerous brain inflammation seen in clinical trials.

Summary of Key Points

Drugs are agents that bind to a target to increase or slow down the activity of the latter.

Most drugs are small molecules that can be taken by the mouth, but many other products are being introduced to the clinic. These are:

Proteins
Nucleic acids
Gene therapies
Cell therapies
Vaccines

References

Akdim F et al (2010) Effect of mipomersen, an apolipoprotein B synthesis inhibitor, on low density lipoprotein cholesterol in patients with familial hypercholesterolemia. Am J Cardiol 105:1413–1419

Aroroa A et al (2007) Needle-free delivery of macromolecules across the skin by nanoliter-volume pulsed microjets. Proc Natl Acad Sci USA 104:4255–4260

Broad Institute CRISPR timeline (2019) https://www.broadinstitute.org/what-broad/areas-focus/project-spotlight/crispr-timeline. Accessed 11 Dec 2019

Conry RM et al (2018) Talimogene laherparepvec: first in class oncolytic virotherapy. Hum Vaccin Immunother 14:839–846

CRISPR Therapeutics (2019) http://www.crisprtx.com/programs/hemoglobinopathies. Accessed 11 Dec 2019

DNA Science blog (2019) https://blogs.plos.org/dnascience/2019/11/07/fda-scrutinizes-fecal-transplants/. Accessed 16 Dec 2019

Drugs@FDA (2020). https://www.accessdata.fda.gov/scripts/cder/daf/index.cfm?event=glossary.page. Accessed 6 Dec 2019

Hobernik D, Bros M (2018) DNA vaccines-how far from clinical use? Int J Mol Sci 19:3605

Janssen HLA et al (2013) Treatment of HCV infection by targeting microRNA. N Engl J Med 368:1685–1694

Jimenez M et al (2019) Microbial therapeutics: new opportunities for drug delivery. J Exp Med 216:1005–1009

Montazersaheb S et al (2018) Potential of peptide nucleic acids in future therapeutic applications. Adv Pharm Bull 8:551–563

Nimjee SM et al (2017) Aptamers as therapeutics. Annu Rev Pharmacol Toxicol 57:61–79

Novartis (2019) https://www.hcp.novartis.com/products/kymriah/acute-lymphoblastic-leukemia-children/mechanism-of-action/. Accessed 13 Dec 2019

Pardi N et al (2018) mRNA vaccines – a new era in vaccinology. Nat Rev Drug Disc. https://doi.org/10.1038/nrd.2017.243

Petrova et al (2013) Structure-functions relations in small interfering RNAs. In: Practical applications in biomedical engineering. https://doi.org/10.5772/53945

Prime editing (2019) https://www.the-scientist.com/news-opinion/new-prime-editing-method-makes-only-single-stranded-dna-cuts-66608. Accessed 11 Dec 2019

Qasim W et al (2017) Molecular remission of infant B-ALL after infusion of universal TALEN gene-edited CAR T cells. Sci Transl Med 9:374

Qu L et al (2019) Leveraging endogenous ADAR for programmable editing on RNA. https://doi.org/10.1101/605972. Accessed 12 Dec 2019

Sangamo (2019) http://www.sangamo.com. Accessed 10 Dec 2019

Sneader W (2005) Drug discovery a history. Wiley, Chichester

The Journal of Gene Medicine Clinical Trial site (2020) http://www.abedia.com/wiley/index.html. Accessed 10 Dec 2019

Vickery HB (1950) The origin of the word protein. Yale J Biol Med 22:387–393

Chapter 3
Background to Chemistry of Small and Large Molecules

Abstract What are small and large drug molecules exactly? How do they physically interact with their protein targets to activate or inhibit them? How are drugs named?

The answers to these questions could take up several books and degree level courses in chemistry and biochemistry. Fortunately, the points that a nonspecialist really needs to know can be simplified without losing their essential meaning. The aim of this chapter is to give readers a guide to the fundamental chemistry of drugs and their targets. Those who are familiar with elementary chemistry will be tempted to skip this chapter altogether. Although this is understandable, they need to be aware that some of the material on drug size and nomenclature may not have been covered in school or college chemistry.

3.1 Introduction

"Better Living through Chemistry." This slogan, based on a DuPont Corporation strap line, was sometimes seen on T-shirts in the latter part of the twentieth century. It can of course be taken either at face value or as a piece of irony, if you feel that chemicals have done more harm to humanity than good. Since this is a book on drug discovery, it can be safely assumed that the author believes that on balance, chemistry has proven to be a benefit to humanity rather than the opposite. Chemistry is defined as "the science of substances: their structure, their properties, and the reactions that change them into other substances." This 1947 definition by the double Nobel Laureate Linus Pauling is a useful starting point. Experimental chemistry has been performed for thousands of years and was developed in the Arab world (the word chemist is of Arabic origin) before being adopted in Europe and developed into a highly sophisticated branch of science.

3.1.1 Elements, Atoms, Molecules, and Compounds

Hydrogen is a colorless odorless gas that given enough time turns into people

© The Editor(s) (if applicable) and The Author(s), under exclusive license to
Springer Nature Switzerland AG 2020
E. D. Zanders, *The Science and Business of Drug Discovery*,
https://doi.org/10.1007/978-3-030-57814-5_3

This quote by the late cosmologist Edward Harrison (Wiley 1995) sums up the situation quite well. Hydrogen is the simplest element that formed soon after the "Big Bang" that brought the universe into existence 13.8 billion years ago. An element is a pure chemical substance made up of one kind of atom. During the evolution of the universe, hydrogen was converted in the process of nuclear fusion to another element, helium. Further nuclear reactions created the elements that exist in stable forms on earth, for example, iron, carbon, and oxygen. These elements were originally blown out of stars in giant supernova explosions, thereby producing clouds of material that ultimately coalesced into planets. Since we too are made up of these elements, this means, rather romantically, that we are ultimately derived from stardust.

There are 94 stable elements found on earth and another 24 produced artificially under extreme conditions, such as those created by nuclear reactions. Out of all these, only a few are used by living organisms (and by drug discovery chemists).

The famous Periodic Table of the Elements was compiled in 1869 by the Russian chemist Dimitri Mendeleev[1] as a means of classifying the elements into groups with shared properties (Fig. 3.1). The names of the elements are abbreviated to one or two letters. For example, hydrogen is H and beryllium is Be. Many abbreviations betray the original name of the substance in antiquity. Potassium, for example, is written K for *kalium* and sodium, Na for *natrium*.

H																		He
Li	Be												B	C	N	O	F	Ne
Na	Mg												Al	Si	P	S	Cl	Ar
K	Ca	Sc	Ti	V	Cr	Mn	Fe	Co	Ni	Cu	Zn	Ga	Ge	As	Se	Br	Kr
Rb	Sr	Y	Zr	Nb	Mo	Tc	Ru	Rh	Pd	Ag	Cd	In	Sn	Sb	Te	I	Xe
Cs	Ba	La	Hf	Ta	W	Re	Os	Ir	Pt	Au	Hg	Tl	Pb	Bi	Po	At	Rn
Fr	Ra	Ac	Rf	Db	Sg	Bh	Hs	Mt									

Ce	Pr	Nd	Pm	Sm	Eu	Gd	Tb	Dy	Ho	Er	Tm	Yb	Lu
Th	Pa	U	Np	Pu	Am	Cm	Bk	Cf	Es	Fm	Md	No	Lr

Fig. 3.1 The periodic table of the elements. A simplified representation of the famous table compiled by Mendeleev which shows just the abbreviation used for each element. Elements are grouped together in the table according to their chemical properties. The section below the main table represents the so-called "lanthanide" and "actinide" group of elements, none of which are naturally occurring in living organisms

---

[1] Element 118 is the latest at the time of writing and named after another Russian, Yuri Oganessian.

3.1.1.1 Atoms and Molecules

Each element is made up of the basic building block of matter, the atom. There are several ways of envisaging what an atom really is, depending on the depth of understanding required for a given scientific problem. For the purposes of this basic introduction, we can think of the atom as a billiard ball without any internal structure.

Molecules are combinations of atoms ranging from two to almost any number. The atoms may be part of the same element or exist as combinations of different elements. For example, under the conditions of atmospheric pressure and temperature found on earth, the elements nitrogen, hydrogen, and oxygen are molecules consisting of two atoms, while argon is present as a single atom (written as N_2, H_2, O_2, and Ar, respectively).

3.1.1.2 Compounds

Compounds consist of atoms of two or more different elements bound together. They have properties that are different from their component elements. Figure 3.2 shows the example of the simple compound water, made from the elements hydrogen and oxygen. Water is a liquid at room temperature, and its properties are clearly different from the gases hydrogen and oxygen.

The number of compounds that can theoretically be produced by the combination of the main elements found in nature (carbon, hydrogen, oxygen, nitrogen, etc.) is almost infinite (see Chap. 7). In drug discovery, the word "compound" is often used to describe a small molecule drug, for example:

this compound is effective for treating diabetes.

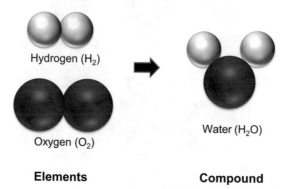

Elements   **Compound**

Fig. 3.2 Elements combine in chemical reactions to produce compounds with quite different physical and chemical properties. This example is an illustration of water being produced by the reaction of hydrogen with oxygen. Note that since both hydrogen and oxygen are made up of two atoms, the numbers here don't add up; the chemical reaction is really $2H_2 + O_2 = 2H_2O$. More detail about basic chemical reactions can be found in any chemistry textbook or online educational resource

Alternatively, the word "molecule" is used, to give:

this molecule is effective for treating diabetes.

Strictly speaking, a molecule could be an element and not a compound (e.g., nitrogen), but this is unlikely to be the case with drug compounds as they are almost always built from at least three different elements.

3.1.1.3 Simple Compounds and Molecular Weights

The "small" in small molecules (and "large" in large ones) refers to their "molecular weights." This is explained as follows: a simple molecule such as ethanol (ethyl alcohol or just alcohol) is a compound made up of two carbons, six hydrogens, and a single oxygen atom. Each element, from hydrogen (the lightest) onwards, has a value called the atomic weight which is normally written underneath the symbol as it appears in the periodic table. The weights are based on a system in which hydrogen has a value of 1, carbon 12, oxygen 16, and so on. The atomic weights can be added together to produce the molecular weight. In the case of ethanol, the molecular weight (MW) is 46 (Fig. 3.3).

Molecular weights are often abbreviated as MW and are expressed as units called daltons, after the English chemist who devised the system in the early 1800s.

Even though ethanol has drug-like properties, its molecular weight is quite modest. Most small molecule drugs range from 200 to about 500 daltons. If the value is too small, the compound will not have enough variety (chemical diversity) to ensure

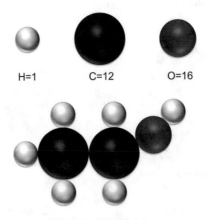

Ethanol: MW=46 daltons

Fig. 3.3 Calculating the molecular weight of a compound. Example shown is the small molecule ethanol with a molecular weight of 46 produced by adding up the atomic weights of 2x carbon, 6x hydrogen, and 1x oxygen atoms. Note that the size of each sphere relates to the element's radius not atomic weight. Molecular weights of natural and synthetic molecules range from tens to millions

strong and selective binding to a target. If the value is much greater than 500, it will be too large to be orally active (see Chap. 2).

Notes

1. The true weight of an atom is so small that it would be inconvenient to use these values. A carbon atom, for example, weighs approximately two hundred thousand million million millionth of a gram!
2. In reality, each element is a mixture of isotopes with slightly different atomic weights, so the values are not exact multiples of 1. Isotopes are important in several branches of chemistry and are powerful tools for labeling molecules to investigate biological processes.
3. Molecules are physically very small. From the example above, 46 grams of ethanol (about a large whisky) contains approximately 6×10^{23} molecules.

3.1.2 Organic Chemistry

Life is based on carbon (C) and several other elements including hydrogen, oxygen, nitrogen, sulfur, and various metals. This is because carbon can readily combine with itself and other elements in a wide range of configurations such as chains and rings. This versatility has been exploited by nature and by human beings for many different applications, such as polymers, dyestuffs, and of course drugs. The study of carbon compounds is known as organic chemistry. The composition of an adult human in terms of the percentages of the different elements is shown in Table 3.1. Oxygen is very abundant, since about 70% of the human body is made up of water. Although the percentage of calcium is quite small, this figure still represents a weight of about one kilogram per person, much of which is in the skeleton, although this metal is also vital for cell function.

3.1.2.1 Writing Down Structures

Simple two-dimensional representations of molecules have their uses and are easy to write down on paper. However, these diagrams do not readily convey one of the essential properties of molecules, which is shape. Drug molecules and their targets have complex three-dimensional shapes that must be compatible if they are to selectively interact with each other. Different combinations of atoms in compounds have well-defined shapes formed by the links (bonds) that hold them together. Carbon, for example, is often found as a tetrahedron in association with hydrogen or certain other elements. The compound methane is illustrated below in the simple 2D representation and in a 3D computer-generated model showing the bonds between the atoms as tubes (Fig. 3.4).

Table 3.1 Percentage of elements in an adult human

Name	Abbreviation	Percentage
Oxygen	O	61
Carbon	C	23
Hydrogen	H	10
Nitrogen	N	2.6
Calcium	Ca	1.4
Phosphorus	P	1.1
Potassium	K	0.2
Sulfur	S	0.2
Sodium	Na	0.14
Chlorine	Cl	0.12
Magnesium	Mg	0.027
Iron	Fe	0.006
Copper	Cu	Trace
Molybdenum	Mo	–
Zinc	Zn	–
Iodine	I	–

Data taken from Emsley (2001) with the kind permission of Oxford University Press

Fig. 3.4 Three-dimensional representation of methane molecule (CH_4) using a computer-generated "ball and stick" model (on right). The shape corresponds to a tetrahedron

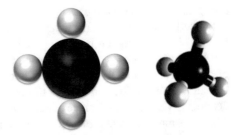

It is obviously impractical to write down chemical structures on paper using either of the above illustrations, so the next section describes how to think like a chemist and interpret the 2D structures that these scientists use on a routine basis. It is possible to make an analogy between chemists and musicians in the following way: a number of great musicians have delivered a perfect performance on their chosen instruments without ever having played the piece before; they learned everything they needed to know just by reading the score. The score is a two-dimensional representation of musical notes that follows a series of accepted rules so that the trained musician knows exactly how the piece will sound. The same principle applies in chemistry. The trained chemist will look at a molecular structure written on paper and be able to identify many of the key properties of the molecule in question. He or she will also be able to name the compound using an internationally accepted set of rules for chemical nomenclature. The following section gives a simple introduction to chemical formulae and nomenclature of the organic molecules that make up most drugs.

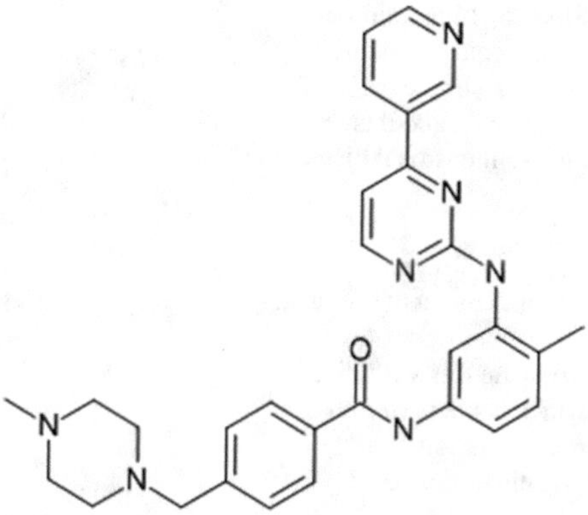

Fig. 3.5 Two-dimensional
structure of a small
molecule drug imatinib
used to treat cancer. This
illustrates the standard
format used by chemists to
write down structures

The structure of the anticancer drug imatinib (trade name Gleevec®) may look quite fearsome to the uninitiated (Fig. 3.5). It can, however, be better understood by learning some simple rules about how carbon and other elements join together in fixed proportions and how complex organic molecules are made up of smaller "functional groups."

Where Are the Carbon Atoms?

It is much easier to draw structures without having to place every carbon and hydrogen atom in the molecule. The above diagram has rings and lines joining them, so where there is no letter specified (like the Ns, nitrogen), it is assumed that a carbon atom is present.

Why Are There Two Lines in Places and No Hydrogen Atoms Displayed?

As with the carbon atoms, hydrogens are left out to avoid cluttering the picture. The number of hydrogens at each position will vary because of the maximum number of bonds that each carbon atom can make. This introduces the concept of "valency" developed in the nineteenth century. The valency of an element is literally its combining power with other elements. For example, carbon has a valency of 4, nitrogen 3, oxygen 2, and hydrogen 1; this means that one carbon atom can combine with a maximum of 4 hydrogen atoms. Some simple examples are shown in Fig. 3.6. Every carbon atom uses 4 bonds, even if there are not 4 recipient atoms. This is illustrated in examples 3 and 4 where double or triple bonds form to make up the valency of 4.

The above two points are reinforced in Fig. 3.7, where imatinib is now displayed with all the carbon and hydrogen atoms. The positions of single and double bonds determine how many hydrogen atoms are attached to each carbon atom, so they can convey where these atoms are situated in the molecule without creating a cluttered diagram like this example.

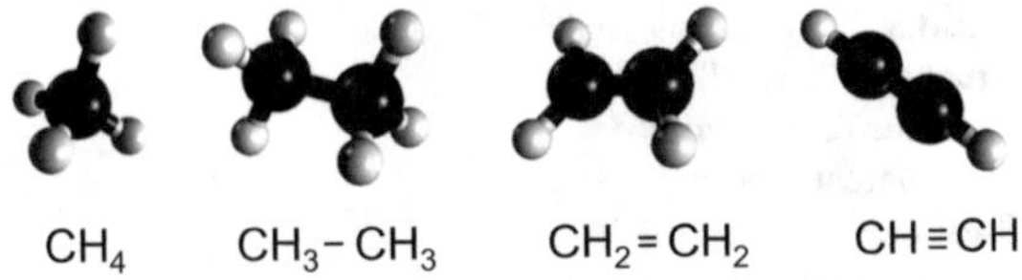

Fig. 3.6 Illustration of single, double, and triple bonds in simple carbon compounds and their three-dimensional display

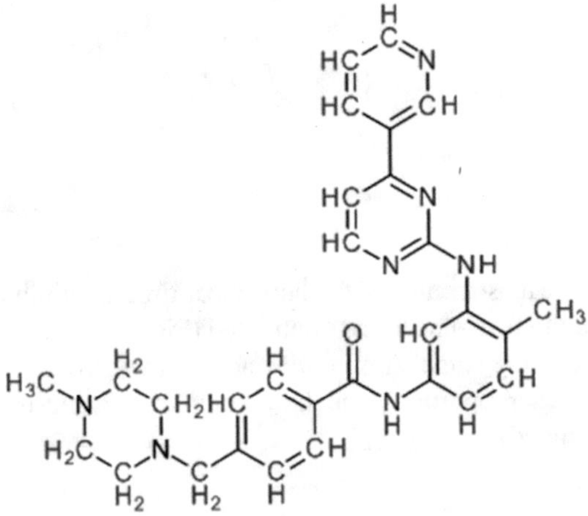

CH_4 CH_3-CH_3 $CH_2=CH_2$ $CH \equiv CH$

Fig. 3.7 Structure of imatinib molecule showing all the hydrogen atoms. Compare this with Fig. 3.5 which is the preferred representation of an organic molecule

3.1.2.2   Working Out the Shape of a Molecule

Shape is an important factor governing the precise interaction between a drug and its target, and in many cases a substantial part of this shape is formed by rings of atoms. There are two general types of ring, alicyclic, and aromatic, which are exemplified in the compounds cyclohexane and benzene shown in Fig. 3.8. The original characteristic of aromatic compounds was their odor, hence the name. According to modern chemical theory, aromatic compounds are defined by the configuration of alternating single and double bonds that force the ring into a flat shape. Alicyclic compounds may have some double bonds, but they do not alternate in the same way, and the rings are bent or "puckered." This "conformation" of the molecule is apparent to the chemist just by looking at the disposition of single and double bonds on the structure drawing.

3.1.2.3   Rings and Other Functional Groups

Many organic compounds and drugs contain rings of different shapes and sizes. The ring backbone may be constructed entirely out of carbon, or else mixed in with other elements such as nitrogen, oxygen, or sulfur in which case they are called heterocycles. Some examples are shown in Fig. 3.9.

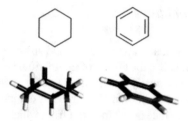

Fig. 3.8 Molecular structures of cyclohexane (left) and benzene (right). The alternating single and double bonds in benzene force the ring into a flat shape shown using the stick display produced by the computer graphics. The disposition of double bonds in a ring indicates to the chemist whether the ring is flat (planar) or bent (puckered). These shapes are important features in drug molecules and their target proteins because they help to determine binding strength and selectivity

Fig. 3.9 Examples of heterocycles used in organic molecules, including drugs

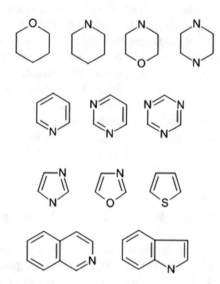

These structures (and many other groups of atoms that are not necessarily ring shaped) are "functional groups" linked together in more complex molecules. These groups are rather like modules or building bricks that can be taken "off the shelf" and assembled in different ways to create a huge variety of different structures. A knowledge of these groups allows the medicinal chemist to create new molecules (or modify existing ones) to produce drug compounds that interact strongly with their target (see Chap. 7). The names of certain atoms specifically associated with functional groups follow the Hantzsch-Widman nomenclature created by two chemists in the 1880s (Heterocyclic Chemistry 2020); for example, thia means sulfur; aza, nitrogen; oxo, oxygen; and chloro, chlorine. A knowledge of the names of these groups helps to understand how complex molecules like drugs are named.

3.1.3 Drug Nomenclature

Naming conventions in science are often hit and miss because most objects are named by the discoverer, hence the efforts of international bodies to create an agreed nomenclature that is linguistically consistent. Organic chemistry has been in existence for roughly 200 years, and many hundreds of compounds have been named, often based on their appearance or the material from which they were first extracted, piperidine from pepper, for example. Small molecule drugs are named at four levels: their formal chemical name, generic name, brand or trade name, and position within the Anatomic and Therapeutic Classification System. The formal chemical name follows a set of rules laid down by IUPAC, the International Union of Pure and Applied Chemistry (IUPAC 2020). Every functional group of a compound is named, and its relative position identified, by a numbering system. In many ways it is like compound nouns in the German language. These can be made up of a string of individual small words to form an impressively long one. "Donaudampfschiffahrtselektrizitätenhauptbetriebswerkbauunterbeamtengesellschaft," for example, translates as "Association for Subordinate Officials of the Head Office Management of the Danube Steamboat Electrical Services."

The IUPAC name for imatinib is 4-[(4-methylpiperazin-1-yl) methyl]-N-[4-methyl-3-[(4-pyridin-3-ylpyrimidin-2-yl) amino] phenyl]benzamide. This is not a particularly good starting point for the uninitiated, so the following compound (a common painkiller) will be used as an example of the different levels of drug nomenclature.

3.1.3.1 IUPAC Name: N-(4-hydroxyphenyl) Acetamide

This molecule is made up of three groups: hydroxyl and phenyl which are linked to acetamide (Fig. 3.10). The hydroxyl group is joined to the phenyl group at position 4. The acetamide is linked to the resulting hydroxyphenyl group via the nitrogen N. Note that one of the hydrogen atoms on acetamide is lost after it has been joined to the H at position 1.

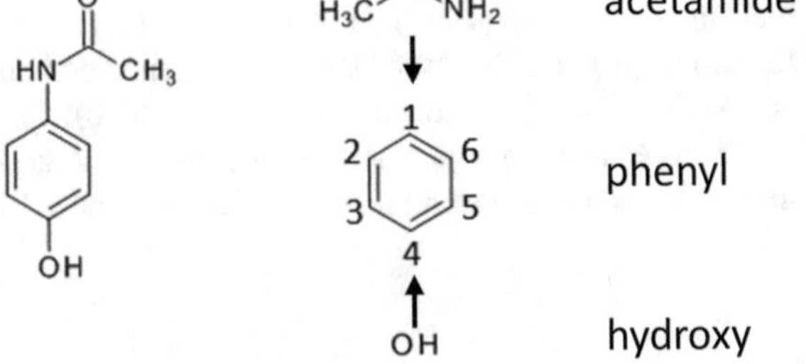

Fig. 3.10 Layout of functional groups (acetamide, phenyl, hydroxyl) and numbering system for N-(4-hydroxyphenyl) acetamide (left)

3.1.3.2   Generic Name

Since the IUPAC nomenclature is too unwieldy for general use, there are international organizations that agree on a manageable name that is not proprietary. This may be the International Nonproprietary Name (INN) (WHO guidelines for INN 1997) or the United States Adopted Name (USAN Council 2020), the former being under the auspices of the World Health Organization (WHO). The INN for N-(4-hydroxyphenyl) acetamide is "paracetamol," and the USAN is "acetaminophen."

3.1.3.3   Proprietary or Trade Name

Here we leave the world of science and medicine and enter the world of creative media. These names are proprietary to the company that sells the product and are generated in the same way as for any other consumer product. There are many trade names for paracetamol, including Calpol, Panadol, and Tylenol.

3.1.3.4   ATC Classification

The Anatomical Therapeutic Chemical (ATC) classification system was set up by the WHO as a means of coordinating data on drug use internationally. The ATC consists of a hierarchy from anatomical group and therapeutic class down to the chemical substance itself (ATC Structure and Principles 2020). The acetaminophen example given here has the classification NO2BE, a code for nervous system, analgesics, other analgesics, and antipyretics, and anilides.

3.1.4   Hydrates and Salts

Drug compounds are often named in two parts, for example, amoxicillin hydrate, ranitidine hydrochloride, sildenafil citrate, and imatinib mesylate. The first word is the drug itself and the second its hydrate or salt form. Hydrates are compounds that are associated with one or more molecules of water. This may help to stabilize them in a crystalline form. If the water is removed, they become anhydrous compounds or anhydrates.

Salt forms are often used in small molecule drug development as they can improve the stability of the parent compound in the different chemical environments of the body. They are also important for manufacturing, so the selection of appropriate salt forms is a key part of the formulation process (covered in Chap. 10). There may also be a strong financial incentive to investigate the development of alternative salt forms as these can extend the patent life of a drug.

Salts are formed by the chemical reaction of an acid and a base to form a neutral compound. Although a detailed explanation of this phenomenon is outside the scope of this book, mention will be made of the term "pH" which is a measure of the degree of acidity or alkalinity (basicity) of a substance. The pH scale ranges from 0 to 14, where 0 is very acidic and 14 is very basic. Strong hydrochloric acid is very acidic, and ammonia is very basic. A neutral substance (e.g., water) has a pH of 7. Since the pH scale is logarithmic, each number represents 10 times more than the other, so a solution at pH 0 is 1 million times more acidic than one at pH 6.

Drugs can either be basic in character and neutralized with an acid or vice versa. Some of the acids used to neutralize the drug are inorganic, i.e., without any carbon atoms. The majority of these are hydrochlorides, sulfates, and phosphates based on hydrochloric, sulfuric, and phosphoric acid, respectively. Alternatively, organic acids (such as acetic acid) are used. Examples of acids commonly used to produce salt forms of drugs are shown in Figure 3.11. The functional group in common is the carboxyl group COOH (except for the SOO in the mesylate) which provides the acidity. Where a base has been used to neutralize an acidic drug, most of the resulting salts contain a metallic element such as sodium, calcium, or potassium.

3.1.5 The Chemistry of Proteins

Proteins have already been introduced in Chap. 2 as injectable drugs, but they are also central to small molecule drug discovery as they form the physical basis of most drug targets. This section introduces the chemistry of proteins and describes the properties that make them central to life processes and drug discovery.

Unlike small molecules of around 500 daltons molecular weight, proteins are classed as macromolecules, with sizes ranging from about 6,000 to over 1,000,000 daltons. Proteins are biological polymers that can be produced in an almost infinite number of chemical forms, so to understand why this is the case, it is helpful to first look at polymers in general.

Everyone is familiar with polymers in the form of plastics, the word polymer deriving from the Greek for "many things." A simple polymer like polythene consists of a chain of identical chemical units or monomers. In this case, the monomer

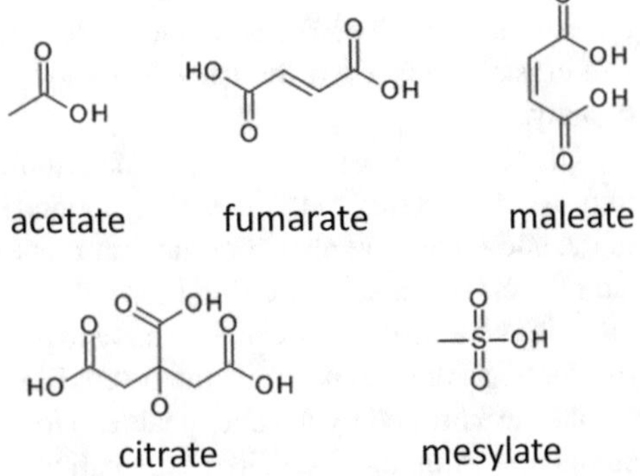

Fig. 3.11 Structures of the acids that are commonly used to produce salt forms of drugs (hydrochloric acid is also used but not shown here)

acetate   fumarate   maleate

citrate   mesylate

is ethylene, hence the name polyethylene, otherwise shortened to polythene. Nylon, on the other hand, is made up of two different monomers linked together in a chain. This chemical linkage is created via the amide functional group; hence nylon is sometimes known as a polyamide. Proteins are naturally occurring polymers made from about twenty monomers called amino acids. Unlike synthetic plastics, the chain length is precisely defined, and the order of amino acids can be varied in any permutation.

This is summarized in Fig. 3.12 where the amino acids are indicated as numbers for simplicity.

The 20 amino acids that occur naturally in most proteins are, of course, more than numbers. They are small molecules with molecular weights around 100–200 with two functional groups that allow them to join in a chain. These are the amino and carboxylic (a type of acid) groups that give this class of compounds its name. When this linkage occurs in a protein, it is called a peptide bond. When a chain consists of two or more amino acids, up to about 40, it is known as a peptide, although the exact upper limit is rather imprecise. After that it is more correct to call it a polypeptide or protein.

Although the names of most amino acids end in the letters "ine," individual compounds are named in same haphazard way as other organic compounds. Asparagine, for example, was so named because it was first isolated from asparagus. Incidentally, this amino acid is responsible for the smell of urine passed after eating this vegetable.

Each amino acid is given a three letter or single letter abbreviation according to use. The latter is used in protein databases where the sequence[2] is written in the

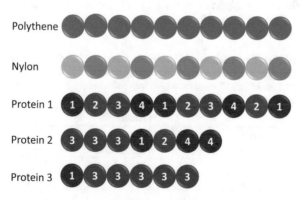

Fig. 3.12 Schematic example of different polymers. Chemically unique monomers are shown as different shaded discs. Synthetic plastics like polyethylene (polythene) and nylon are made up from one and two different monomers, respectively. Proteins are built up from around 20 different monomers called amino acids. The three proteins here use amino acids 1, 2, 3, or 4 in different permutations. In reality, the protein chain will be much longer, and the amino acids given specific names (see below)

---

[2] The use of the words "sequence" and "sequencing" is the same for proteins as it is for nucleic acids, but in the latter case, the units are nucleotides and not amino acids.

single letter code for clarity and for reading by computers (see bioinformatics, Chap. 6).

Table 3.2 lists the common amino acids by name and abbreviation

Despite having amino and carboxyl functional groups in common, each amino acid is chemically quite unique; this means that the proteins formed from their joining up (polymerization) show great diversity in their chemical structure and therefore their biological function. Figure 3.13 shows the structure of a heptapeptide, i.e., a peptide made up of seven amino acids[3]. Notice the different rings and chains that grow off the string of atoms that is formed by peptide bonds. It is the side chains that impart different structures to the peptides and proteins formed when the amino acids are linked together.

3.1.5.1   Molecular Origami

Chains of amino acids are known as the primary sequence of a protein and can be written down as a string of letters, for example (single letter code), "GHISSTWLSTVVN." The structures formed when the protein chain folds up according to the chemical structures of the constituent amino acids are called the secondary and tertiary structures. The quaternary structure relates to association of more than one polypeptide chain. The example of the heptapeptide in Fig. 3.13 introduces the idea of complex structures in protein chains. This is one of the most important areas in biochemistry, as it determines how proteins function in living cells and whether they are suitable drug targets for small molecules or other proteins. Once the protein chain is formed in cells, it folds into a precisely defined conformation, although it is still unknown how this happens as there are literally

Table 3.2 List of 19 common amino acids and their 3-letter and single-letter codes. These codes are used to illustrate protein sequences in the same way as A,C,G,T are used for nucleic acids. Amino acids not shown include cystine (two cysteines linked together) and hydroxyproline, which is present in connective tissue proteins

Name			Name		
Alanine	Ala	A	Lysine	Lys	K
Aspartic acid	Asp	D	Methionine	Met	M
Asparagine	Asn	N	Phenylalanine	Phe	F
Cysteine	Cys	C	Proline	Pro	P
Glycine	Gly	G	Serine	Ser	S
Glutamic acid	Glu	E	Threonine	Thr	T
Glutamine	Gln	Q	Tryptophan	Trp	W
Histidine	His	H	Tyrosine	Tyr	Y
Isoleucine	Ileu	I	Valine	Val	V
Leucine	Leu	L			

[3] The length of small peptides is indicated by the Greek number prefix, i.e., di, tri, tetra, penta, hexa, hepta, octa, nona, and decapeptide for 2–10 amino acids.

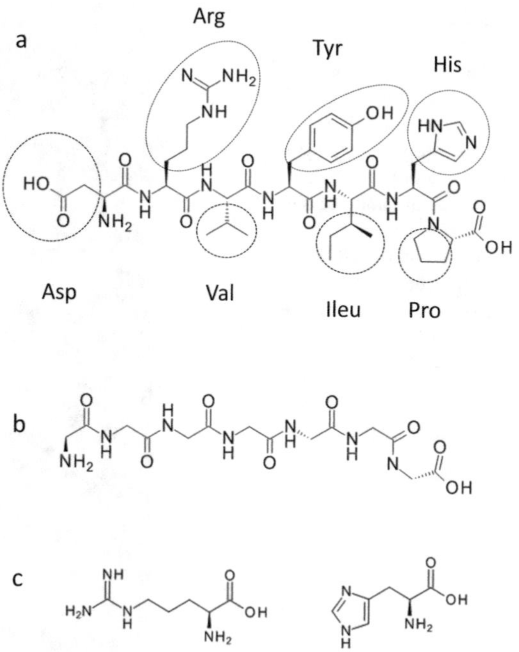

Fig. 3.13 Structure of the heptapeptide Asp Arg Val Tyr Ileu His Pro (DRVYIHP in single letter code). (**a**) Structure with side chains of the individual amino acids highlighted. (**b**) The same diagram as (**a**) but minus the side chains to show the peptide structure that is common to all peptides and proteins. (**c**) Examples of unlinked amino acids, arginine, and histidine

trillions of possible structures. Proteins are either globular or fibrous according to their overall shape. Globular proteins perform catalytic functions within cells (see enzymes, Chap. 6) and many other roles. They tend not be structural, however, so fibrous proteins are instead employed in tissues such as hair, nails, and skin where their greater mechanical strength is required. Structures for both types are shown in Fig. 3.14.

More details about proteins and their functions in drug discovery will be presented in later chapters. Meanwhile, this chapter will continue with a description of large molecules other than proteins and conclude with a section on the way in which drugs bind to their targets, focusing on the interactions between small molecules and proteins.

3.1.6 Large Molecules Other Than Proteins

3.1.6.1 Nucleic Acids

Deoxyribonucleic acid and ribonucleic acid (DNA and RNA) have already been introduced in Chap. 2. So far, the nucleotide building blocks that make up these molecules have just been given the letters A C G T and U, but the full names of the four nucleotides used to build a DNA molecule are deoxyadenosine triphosphate (dATP), deoxythymidine triphosphate (dTTP), deoxyguanosine triphosphate

Fig. 3.14 Three-dimensional structures of collagen (left) and pepsin (right). Computer-generated wire frame models showing extended fibrous nature of collagen and globular nature of the enzyme pepsin. Amino acid side chains can be seen protruding from the collagen structure which is a triple helix made up of three protein chains

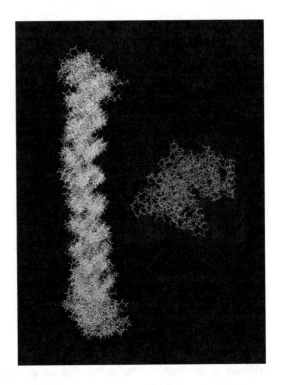

(dGTP), and deoxycytosine triphosphate (dCTP). These are small molecules consisting of a nucleobase linked to deoxyribose linked to triphosphate (Fig. 3.15). Therefore DNA is called deoxyribonucleic acid (the acid part is provided by the phosphate group). RNA is made up of adenosine triphosphate (ATP), CTP, GTP, and uridine triphosphate (UTP) instead of TTP. In RNA the deoxyribose is replaced by ribose, hence ribonucleic acid.

The structure in Fig. 3.16 shows the four nucleotides of DNA joined in a chain. The principle is like that of the peptide chain being built up from amino acids, although the phosphate-deoxyribose linkage is chemically different. The nucleobases protruding from of the DNA chain allow two chains to link together to form a double helix. The figure shows the three-dimensional structure of part of a double helix where bases from each strand of DNA are aligned together and form weak hydrogen bonds (see later in chapter). From the earlier section on ring structures, it will be noticed that the rings in the nucleobases are completely flat because of the disposition of single and double bonds.

The synthesis and manipulation of nucleic acids is a fundamental part of biotechnology and modern drug discovery, so this important class of large molecules will be revisited in later chapters.

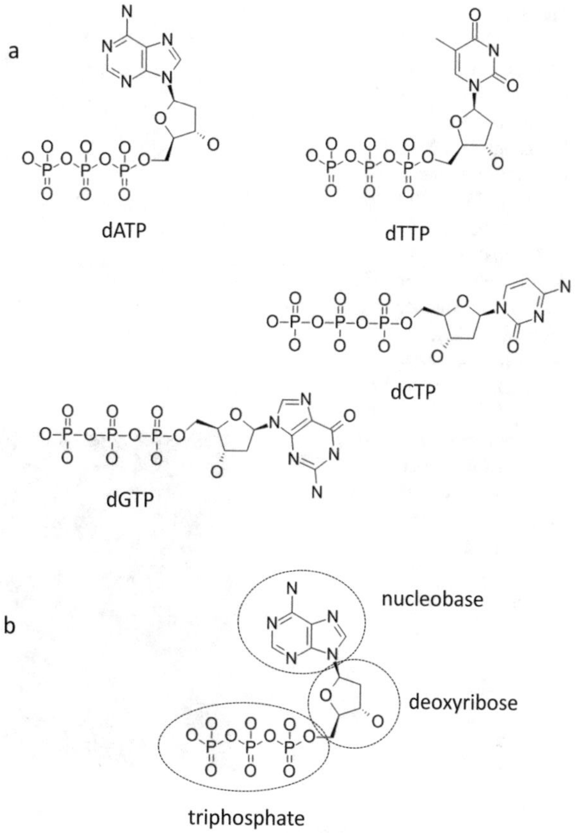

Fig. 3.15 (a) Structures of the nucleotides used to produce DNA with hydrogen atoms omitted. The deoxyribose and phosphate groups are the same for each compound, the difference being the nucleobases whose initials A, C, G, and T are used when writing down a DNA sequence. (b) The nomenclature of the different chemical groups that make up DNA

dATP

dTTP

dCTP

dGTP

nucleobase

deoxyribose

triphosphate

3.1.6.2   Lipids

The lipid family is extensive and diverse and is not as readily categorized as proteins and nucleic acids. Nevertheless, the basic principles of lipid structure are straight-forward; the molecules consist of an extensive water-insoluble portion which dissolves readily in an organic solvent like gasoline but not in water (unless it is a natural detergent-like molecule). Lipid molecules form the membranes that enclose cells, along with specialized proteins, and small lipid molecules are important signaling molecules that allow communication between cells. Lipids are important for drug development for several reasons: cholesterol, for example, is a lipid that has been implicated in coronary heart disease, so drugs that reduce its level in the blood are of great importance. Other lipids have been used for the delivery of drugs into cells or modified to become drugs themselves. Some of these points are covered in later chapters. Representative examples of different lipids are shown in Fig. 3.17. Note that their molecular size is modest compared with proteins and nucleic acids, but they can associate with each other to form much larger "complexes," as, for example, in the cell membrane.

Fig. 3.16 (**a**) Linkage of three nucleotides (dATP, dTTP, and dTTP) to illustrate the way DNA strands are built up. (**b**) Backbone of DNA strand which is common to the whole molecule. (**c**) Three-dimensional structure of part of DNA double helix showing the nucleobases stacked in the same plane. The arrow points to a nucleobase from each strand linking together in space to hold the structure together. This reversible linkage forms the basis of hybridization of nucleic acid strands to each other and is critical for normal gene function as well as biotechnology

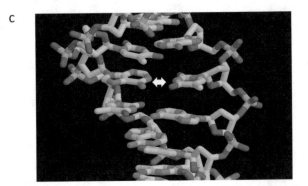

3.1.6.3 Carbohydrates

Like lipids, carbohydrates are varied and complex molecules with a wide range of molecular weights. They are built up from units called saccharides, which are commonly known as sugars. Glucose consists of one saccharide molecule, while sucrose has two, so these compounds are known as monosaccharides and disaccharides, respectively. Large molecule carbohydrates like cellulose are polysaccharides, by analogy with polypeptides and polynucleotides. Carbohydrates not only provide the fuel that feeds the energy producing systems in cells but also combine with proteins to provide structural elements in the body, like cartilage and connective tissue. These protein-carbohydrate associations (glycoproteins) are expressed on the surfaces of cells to create adhesive surfaces that allow cells to physically interact with each other in biological tissues. This is important in terms of drug discovery for several reasons. Firstly, cell interactions via glycoproteins are the hallmark of inflammation and viral infection, so compounds which inhibit the interactions between carbohydrates on different cells (or between a virus and a cell) may be useful drugs for these conditions. Secondly, carbohydrate molecules are sometimes necessary for optimal activity of protein drugs produced using recombinant DNA technology. This is covered later in Chap. 8.

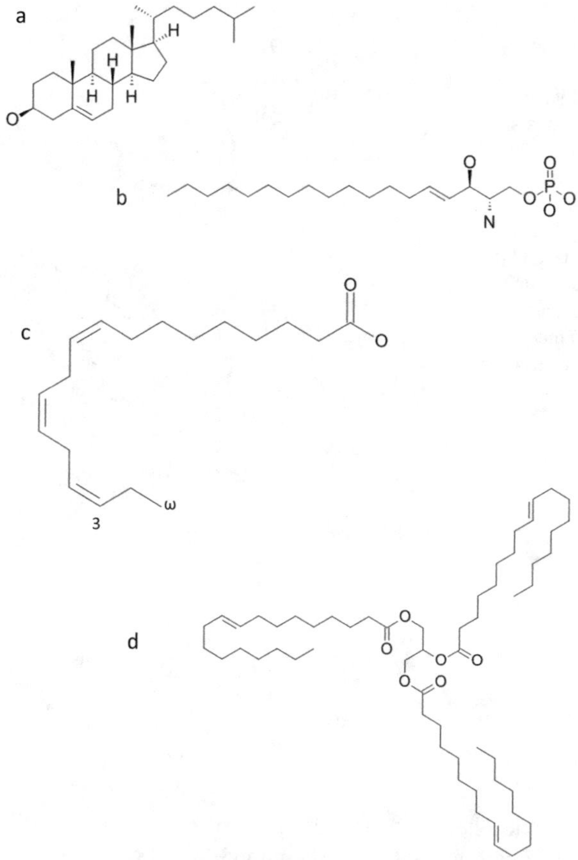

Fig. 3.17 Examples of different lipid molecules with most of the hydrogen atoms omitted. (**a**) Cholesterol; note the predominance of carbon atoms that confers solubility in organic solvents. The complex ring structures are formed from open chains that curve round on each other. (**b**) Sphingosine-1 phosphate; a complex lipid involved in signaling between cells. (**c**) α-Linolenic acid; a dietary lipid member of the omega-3 fatty acid family promoted as a supplement for improved cardiac health. The naming derives from the double (unsaturated) bond 3 carbon atoms away from the omega (ω) carbon atom at the end of the chain. (**d**) A triglyceride; this is a lipid formed between three fatty acid chains (hence "tri") and a smaller glycerol molecule. These large water insoluble structures are commonly known as fats

The nomenclature of carbohydrates is quite involved, particularly the terms used to describe their three-dimensional configuration in space. Glucose, for example, is also called α-D-glucopyranose, the α-D referring to the three-dimensional configuration of the molecule in space, a property known as "chirality" (see Chap. 7). The word pyranose refers to the size of the ring (six-membered) as opposed to the five-membered furanose ring found in other sugars like sucrose and is the reason why sugars are also classified as pentoses and hexoses. Further detail can be found in any

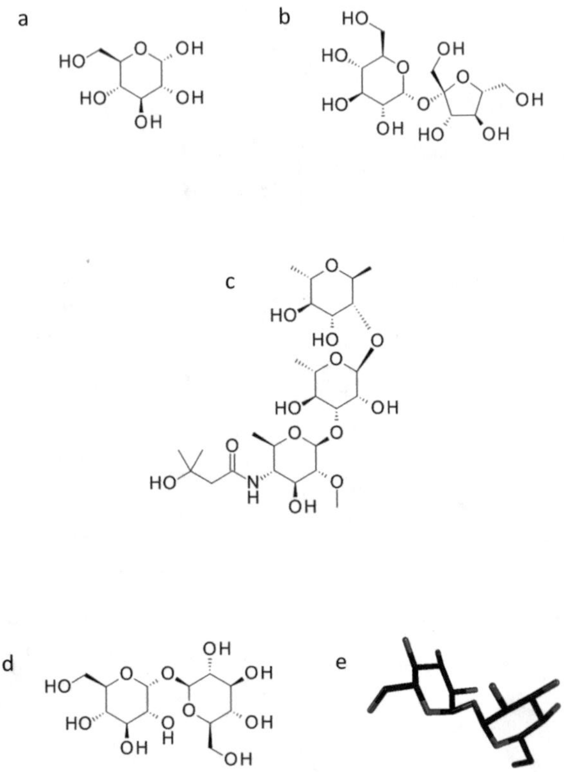

Fig. 3.18 Examples of
carbohydrates (**a**) glucose,
a monosaccharide; (**b**)
sucrose, a disaccharide;
notice the joining of a six
membered and five
membered ring via an
oxygen atom. (**c**) A
complex carbohydrate of
the type found attached to
proteins on the cell surface.
(**d**) The disaccharide
trehalose and (**e**) the
three-dimensional structure
of (**d**)

textbook of organic chemistry or biochemistry or online educational resources.
Some examples of carbohydrates are shown in Fig. 3.18.

3.1.7   Drug-Protein Interactions

3.1.7.1   Visualizing Interactions in the Computer

Computer-generated 3D images of proteins and small molecules have transformed
the way in which drug-protein interactions can be visualized. Several examples
have already been presented in the figures using the ball and stick or stick representations of atoms and bonds; others, including wireframe, spacefill, and ribbon
displays, are useful where the images are complex. An example is the visualization
of the small molecule imatinib bound to its target protein which is activated in certain cancers. The drug works by locking the target protein in a fixed position thereby
preventing its normal function as a signaling molecule (Fig. 3.19).

The imatinib structure fits into a pocket inside the protein formed by the folded
amino acid side chains. The shape and configuration of the compound will ensure

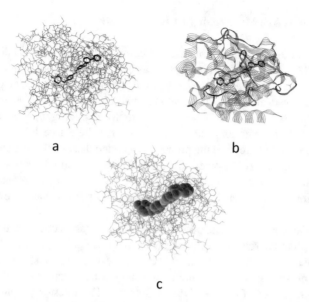

Fig. 3.19 Computer-generated models of the small molecule drug imatinib (Gleevec®) binding to its drug target protein. (**a**) Protein displayed as wireframe with imatinib buried within its surface and visualized as a stick model. (**b**) Protein rendered as a ribbon diagram that shows the way this polypeptide chain forms a helix. (**c**) Drug molecule displayed as a surface model with each atom denoted by a sphere. With no pun intended, these are "spheres of influence" that define the limits of interaction between the drug and the protein side chains

that it is able to neatly access the pocket, but these features are not enough for the drug to influence the protein. It must also bind into the protein pocket with enough strength to hold it in position, but not so strongly that it is impossible to remove it[4]. This is where other features of the molecules come into play.

3.1.7.2   Electricity and Fat

These may seem strange subjects to introduce, but in fact they are related to chemical bonding. This topic is part of the basic chemistry of atoms and molecules but has not been mentioned up to now to avoid unnecessary complication. Unfortunately, it is impossible to understand drug binding without at least a basic idea of how molecules interact with each other. These interactions involve electrical charges and attractions to fatty molecules.

---

[4] This is not true for all drugs; some work by permanently blocking an active site on the target and are called irreversible (as opposed to reversible) inhibitors.

3.1.7.3 Atoms Are More than Just Billiard Balls

Our understanding of atomic structure has come a long way since Dalton's time in the early 1800s. Instead of just being seen as billiard balls, modern quantum theory describes atoms as collections of subatomic particles whose exact position can never be observed directly. The subatomic particles created in the first moments after the universe was born, later merged to form hydrogen, the first element. Modern particle accelerators (atom smashers) like the Large Hadron Collider at CERN are designed to reverse the process to recreate these early particles. While all of this is fascinating from a scientific point of view, the simple notion of atoms as combinations of two charged particles, protons and electrons, is enough for a basic understanding of chemical bonding using the illustrations that follow to convey the point[5].

Protons carry a positive charge and electrons a negative one. Since like charges repel and unlike charges attract (same with magnetic poles), atoms with different charges will attract or repel each other accordingly. Figure 3.20 shows an atom with eight protons shown with + symbols surrounded by electrons (− symbol). The atomic diagrams are of course highly schematic. The interior (the nucleus) is so small relative to the total size of the atom that the physicist Ernest Rutherford described it as the "fly in the cathedral." The electrically neutral atom in this example has eight electrons to balance the charge. The positively charged atom has lost one electron and the negative atom has gained one. Without this free movement of electrons in and out of atoms, we would not have electricity, which of course in our world is unthinkable.

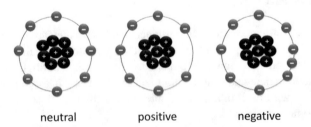

neutral   positive   negative

Fig. 3.20 Simple diagram of atom showing protons in the nucleus (+ sign) orbited by electrons (− sign). A neutral atom has an equal number of protons and electrons; when one electron is missing, the atom is positively charged and if an electron has been gained, it is negatively charged. This charge is one of the fundamental ways in which atoms are formed

---

[5] Any detailed study of this area must cover the quantum mechanical theory of molecular orbitals, which is beyond the scope of this book.

Covalent and Ionic Bonds

The strength of the bond between different atoms in a molecule varies in strength according to the disposition of electrons. Most of the bonds that make up a drug molecule, for example, are strong covalent bonds where the atoms link closely by sharing electrons. These bonds can be so strong that they require high temperatures or reactive chemicals to break them. Ionic bonds are formed when a positively charged atom links with a negatively charged one by electrical attraction. Ionic bonding is so named because charged atoms are called ions. Bonds of this type are found in salts like sodium chloride, where positive sodium ions interact with negative chloride ions. Figure 3.21 shows a diagram of covalent and ionic bonding.

Hydrogen Bonding

Hydrogen bonding occurs between hydrogen and a negatively charged atom such as oxygen or nitrogen. Because less energy is required to break hydrogen bonds compared with covalent or ionic bonds, they are useful for holding molecular structures in specific conformations without locking them too tightly. The two helices of DNA

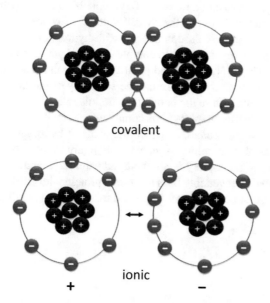

Fig. 3.21 Illustration of two main types of strong electrical bonding between atoms. Covalent bonding involves the sharing of electrons between the two atoms to produce a strong bond. All the compounds and large molecules illustrated in this chapter are formed by covalent bonding between their constituent atoms. Ionic bonding is a strong interaction between a positively charged and a negatively charged atom. These bonds are more easily broken and are used to create the salt forms and hydrates of drugs along with some interactions between drug molecules and their protein targets

are held together by hydrogen bonds, as are protein chains and many other molecules, including water. Most interactions between a drug and its target occur through the formation of hydrogen bonds between amino acids in the protein binding pocket and atoms within the drug itself.

Hydrophobic Interactions

This is where the fat comes in. Many organic molecules like the hydrocarbons found in oil do not dissolve in water. Water is a polar solvent, that is, a substance bearing an electrical charge, but oil has no charge and therefore cannot interact with other molecules through ionic binding. Oily molecules are classed as hydrophobic (literally "water hating"), while salts and many other compounds are hydrophilic ("water loving"). The opposite terminology is often used in drug discovery: oily compounds are lipophilic (fat loving) and possess the property of lipophilicity. Oily molecules prefer to associate with each other through hydrophobic interactions, which are part of a more general group known as van der Waals forces. Named after a Dutch scientist, these forces are weaker than covalent, ionic, or hydrogen bonds and operate at close distances between atoms. Despite this weak binding activity, they are important for holding molecular structures together.

Drug molecules consist of a mixture of hydrophobic and hydrophilic groups, the proportion of which determines the overall amount of charge or polarity of each molecule. The hydrophobic groups interact with hydrophobic amino acids in the binding pocket of the target protein to complement the hydrogen bonding elsewhere in the molecule. The lipophilicity and polarity of compounds are vitally important for drug development as they determine the solubility of the drug in the blood as well as how it is dealt with in the body after administration.

The final figure in this chapter shows computer-generated images of a small molecule bound to a target protein, in which the areas of hydrogen and hydrophobic bonding are marked with colored spheres. This figure represents the fundamental essence of what drug discovery is all about, namely, the search for molecules that strongly and (usually) reversibly bind to a target protein with exquisite selectivity in order to change the function of that target and treat disease (Fig. 3.22).

Fig. 3.22 A small drug molecule (green stick model) interacting with a protein target (wire frame). (**a**) Points in space where hydrogen bonding can occur between the drug and protein marked with yellow, white, and purple spheres. (**b**) The same interaction, but with points of hydrophobic bonding marked with red spheres. As well as illustrating how drugs can bind to their target, this information is useful for computer-aided drug design (Chap. 7). Images courtesy of Dr Philip Dean, Cambridge, UK

a

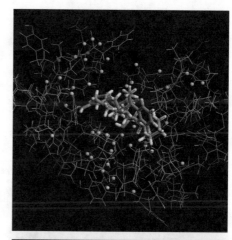

b

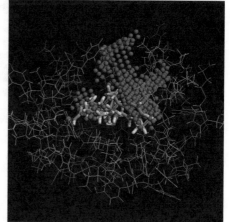

Summary of Key Points

All matter is based on elements that are built from atoms or molecules.

Two or more elements bind together to form compounds.

The chemistry of life (and drugs) is based on the element carbon.

Molecular weights are based on a value of 1 for the lightest element, hydrogen. Small molecule drugs have molecular weights from 200 to 500.

Proteins and other macromolecules are large molecules with molecular weights from thousands to millions.

Organic compounds are built from smaller functional groups with defined three-dimensional shapes.

Drugs are named at four levels:

Organic compounds (IUPAC rules)

Generic names (INN/USAN rules)
Proprietary/trade names
ATC Classification

Hydrates and salts of drug compounds can improve their medicinal properties.

Salts are neutral combinations of acids and bases. Degrees of acidity and basicity are measured using the pH scale.

Proteins are polymers built from around 20 amino acids and fold into defined three-dimensional shapes including those that are that are globular or fibrous.

Proteins have defined functions in life processes and are the targets for most drugs, as well as being drugs themselves.

Lipids and carbohydrates are important classes of organic molecules which are used in different aspects of drug development, including targets, drugs, and delivery agents

Atoms are bound together by chemical bonds with different strengths:

Covalent > Ionic > Hydrogen bonding > Hydrophobic interactions

References

ATC Structure and Principles (2020) https://www.whocc.no/atc/structure_and_principles/. Accessed 18 Jan 2020

Emsley J (2001) Nature's Building Blocks An A-Z Guide to the Elements. Oxford University Press, Oxford

Heterocyclic Chemistry. https://www2.chemistry.msu.edu/faculty/reusch/VirtTxtJml/heterocy.htm. Accessed 18 Jan 2020.

IUPAC (2020) http://www.iupac.org/. Accessed 18 Jan 2020

United States Adopted Names Council (2020) https://www.ama-assn.org/about/united-states-adopted-names/usan-council. Accessed 18 Jan 2020

WHO guidelines for INN (1997) http://apps.who.int/medicinedocs/en/d/Jh1806e/1.html. Accessed 18 Jan 2020

Wiley JP (1995) Smithsonian Magazine. https://www.smithsonianmag.com/science-nature/phenomena-comment-notes-2-106257735/. Accessed 18 Jan 2020

Chapter 4
Laying the Foundations: Drug Discovery from Antiquity to the Twenty-First Century

Abstract This chapter will give a brief overview of the history of drug discovery from antiquity up to the early decades of the twentieth century and then will describe the "golden age" of drug discovery when most of the medicines in current use were developed. This latter era saw the introduction of pharmacology as a distinct discipline of biological science, along with major advances in synthetic chemistry. The basic concepts and terminology of pharmacology will be explained and examples given of how the discipline has been used to develop medicines to treat diseases such as asthma and duodenal ulcers. New biological technologies began to make an impact on mainstream pharmaceutical research in the 1990s, foreseeing the wider use of large molecules as medicines. The transition from traditional pharmacology to twenty-first century drug discovery will be discussed.

4.1 Introduction

"You have to know the past to understand the present." This quote from the astronomer Carl Sagan (1980) was never as true as with the science of drug discovery. The "present," as much of this book will show, is the technology-driven push towards making drug discovery as efficient a process as possible. The "past" is the accumulated knowledge of medicine, biology, and chemistry that has underpinned pharmaceutical research over the years. Science, although rigorously objective in outlook, is subject to trends and fashion because it is conducted by human beings. This means that "what goes around comes around again" just like clothing or music styles. Old ways of looking at chemistry or biology are reviewed in the light of new ideas and technologies. Alternatively, new uses are found for old drugs, to the benefit of both the patient and the manufacturer.

E. D. Zanders, *The Science and Business of Drug Discovery*,
https://doi.org/10.1007/978-3-030-57814-5_4

4.1.1   Medicines from Antiquity

We often view the past from the perspective of the present, and this is also true when considering the use of medicines. It is difficult for those of us with a modern way of thinking to avoid at least a hint of bemused condescension when learning about the bizarre rituals and often disgusting material administered to sick people from ancient times, even up to the present day[1]. However, the inner voice of reason tells us that something good must have come out of literally thousands of years of trial and error with materials derived from plants and animals. The numbers may be low, but the products, like alcohol and the pain killing opiates, continue to have a major impact upon human society.

4.1.1.1   Greek and Arabian Influences

The Ancient Greeks were the first to approach medical practice and drug therapy in a rational way. Aristotle sought reasoned explanations for physical phenomena, although he never troubled himself with experiments to see if the theory stood up to reality. It was left to major figures like Hippocrates (around 400 BC) and Galen (around 150 CE) to turn speculation into experiment by making systematic observations of the effects of medicinal extracts on their patients. Later generations who followed these prominent teachers had an enormous influence on drug discovery right up to the nineteenth century. Unfortunately, the medical theories of the Ancient Greeks were based on false premises. Their idea that diseases arose due to imbalances in the four humors of the body while, at least providing a hypothesis to explore, excluded the notion that disease could be due to external causes. Another obstacle was the idea that a disease is specific to each individual patient, so that the required therapy must be prepared to order. This holistic approach to medicine makes it impossible to evaluate treatments in a truly objective way using controlled trials on a statistically significant number of subjects. There is also the issue of polypharmacy, in which more than one medicine must be administered to have an effect. This means that it is impossible to identify a single agent that can be administered in standardized form with the confidence that it will work in most patients. Although these ideas were superseded by the discoveries of single agents that could treat large numbers of patients, they foreshadowed important issues in modern drug discovery. For example, not all patients with a given disease will respond to the same medicine; understanding and dealing with this fact is now possible through the science of pharmacogenetics, where the genetic makeup of patients influences their responses to medicines (covered in Chap. 14). Another example is the use of combination therapies to treat illnesses. These therapies consist of more than one

---

[1]For example, a medieval asthma treatment: take 20 live woodlice, 6 grains saffron powder, 4 grains tincture of amber, and 2oz pennyroyal water; strain and add 1oz syrup of balsam. Take 3 spoonfuls when the cough is troublesome.

purified medicine administered at the same time in single or multiple tablets. A good example of combination therapies are the drugs used to control AIDS; these consist of as many as three separate compounds which simultaneously attack individual targets in the HIV virus. These are modern developments on an ancient theme, perhaps echoing Carl Sagan's quote at the beginning of the chapter.

The first millennium CE saw the center of gravity of medical and pharmaceutical study move towards the Arab world. Abu Bakr Al-Razi was a major figure in Baghdad who suggested that medicines should be tested on animals before administration to humans. Later figures, such as al-Zahrawi (936–1013 CE), produced comprehensive texts that laid out systematic procedures for extracting, storing, and using medicinal preparations from a large variety of plant and animal sources. This knowledge, along with the Greek approach to understanding disease, was carried from the Arab world to Europe via Salerno in the eleventh century. Due to religious and other factors, pharmaceutical thinking remained static for hundreds of years and was confined to the teachings of Galen and the Arab physicians. Some progress was made, nonetheless. The preparation of medicinal products became the responsibility of specialists called apothecaries (now pharmacists), who devised a series of weights and measures for quantifying small quantities of medicines. The term "apothecary" is derived from *apotheca*, a storage place for wine, spices, and herbs.

4.1.1.2 Asian Medicine

Records of ancient Indian medicine (Ayurveda) are contained in one of the Vedas that constitute the written basis of Hinduism. The basic concepts behind Indian medicine and that of Galen are similar in that disease is caused by an imbalance between "humors" in the body. Therapy is directed towards correcting the imbalance by compensating for a loss of humor (no pun intended) or antagonizing an overactive humor.

Chinese herbal medicine is very much in evidence today, although it has its origins in the first millennium CE. As with medicine from the other civilizations, it relies upon correcting an imbalance of humors, in this case the well-known balance of *yin* and *yang*. Chinese herbal remedies have received a great deal of attention from modern pharmaceutical scientists because there is a spirit of cooperation between them and the herbalists who treat disease in modern Western settings. This does not of course mean that the scientist is not exasperated by the fact that the exact composition of a herbal mixture is changed for each patient or that some of the treatment might be presented in an apparently bizarre way. However, most scientists in biopharmaceutical companies realize that it is foolish to dismiss Chinese herbal medicine out of hand. I have had personal experience with herbal medicines used to treat skin inflammation in children. These were concocted by a practitioner of Chinese medicine in London and used with apparent success by a major children's hospital using Western criteria for measuring clinical outcomes. As I was working on inflammation biology at the time, this seemed promising enough to investigate further with the aim of extracting the active ingredient(s) in purified form. We therefore dispatched a member of the project team to Chinese stores in Soho, London, to collect samples according to the various recipes given by the herbalist. Rather than try to use individual mixtures according to Chinese practice, we just pooled everything and extracted the material with solvents to produce an uncharacterized mixture of chemicals to be tested against various inflammation targets. There appeared to be some anti-inflammatory activity, so each individual herb was extracted and tested in turn[2]. One of these produced the effect we were seeing with the whole mixture, and the active compound was subsequently purified and identified. It turned out to be a known anti-inflammatory compound, although one that had modest activity. We decided to terminate the project based on the conclusion that this compound was not interesting enough to be pursued further. There was still the possibility that more than one herb contained active compounds that administered in combination would give a greater benefit than each one in isolation. This is called an additive effect, when the total activity is the sum of the activity of each individual compound. If the total effect is greater than the sum, it is known as a synergistic effect. Nevertheless, the prospect of testing and purifying many combinations of herb extracts was just too daunting. Even if there were a series of novel compounds that had to be used together, the obstacles to developing them as a combination therapy would have been far too great. Although this outcome was not positive, there is a great deal of research into Chinese medicine by conventional scientists, including, of course, those at Chinese companies and universities. The rapid expansion of the Chinese biopharmaceutical industry will no doubt ensure that this research continues apace.

4.1.2 The Dawn of Rational Drug Discovery

For centuries after the fall of the ancient empires, doctors still relied almost exclusively upon crude medicines isolated from naturally occurring sources. Some genuinely effective treatments were discovered, including cinchona bark brought to Europe from South America in the early seventeenth century. Extracts of the bark were seen to be effective in controlling fever, but it was not until the early nineteenth century that the active principle quinine was isolated and used as a pure compound for malaria. Despite the successes with herbal extracts, the theoretical foundations of drug discovery were still based to a large extent upon the old ideas of correcting imbalances in the body. Slowly, however, these gave way to new theories based on the application of the scientific method. This method is summed up as follows: perform experiments to produce data upon which to form a hypothesis and then repeat the process by performing further experiments to confirm, reject, or modify the hypothesis and so on. The scientist must be as objective as possible by interpreting

---

[2] Details of how these experiments are performed will be given in later chapters as they are central to modern drug screening.

the experiments in the light of observations rather than preconceived prejudices – the experiment does not lie (usually).

One of the first examples of this approach was the first controlled clinical trial undertaken in 1747 by the British naval surgeon James Lind. He wanted to find agents to treat scurvy, a disease that was rife in the British navy. Twelve patients were assigned an identical diet along with selected agents such as sea water, dilute sulfuric acid (!), and citrus fruit. The sailors who had eaten oranges and lemons were rapidly cured, while the others were unchanged. Although this was a dramatic example of the application of a scientific method, it was not until 50 years later that the British Admiralty made the supply of lemons or limes mandatory on its ships.

4.1.2.1 The Move Towards Pure Drugs

A major figure in the history of drug discovery was the Swiss physician Paracelsus (1493–1541). He was the one of the first to advocate the use of pure chemical compounds as medicines rather than the herbal mixtures used in previous centuries. Although his ideas of disease were wildly off the mark, his philosophy of the quintessence of nature led him to administer drugs singly rather than in combination. This was a significant turning point towards modern practice. Since Paracelsus and his followers were operating long before the era of organic and natural product chemistry, they used several *inorganic* compounds, mainly minerals containing specific metals (mercury, e.g., was used to treat syphilis). The use of arsenic, another toxic metal, has a long history going back perhaps 5000 years in Chinese medicine. Paracelsus recognized it as a treatment for cancer, and in fact the compound arsenic trioxide, a component of traditional Chinese medicine, has been approved for the treatment of leukemia. The detailed mechanism of action of this compound has only just been unraveled, but it is interesting to note that historical treatments, like this example, can be put on a firm scientific foundation.

4.1.2.2 Natural Products

As the name suggests, these are compounds that have been isolated from natural sources, i.e., plant, animal, or microbial. Approximately fifty percent of current medicines have their origins as natural products, so they are of intense interest to the biopharmaceutical industry. It is worth stating at the outset that natural products are no different in chemical terms from those made purely synthetically in the laboratory. In the early years of chemistry, it was thought that natural products had a special property (vitalism) just because they were found in living things (shades of this today?). This was disproved in 1828 by Wöhler, who showed that urea excreted by animals could also be synthesized in the laboratory to producing a compound that was identical to the naturally occurring substance.

Natural product chemistry advanced through the development of techniques to extract substances from plants or other material and identifying the chemical

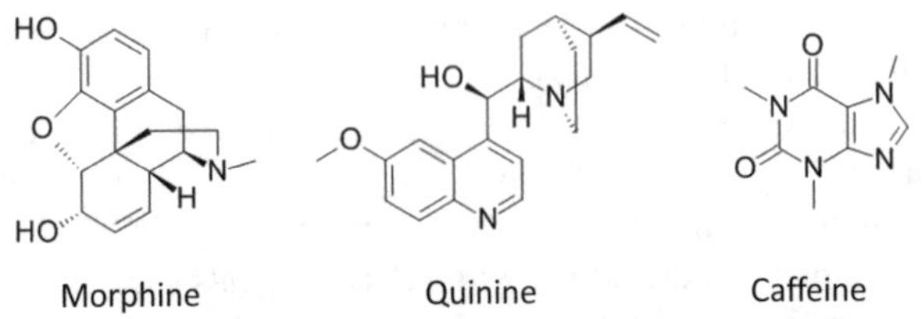

Morphine   Quinine   Caffeine

Fig. 4.1 Structures of common alkaloids. These heterocyclic compounds are natural products from poppies, cinchona bark, and coffee, respectively. They are extracted from plant material using alkaline solutions, hence the name

structures of the biologically active molecules. Solvent extraction was introduced (and is still used today) as a means of dissolving compounds in liquids that could be later evaporated to reveal crystalline substances. The act of making a cup of tea is not dissimilar to the first stage: add plant material to boiling water, infuse for a while to extract the active substances, and then filter out the insoluble material using a tea strainer or the paper of a tea bag. By using a variety of solvents (e.g., acids, alkalis, and alcohols), chemists were able to identify many different plant substances, one of the first of which was morphine. This was extracted from poppies and appeared to be alkaline in nature as opposed to the many acidic substances that had been isolated from plants up to that point. Later, the term alkaloid was applied to natural products of this type, which are of fundamental importance in small molecule drug development. Many of the heterocycles that make up drug molecules (described in Chap. 3) were first identified in alkaloids. Other important chemical features of natural products will be discussed in the chapter on medicinal chemistry (Chap. 7). These discoveries advanced drug discovery enormously, but many natural products are too complex to make in the laboratory, so drug developers must rely upon an abundant source of the biological starting material. Morphine, for example, is a complex compound with a difficult synthesis that was only achieved in 1950. Figure 4.1 shows the structure of morphine and some other common alkaloids.

4.1.3   Nineteenth-Century Synthetic Chemistry

Despite the rapid progress in drug discovery in the nineteenth century brought about by natural product chemistry, the problems of production at scale meant that ways would have to be found to produce purely synthetic drugs. The opportunity came in the form of an ugly waste product of the gas industry, coal tar. This material provided a wealth of organic compounds including aniline, which was used to make synthetic dyes, the first, mauveine, being invented by William Perkin in 1856. I recall seeing an example of nineteenth-century mauve clothing exhibited in the main reception area of the pharmaceutical company I was working for, presumably displayed to remind visitors about the origins of the business in synthetic chemistry.

The discovery of a whole range of synthetic dyes based on aniline and other coal-derived compounds helped to fuel the massive growth of the chemical industry, particularly in Germany. The advances in organic chemistry that accompanied this industrial activity allowed chemists to make a wide variety of compounds, not necessarily dyes, which could be tested for medicinal activity. A classic example is the development of aspirin by F. Bayer and Company in Germany. This has its origins in the extracts of willow bark known to have antipyretic (fever-reducing) properties. The active compounds were called salicylates based on *Salix*, the botanical name for willow. These compounds are based on phenol which was purified from coal tar and later converted to salicylic acid. This acid has antipyretic properties but is corrosive and therefore irritating to the stomach when swallowed. Felix Hoffmann was involved with attempts to reduce the irritant effects by masking the phenol group. His compound, acetylsalicylic acid, was tested in 1897 and found to be well tolerated with antipyretic (fever-reducing) and analgesic (pain-killing) properties. It was marketed as Aspirin® and turned out to be a remarkably successful drug for fever and inflammation, as well for heart disease due to its activity on blood platelets. Despite this success, it is sobering to think that aspirin would not be approved as a drug in our current regulatory environment. The compound blocks the production of small molecule prostaglandins in the digestive system which act to protect against stomach and duodenal ulcers. Long-term administration of high doses of aspirin to patients leads to a high incidence of ulcers, which can lead to fatal internal bleeding. This has led to the development of safer aspirin analogues (such as diclofenac) that have a reduced side effect profile.

4.1.4 Interactions Between Synthetic Compounds and Biological Tissues

The staining of biological materials with dyes has been observed since antiquity and is routinely employed with clinical samples for diagnosis in hospital laboratories. The German scientist Paul Ehrlich recognized the connection between the chemical composition of different dyes and the types of animal tissue to which they bound. He developed the method of "vital staining" in which live animals were injected with dyes and then sacrificed, before examining their different tissues under the microscope. In 1885 he showed that one of these dyes, methylene blue, stained nerve cells while leaving most other tissues unaffected. Two consequences followed from these investigations, one being that the dyes could be used as medicines. The results from the methylene blue staining implied that the compound could act as a pain killer because it was bound to nerve cells. Ehrlich did indeed find that this was the case when tested on patients, although the dye was too toxic on the kidneys (nephrotoxic) to be clinically useful. The other important point was that small molecules could bind selectively to biological structures; the era of identifying drug targets had begun.

4.1.4.1 Introducing Pharmacology

Pharmacology[3] (Greek *pharmakon*, drug) is a scientific discipline concerned with the study of how drugs and other chemicals interact with the body. Although pharmacology has been undertaken in one form or other for many hundreds of years, the first department of that name was established in 1847 in Estonia by Rudolf Buchheim. His former student Oswald Schmiedeberg moved to Strasbourg to set up an Institute of Pharmacology that was to prove highly influential in training generations of drug discovery scientists. Later in the nineteenth century, the first hints of structure activity relationships between drugs and physiological responses were obtained because of advances in organic chemistry. For example, Alexander Crum Brown and Thomas Fraser presented a paper to the Royal Society of Edinburgh in 1868 entitled "On the Connection between Chemical Constitution and Physiological Action; with Special Reference to the Physiological Action of the Salts of the Ammonium Bases Derived from Strychnia, Brucia, Thebata, Codeia, Morphia, and Nicotia." They showed that whatever the normal effect of these alkaloids, the change of a tertiary nitrogen atom to the quaternary form invariably produced a curare-like paralyzing action, thus providing the opportunity for making novel agents.

Theories of drug action had to fit the experimental observations made by pioneering pharmacologists, microbiologists, and chemists. One important aspect of this related to the idea of affinity between a drug and the cells and tissues of the body. The title of Goethe's 1809 novel about understanding human relationships Wahlverwandtschaften (Elective Affinities) was applied to pharmacology by Friedrich Sobernheim in terms of specific elective affinities. Another key aspect of drug action was that disease results from alterations in cellular structure and activity, an idea published by Rudolf Virchow in 1858. This observation, coupled with Ehrlich's data showing that dyes would selectively bind to specific cell types and structures, created the groundwork for a receptor theory of drug action.

The Receptor Theory of Drug Action

Receptors (Latin *receptor*, receiver or harborer) are the central focus of pharmacology, as they constitute the physical entities to which drugs bind in the body. In 1900 Ehrlich and his collaborator Julius Morgenroth introduced the term for the first time: "For the sake of brevity, that combining group of the protoplasmic molecule to which the introduced group is anchored will hereafter be termed receptor."

As noted by Prüll et al. (2009), the receptor theory was not immediately accepted, despite the support of Sir Arthur Conan Doyle, formerly an ophthalmologist, but better known as the creator of Sherlock Holmes; this support was reciprocated as Ehrlich was a great fan of detective stories. The most prominent alternative to a

---

[3] Pharmacology is not the same as pharmacy, which is concerned with the preparation and administration of medicines.

chemical receptor theory was the idea that the physical properties of molecules and target tissues dictated drug action. This viewpoint, held notably by Walther Straub in Germany, was part of a major controversy in pharmacology, amazingly until as late as the 1940s.

Agents that bind to receptors are called ligands (Latin *ligare*, to bind), so ligand-receptor interactions are highly significant. The terms low molecular weight ligand or high molecular weight ligand are sometimes used to refer to small and large molecules (e.g., proteins) that bind to their receptors.

The receptor is the drug target, which in most cases, is a protein. Not all drug targets are classed as receptors, however. Enzymes, for example, are part of a separate target class of proteins that catalyze biochemical reactions; these targets have been used to develop an impressive range of drugs to treat AIDS, high blood pressure, and many other conditions. This chapter will concentrate more on receptors, with other target classes being covered in due course.

Physiology and Cellular Communication

This chapter now moves towards the biology of health and disease to explain how pharmacology works in practice. The whole point of drug discovery of course is to offer the medical and veterinary profession the tools to cure a disease (or at least to alleviate its symptoms). To understand disease, we have to understand physiology, the healthy background from which pathology arises, either through infection or internal changes in the body. Physiology is the science of how the body functions by regulating fundamental processes such as respiration, nervous transmission, excretion, and digestion. In mammals, these work together in homeostasis, where changes in the body are automatically brought back to a normal state, for example, in the response to external heat where the skin is cooled by the evaporation of sweat. These processes are controlled through feedback mechanisms that operate in a similar way to the thermostat on a central heating boiler that maintains a constant temperature in the home. In biology, feedback regulation occurs through the actions of both large and small molecules to regulate cell function, so this area is of great interest to cell biologists as well as physiologists. Scientists in each discipline find it easier to envisage these regulatory pathways as being comprised of familiar objects from engineering such as wires, switches, wheels, and pulleys. They are highly sophisticated self-organizing machines despite the tissue they reside in having the appearance of a wet gooey mass. Even the simplest organisms have an awesome complexity, something I briefly reflect upon as I swat yet another fly in the kitchen.

Disease occurs when these regulatory systems are disrupted in some way and cause an imbalance. This is a modern take on the ancient doctrine of humors, where disease was thought to be caused by imbalances in phlegm, yellow bile, black bile, and blood that were associated with phlegmatic, choleric, melancholic, and sanguine temperaments, respectively. In modern terms, a viral or bacterial infection provokes an immune response in the body which is orchestrated by many cells and chemicals that were previously at low levels in the healthy person. The result is (hopefully) the clearance of the infection, but unfortunately the symptoms of fever and muscle pain are part of the cure. Disease may also arise through defects caused by genetic mutation, for example, in conditions where proteins involved in the clotting of blood are inactivated, thus leading to hemophilia, or uncontrolled bleeding.

The body is organized at different anatomical levels, with organs being the largest, followed by tissues, cells, and then subcellular components. As a generalization, pharmacology has traditionally been concerned with whole organs and tissues, while cell and molecular biology operates at the level of cells and the molecules that comprise them.

4.1.4.2 Chemical Signals in the Body

Communication between different parts of the body occurs through both the transmission of chemical signals in the blood or other body fluids and electrical signals conducted through cells like current through a wire. The molecules that drive these chemical and electrical signals are used by pharmacologists as starting points in the design of small molecules to interact with drug target. The different classes of chemical signal are as follows.

Hormones

This large family of compounds is comprised of small molecules, peptides, and proteins that circulate in the blood to communicate with target organs. They are produced by specialized organs such as the pancreas (insulin), adrenal glands (adrenaline), and the testes (testosterone). These compounds have multiple effects on the body through their interaction with receptors on different cells within the target organ. Adrenaline (epinephrine in US terminology), for example, is a small molecule hormone that is secreted into the bloodstream by small glands situated above the kidneys. It is well-known as the "fight or flight" hormone that prepares the body for dealing with stress. This is achieved through the enhancement of brain, lung, and circulatory system functions that result from the hormone binding to receptors on some of these target organs. Hormones are also implicated in cancers; for example, some prostate and breast cancers grow in response to male and female sex hormones, respectively.

Cytokines

While the study of hormones has a long history, the identification and characterization of cytokines only really started in the 1970s, although the interferons were discovered in the late 1950s. Cytokines are proteins that are produced by many different cell types, in contrast to the specialized endocrine cells that secrete hormones.

They can operate over short ranges to allow cooperation between cells that cluster together, or even activate the same cells that produced them. The immune system operates through cytokine communication; the feeling of malaise that accompanies a cold is caused by cytokines binding to receptors in the brain to switch on fever mechanisms or to act on muscles to induce fatigue. Extreme cases of immune system activation through cytokines occur in septic shock caused by bacterial infection. The resulting "cytokine storm" leads to a series of extreme reactions including a lowering of blood pressure that can prove fatal.

Cytokines are not restricted to the immune system but are present throughout the body to stimulate the growth of many different cell types in a controlled manner[4]. However, cancer can arise if the cell growth becomes independent of cytokines because of mutations, since the cell is no longer controlled but just keeps on dividing. Anti-cancer drugs that target the mechanisms of cell growth normally controlled by cytokines have been developed by a significant number of biopharmaceutical companies; for this, and other reasons, there is a great deal of interest in the biology of these important molecules and their potential use in drug development.

Neurotransmitters

In contrast to hormones and cytokines, neurotransmitters operate through a physically wired network of nerve cells that can produce extended structures up to a meter long. Electrical impulses are transmitted down nerve cells or neurons linked together in a chain. The linkage is not complete, however, since there is a gap of about a 20 millionth of a millimeter between them (called a synapse after the Greek for "holding together"). Some synapses transmit the nerve impulse via an electrical signal, but others employ small molecules called neurotransmitters. Acetylcholine was the first such molecule to be discovered and was found to transmit signals between nerve cells and muscle. Once acetylcholine has performed its task in transmitting a signal, it is rapidly broken down by the enzyme acetylcholinesterase to terminate the signal. Some nerve gases inhibit this enzyme leading to enhanced nerve activity in the muscle with resulting spasms and death. More subtle inhibition can however be used to compensate for the loss of neurotransmission in diseases such as Alzheimer's where nerve cells have been destroyed. There are other neurotransmitters based on amino acids (glutamate) or hormones (adrenaline/epinephrine, noradrenaline/norepinephrine). Electrical communication in the body occurs through the movement of charged elements such as potassium and calcium through ion channels in many cell types, including nerves. Ion channels will be covered later in Chap. 6. Research into both neurotransmitters and ion channels is important in understanding central nervous system (CNS) diseases as well as the mechanisms of sensing pain, both of which are

---

[4]Cell growth in this context means cell division to produce daughter cells.

Fig. 4.2 Modes of communication between organs, tissues, and cells in the human body. (**a**) Hormones produced by specialized endocrine organs such as the adrenal glands situated above the kidneys. Adrenaline/epinephrine released into the blood and acting on the heart and (indirectly) on the brain to increase responsiveness to external threats. (**b**) Communication between two white blood cells by means of protein molecules called cytokines. (**c**) Neurotransmitters passing across a synapse between two nerve cells to propagate a signal

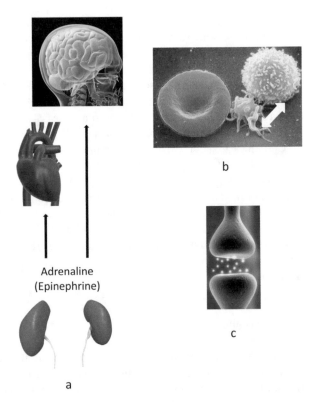

Adrenaline
(Epinephrine)

b

c

a

extremely active areas of drug discovery. The above categories of signaling molecule are summarized in Fig. 4.2.

4.1.4.3 Experimental or Classical Pharmacology

This is a venerable discipline going back to the early years of the twentieth century, hence the slightly pejorative term "steam pharmacology" given to it by biologists caught up with modern cell and molecular biology. However, classical pharmacologists can rightly point to the many important drugs which have been discovered using the techniques of experimental pharmacology. These techniques involve the following actions:

1. Identify the human or animal tissues that provide a model for the disease of interest.
2. Identify a physical response of the organ or tissue and devise a way of measuring it in a reproducible manner.

3. Identify hormones or neurotransmitters involved in the response and using natural products or synthetic compounds to either enhance or suppress the response[5].
4. Select compounds with the highest activity and selectivity for the natural molecules; then test them in whole animals (and ultimately humans) to assess their potential as drugs.

Example: Small Molecule Drugs for Treating Asthma

This classical pharmacology approach is now illustrated using the above procedures to discover drugs to treat the respiratory disease, asthma.

Disease Model, Selection of Tissue, and Test System

Asthma is characterized by an inflammation of the airways in the lungs brought about by allergic reactions to pollen and other environmental stimuli. The main symptom is the shortage of breath caused by the narrowing of the bronchial tubes, thereby reducing the supply of air to the lungs. A drug that counteracts the narrowing will therefore relieve the symptoms (although it would not provide a cure). This focuses attention onto the bronchial tubes since these tissues could be used to test potential drugs. The tubes are made up of smooth muscle tissue that contracts and relaxes in the similar way to the skeletal muscle that moves the limbs. This contraction (and relaxation) can be measured in the laboratory using a tissue bath and recording equipment. (Fig. 4.3) Once this is set up, it is then possible to introduce experimental compounds into the apparatus and look for those that relax the bronchial tissue (see Jespersen et al. 2015 for details).

Identify Molecules that Relax the Bronchial Tissue (Bronchodilators)

It was known for many years that adrenaline (epinephrine) caused bronchial smooth muscle to relax. This hormone is used as a drug to counteract the anaphylactic shock brought about by severe allergic reactions. Unfortunately, it is unsuitable for use as a bronchodilator because of its undesirable side effects (e.g., raising blood pressure and causing insomnia) consistent with its role as a "fight or flight" hormone. Adrenaline also has a short half-life[6] because it is rapidly broken down in the body, so symptomatic relief of asthma would be only short-lived. This is where pharmacologists use analogues or derivatives of the natural hormone to see if they have improved properties. A range of such analogues was produced during the first half of the twentieth century, the most relevant one here being isoprenaline (isoproterenol), first synthesized in the late 1930s and used as a bronchodilator for many years.

---

[5] Note that cytokines did not have much impact upon classical pharmacology because little was known about them and the fact that they are proteins and not small molecules.

[6] The half-life is the time taken for 50% of the compound to disappear.

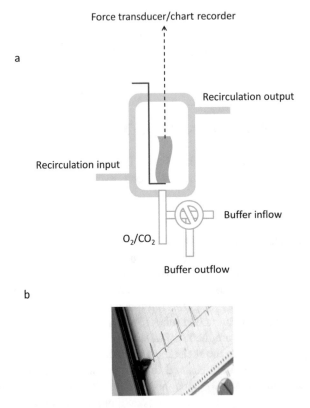

Fig. 4.3 (**a**) Typical pharmacology experiment using a tissue strip suspended in an organ bath. The apparatus consists of a sealed outer chamber filled with circulating water to keep the tissue at body temperature (37 °C). The inner chamber is filled with nutrient solution through the three-way tap to keep the tissue strip in the same state as in the live animal/person. An oxygen/carbon dioxide mixture is also fed into the chamber to maintain viability. Drugs or test compounds are introduced directly into the chamber and any contraction/relaxation of the muscle tissue measured using a force transducer that sends a signal to a chart recorder. (**b**) A chart recorder showing peaks drawn by a pen over a regular time period. The height of the peak and the frequency with which they occur provide numbers that can be used to further analyze the effects of the drug in a quantitative way

Isoprenaline was an improvement on adrenaline, as it did not cause as much elevation of blood pressure through constriction of the blood vessels, but it was rapidly broken down in the body. The problem of the short half-life was solved by altering the parts of the molecule that caused it to be broken down by the body's metabolism (Chap. 11). In addition, compounds were identified (using the tissue bath experiments and other techniques) that had limited effects on heart tissue; in other words, they exhibited greater selectivity for bronchial smooth muscle over heart tissue. One of the most successful of these compounds was salbutamol, discovered by David Jack and others in the 1960s, at Allen & Hanburys a division of Glaxo (now

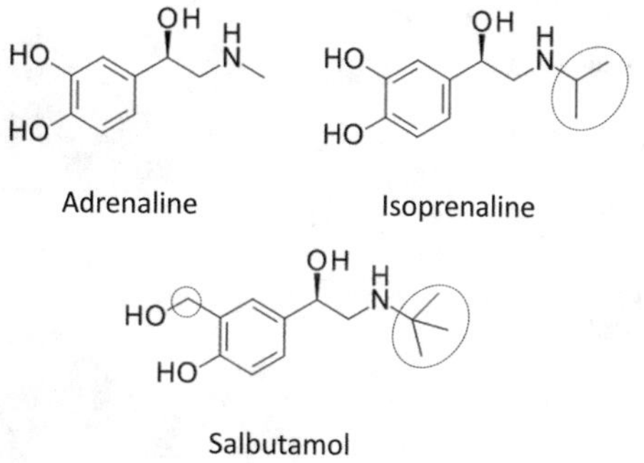

Fig. 4.4 Modification of a natural hormone adrenaline to produce drugs that relax bronchial smooth muscle in the airways. Note the small modification (addition of isopropyl group to adrenaline) in isoprenaline to prevent undesirable entry to the brain. The ring with two OH groups is a catechol group that is rapidly broken down by the body. A simple modification of one of the OHs in salbutamol slows down the breakdown and makes the drug effective for several hours, while the substitution of a t-butyl group (circled right-hand side) results in greater selectivity for lung tissue

GlaxoSmithKline). This was marketed as an inhaled bronchodilator with the trade name Ventolin® and gives symptomatic relief for four hours (Fig. 4.4).

Until recently, drug discovery scientists might taste the compounds they produced, testing for sweetness or other sensations. This certainly helped in making discoveries but had an obvious downside. David Jack was a discovery scientist in this heroic mold, but one night had to phone his doctor because of a massive increase in his heart rate brought on by tasting some adrenaline analogues; luckily he survived to become a research director, but they certainly do not make them like that anymore.

The above example introduced an important facet of experimental pharmacology, namely, the ability to find compounds that bind to receptors and exert an effect on biological tissues in a highly selective manner. By implication, there must be separate receptors for adrenaline on different tissues, in other words, receptor subtypes. This is a fundamental aspect of drug discovery that will be examined later in this chapter.

4.1.4.4 Agonists and Antagonists

The two main categories of receptor binding drugs are agonists and antagonists.

When a natural ligand (e.g., hormone or neurotransmitter) binds to its receptor, the latter is activated to switch on whatever function is associated with it. In the case of adrenaline, this is the relaxation of smooth muscle. The agonist is a molecule that binds to a receptor in the same way as the hormone and triggers the response, but it may have a quite different chemical composition. This is illustrated in the figure where the normal house key fits into the lock, but the agonist is in the form of a hairpin that is used to pick the lock. An antagonist is a molecule that blocks the

Fig. 4.5 Agonist and antagonist drugs illustrated by analogy to a key in a lock. The natural hormone, cytokine, or neurotransmitter is the key; the agonist is a hairpin with the correct shape to allow the lock to be picked yet without the overall appearance of the key; the antagonist is a twig jammed in the lock to prevent the key from getting inside it

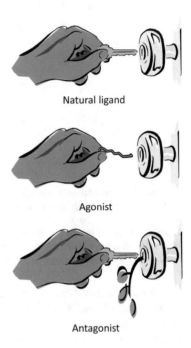

Natural ligand

Agonist

Antagonist

binding of the natural ligand to the receptor, thereby preventing its activation. This is illustrated in the figure by the attempt to put the key into the lock having previously jammed it with a twig by mistake (perhaps while under the influence of one of the oldest drugs?). The names agonist and antagonist are derived from the Greek *agonistes* "contestant" and *antagonistes* "opponent," respectively. This is quite appropriate since the agonist competes for its place on the receptor with the natural ligand and will displace it because it binds more tightly. The antagonist must also bind more tightly than the natural ligand, but in doing so will oppose its action. There are further pharmacological terms such as partial agonist or inverse agonist that will be explained in any basic pharmacology resource and to avoid further complication will not be discussed here any further (Fig. 4.5).

4.1.4.5　Measuring Drug Efficacy and Selectivity

Pharmacology is not just a qualitative (descriptive) discipline; it is also quantitative, because the action of drugs on their targets can be described in mathematical terms. The simple weights and measures used in drug discovery are introduced in this section along with the concepts of affinity and efficacy.

Table 4.1 Units of
measurement used in drug
discovery. Because drug
molecules are highly potent,
the quantities used for
experiments are small
fractions of a gram

Name	Abbreviation	Numerical value
Milligram	mg	1 thousandth gram
Microgram	µg	1 millionth gram
Nanogram	ng	1 billionth gram
Picogram	pg	1 trillionth gram
Femtogram	fg	1 thousand trillionth gram

Weights, Concentrations, and Molarities

Consider the fact that many small molecule drugs are administered in doses as small as 10 milligrams to a person weighing 70 kg. This represents one seven-millionth of the weight of an average sized person, so the drug must clearly be very potent. Drugs are often used in extremely low concentrations in laboratory experiments, and this is reflected in the terminology for weights and measures. Table 4.1 shows some examples of weights defined as fractions of a gram[7].

The table above is applicable to volumes as well, so the word "gram" is substituted with "liter" (note "litre" used in UK English).

There are two ways of expressing the concentrations of a drug: the weight per unit of volume and the molarity. In the first case, a solution of drug made up in water (solid compound dissolved in water) might have a concentration of 10 milligrams per milliliter. Laboratory scientists would normally write this as 10 mg/ml and pronounce it "migs per mill" (µg/ml is pronounced "micrograms per mill," not "mikes per mill"). Sometimes the concentration is expressed as mgL^{-1} which is the same as milligrams per liter. The different usage of symbols for weights and measures can be confusing, particularly for technical translators who must work with formal regulatory documents. Guidelines do exist for publication in academic journals as well as for regulatory submissions, examples of which can be found in ICH M5 EWG Units and Measurements Controlled Vocabulary (2005).

The other way of expressing concentrations is through the term molarity. This is the concentration of a compound expressed as a function of its molecular weight. An experiment might be designed to compare the activity of two drugs, for example, a small molecule and an antibody with molecular weights of 500 and 150,000 daltons, respectively. The antibody has 300 times more "activity" than the small molecule on a weight for weight basis, so it is impossible to make a comparison based on weight alone[8]. The solution to this problem is to use a system of measurement that combines the molecular weight and the weight of a compound. The terms mol

---

[7]As an aside, the world reference kilogram weight made of platinum-iridium alloy is stored near Paris; it is no longer its original weight because it is losing too many atoms, so it has been redefined to relate to the energy of photons which can be measured in a precise way.

[8]The terms weight and mass are often used interchangeably. These are only the same when considering the force of gravity on Earth; the weight of a compound with a given mass will be greater on our planet compared with, for example, the moon.

Fig. 4.6 Mol, molarity and molar. In this example, the molecular weight of ethanol is 46, so 1 mol of ethanol weighs 46 grams. A 1 molar (1 M) solution of ethanol in water contains 46 grams of ethanol dissolved in one liter of water

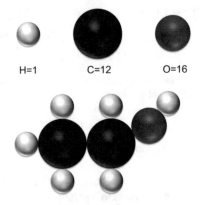

H=1   C=12   O=16

Ethanol: MW=46 daltons

1 mol ethanol = 46 grams
1 molar (M) ethanol = 46 grams/liter

and molar (M) are introduced in Fig. 4.6 using the molecular weight of ethanol as also described in Chap. 3.

Going back to the comparison between the small molecule and antibody, 1 mol of each would weight 500 grams and 150,000 grams (150Kg), respectively. In practice, the amounts used in laboratory experiments are reduced by factors of a thousand or more. Many experimental compounds, for example, are dissolved at millimolar concentrations and still have activity when diluted 1000 times to micromolar (μM) or even 1,000,000 times for nanomolar (nM). Drug discovery scientists like compounds that are active in the low micromolar or nanomolar range, and some drugs, like steroids, are still active at picomolar (pM) concentrations.

Ligand Binding to Receptors

To make meaningful comparisons between natural ligands and synthetic drugs it is necessary to express the strength of their binding to receptors as a number. This number, called the affinity, is calculated using an equation that requires knowledge of the concentrations of free ligand and receptor as well as the concentration of bound ligand. Drug binding of course is not an end in itself; it must lead to a biological response, such as muscle contraction. The degree to which a drug promotes such a response is called the efficacy. It is possible for a drug to bind to only a small fraction of receptors present in a tissue and yet to display a high degree of efficacy. These concepts arise from experiments in classical pharmacology in which compounds are used to provoke a response in isolated tissues or organs that can be measured in the lab. Pharmacology experiments can be set up in different ways, for example, to see if the tissue response increases in proportion to the amount of drug added; alternatively, the time taken to reach a maximum drug response might be measured. These measurements help to determine the rate of association of a drug to its receptor. Ligands do not remain bound to their receptors, but rapidly become detached from them (dissociate) and exist in a balance (equilibrium) between the bound and unbound states. These measurements of rates of association and dissociation are known as the binding kinetics of a ligand to its receptor. The various measurements described here provide information about the number of receptors present on each tissue and their biological properties. Very importantly, the presence of more than one type of receptor for the same ligand can be inferred from these experiments, something that is relevant to the adrenaline agonist (salbutamol) described previously and the histamine antagonists to follow later.

Drug Potency: The EC_{50} and IC_{50}

The EC_{50} of a drug literally means "effective concentration 50%." In a typical pharmacology experiment to test the activity of a drug, there will be a point at which adding more and more of the drug will make no difference to the activity; in other words, the latter will have reached its maximum. The EC_{50} is the amount of added compound (measured in weight or molarity) that gives 50% of the maximum activity.

If the drug is inhibiting the binding of a ligand to its receptor (i.e., is an antagonist), the term IC_{50} is used, meaning inhibitory concentration 50%. The actual way

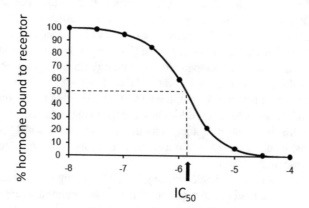

Fig. 4.7 Estimating the IC_{50} for a drug binding to a receptor. Example of a hormone binding to its receptor on a tissue; the amount bound is expressed as a percentage, so 100% is the maximum amount that can be bound. An antagonist drug is added to push the hormone off the tissue (displace it). As the amount of antagonist added is increased, the level of hormone binding drops off until it eventually it is all removed (at the 0% point on the graph). The amount of added competitor drug is expressed in molar units using the shorthand $-8, -7, -6$, etc. which is another way of expressing $10^{-8}, 10^{-7}, 10^{-6}$. To clarify further, 10^{-6} M is one millionth or one micromolar. The IC_{50} is read off the graph as indicated. In this case, the value is roughly 2×10^{-6} M or 2 micromolar

in which the binding is measured will be covered later in the chapter, but in the meantime the basic concept is illustrated in Fig. 4.7.

The EC_{50} and IC_{50} values are referred to in various ways; for example, "compound Z inhibited the binding of adrenaline to its receptor with an IC_{50} of 20 nM." Alternatively, the following phrase might be used when comparing the action of two drugs: "Compound X had an EC_{50} of 30 μM on bronchial smooth muscle relaxation compared with compound Y which had an EC_{50} of 30 nM." Note that the EC_{50} values in the example can be used to evaluate the comparative efficacy of compound X over compound Y and its selectivity for the target receptor on the bronchial tissue. In this example, compound X must be used at a 1000 times higher concentration to give the same effect as compound Y (30 μM is 1000 times greater than 30 nM). In drug discovery terminology, compound Y has 1000-fold selectivity over compound X.

4.1.4.6   Receptor Subtypes and the First Blockbuster Drugs

The earlier section on the development of adrenaline agonists to treat the symptoms of asthma highlighted the fact that adrenaline itself acts on different organs of the body including the lungs and blood vessels. However, agonist drugs like salbutamol are mostly active on the lung. This means that the same ligand (adrenaline) must bind to different subtypes of the receptor and generate responses according to where each subtype is expressed in the body as first demonstrated by the American pharmacologist Raymond Ahlquist. Briefly, he determined the rank order of potency of a series of compounds (including adrenaline) on the excitation or inhibition of various tissues and in the process discovered two classes of adrenergic receptors which he named α- and β-receptors. This work was aided in part by having access to using sophisticated instruments developed from technology developed during the recent World War. Ahlquist's seminal work was published in 1948, but he considered the idea of receptors as a theoretical tool; the later subdivisions of adrenoreceptors into α1, α2, β1, and β2 subtypes caused him anxiety (β3 came much later). He believed that "if there are too many receptors, something is obviously wrong" (Prüll et al. 2009). What he would have made of our current inventory of receptors and subtypes is best left to the imagination.

It remained for Sir James Black, working for the UK company Imperial Chemical Industries (now subsumed into AstraZeneca) to turn the theory of receptor subtypes into new drugs. His work on agents to treat angina pectoris in the late 1950s led to the first "beta-blocker" drugs, pronethalol and propranolol, thus pioneering the exploitation of receptor subtypes that is now routine practice in biopharmaceutical companies.

By the 1960s the receptor theory of drug action was accepted by pharmacologists, but still not understood at the molecular level. D.K. de Jongh's comments written in 1964 sum the situation up in a rather literary manner of its time: "To most of the modern pharmacologists the receptor is like a beautiful but remote lady. He has written her many a letter and quite often she has answered the letters. From these answers the pharmacologist has built himself an image of this fair lady. He cannot, however, truly claim ever to have seen her, although one day he may do so." That day came soon enough after the application of cell and molecular biology to pharmacological problems (see later).

Development of Antihistamine Drugs

The discovery and development of the histamine antagonists is an example of how pharmacology became the most successful drug discovery technology in the years leading up to the modern era of biotechnology. Sir James Black, then at Smith, Kline & French in the UK (now part of GlaxoSmithKline) turned his attention to anti-ulcer drugs based on antagonists of histamine. This small molecule, derived from the amino acid histidine, contributes to the symptoms of allergic reactions such as skin reddening (erythema), running eyes, and nasal congestion. These unwanted symptoms are treated with histamine-binding antagonists, the so-called antihistamines, which are commonly prescribed in the hay fever season. In addition to its role in allergic diseases, histamine promotes the secretion of gastric acid in the stomach. This concentrated solution of hydrochloric acid contributes to the breaking down of protein chains during the initial stages of digestion. Despite the corrosive potential of the acid, the stomach is normally protected from damage; however, this protection sometimes breaks down, leading to peptic ulcer disease in which the stomach lining and duodenum are perforated, with potentially life-threatening consequences. It has been known since the late 1940s that certain histamine analogues promote acid secretion, and by the early 1970s, it was clear that there were two histamine receptors, named H1 and H2 (for the receptors linked to allergy and acid secretion, respectively)[9]. This meant that if compounds could be developed to selectively block H2, they would reduce the amount of acid secreted into the stomach and thereby give enough time for the peptic ulcers to heal. James Black and his colleagues took up the challenge, the first task being to establish a suitable biological test system. There is a general terminology for these tests depending upon the level of tissue being examined. The assay is conducted in vivo if a whole animal is being used to test an experimental compound. If a piece of tissue has been taken out of the animal and is still viable, then it is an ex vivo assay. If the tissue and cells are ground up into fragments and assayed in a test tube (or modern equivalent), this is an in vitro assay: *in vitro veritas*, as the old biochemistry joke goes.

Black and his co-worker set up a low-throughput in vivo assay that measured the secretion of acid into the stomachs of anesthetized rats. The chemists then synthesized over 200 analogues of histamine to try and find one that would inhibit acid production in the assay. It was 4 years after the start of this work in 1964 that one of their compounds, guanylhistamine, was shown to have some modest antagonist

---

[9] In later years, the number of receptors was increased to four by the discovery of the H3 and H4 receptors, neither of which relate to acid secretion.

activity; however, it also stimulated some acid release on its own, a phenomenon known as partial agonism. To eliminate this undesirable effect (sometimes called breeding out the effect), the medicinal chemists identified the part of the molecule responsible for inducing acid secretion and replaced it with a different chemical group (a thiourea). In 1970, the new compound, burimamide, was found also to block acid secretion, albeit with low activity. Again, the compound structure was reviewed, and a new analogue, metiamide, with a tenfold higher activity than burimamide, was produced and tested on 700 patients. A few of these developed a blood disorder, which fortunately was reversible, but which meant that the drug could not be developed further. The thiourea group was shown to be responsible for the toxicity, so metiamide was modified to produce a new compound cimetidine. This was launched as Tagamet® in 1972, creating revenues of millions of dollars for Smith, Kline & French and becoming the first blockbuster drug in the process. It had taken 8 years to produce a compound (cimetidine) with the required activity and features (like oral bioavailability and minimal toxicity) that would make it a good medicine. Despite this, it took a further 4 years before cimetidine reached the marketplace. The work of James Black on histamine antagonists, along with his earlier work on beta blockers, helped to earn him a Nobel Prize in 1988.

The phrase "nothing succeeds like success," to reiterate the point made earlier, is very applicable to drug discovery; the commercial success of Tagamet® ensured that other companies would attempt to develop their own H2 blockers and occupy some of the by now lucrative market for anti-ulcerant drugs. The Glaxo scientists Roy Brittain and David Jack (of salbutamol fame) were aware of burimamide and its lack of oral bioavailability. They therefore produced their own analogues with improved activity until they had a compound with the same thiourea group as the one found in metiamide. Glaxo scientists then replaced the offending group with a non-toxic substitute to produce ranitidine. This was more potent than cimetidine and did not have the side effects of that drug (see also Chap.11). Ranitidine was developed and marketed as Zantac®, becoming the best-selling drug in the world for several years. For Glaxo employees at that time, it really was a golden age of drug discovery (Fig. 4.8).

Several important points arise from the histamine antagonist program:

1. Some understanding of disease is necessary for identifying a molecular target that can be affected by a small molecule drug. In the previous example, peptic ulcer disease was known to result from an excess secretion of acid in the stomach driven by the action of histamine (although other mechanisms operate as well). Classical pharmacology experiments led to the identification of histamine H2 receptors as suitable targets for antagonist drugs.
2. A test system must be put in place to measure the activity of experimental compounds. In the above example, a simple in vivo test was used. The (rate) limiting step in this program was not the assay itself, but the several years taken to synthesize compounds, test them, and then modify active molecules in the laboratory. Although the basic principles of synthesis and testing are the same today, the test (assay) would now be performed on purified cell fractions or even the

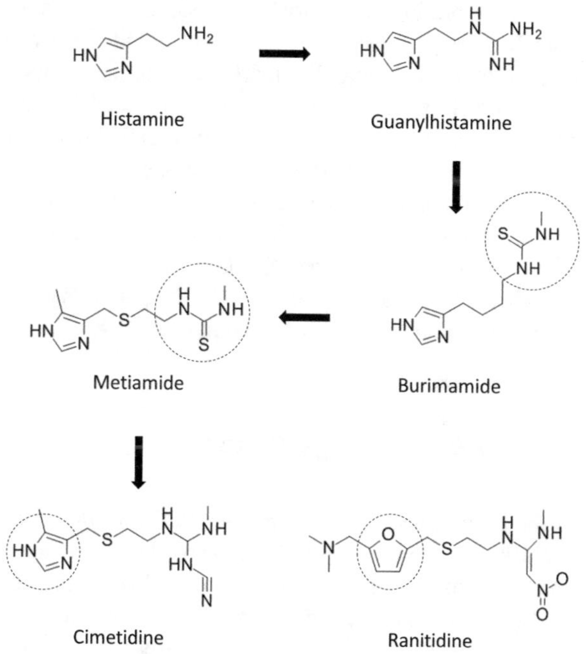

Fig. 4.8 Creation of H2 receptor antagonist drugs based on naturally occurring histamine molecule. Chemical modifications of histamine leading to cimetidine marketed as Tagamet®. The thiourea group that caused toxicity is circled in burimamide and metiamide. A rival H2 antagonist ranitidine (marketed as Zantac®) is shown for comparison with cimetidine. In this molecule the original histamine structure has all but disappeared, with the basic ring structure completely replaced (technically an imidazole group by a furan, circled)

Histamine

Guanylhistamine

Metiamide

Burimamide

Cimetidine

Ranitidine

receptors themselves. Furthermore, thousands of compounds would be submitted for testing rather than just a few hundred.

3. A thorough knowledge of organic chemistry is required to modify drug molecules and to predict the biological consequences of these modifications. The histamine work was a tour de force of medicinal chemistry where functional groups were chosen to impart good solubility in water, reduced brain penetration, and greater binding to the histamine receptor. The scientists also learnt by hard clinical experience that the thiourea group was toxic and had to be avoided in future programs. There are now chemical databases of so-called toxic groups that are routinely searched when designing new molecules.

4. It is possible for competitors to produce their own compounds that act on the same target to treat the same disease. Some areas of drug discovery are so competitive that there may be as many as six different compounds under development for the same target. This is only commercially viable if each company has secure patent protection on their inventions. In the case of histamine H2 antagonists, cimetidine and ranitidine were protected by composition of matter patents because the two compounds are sufficiently unrelated in chemical terms. Patents of course have a limited lifetime, so Zantac® (which was marketed later than Tagamet®) has lost patent protection and is now sold over the counter.

The phrase "me-too drugs" has entered the vocabulary of drug discovery although it often bears negative connotations. "Me-too" implies that each company is marketing a drug to gain profit even though it may not offer any advantage to a patient over an existing product that works by the same mechanism. It is certainly true that many drug discovery projects have been (and still are) modeled on those being undertaken by rival companies and that marketing campaigns will attempt to exploit even the slightest difference between competing products. In the case of the H2 antagonists, however, ranitidine was shown to have eight times more activity than cimetidine and display fewer side effects (Konturek et al. 1980). The physician has the choice of several medicines, even if these work on the same target. In real-world clinical practice these medicines will have different effects on different people (a fact to which I can testify personally). The reasons are not always clear, but the responses (or side effects) may be influenced by the patient's genetic background. This area of pharmacogenetics will be covered in Chap. 14.

4.1.5 Pharmacology Revisited

In the period leading up to the late 1960s, drug discovery was undertaken almost exclusively with live animals, whole organs, or isolated tissues. The cells that make up these structures all express receptors that are targets for the thousands of different ligands found in the human body. Many of these will have no relevance to drug discovery at all, either now or in the future, but natural scientific curiosity, if nothing else, demands that their biochemical properties be fully understood. This can only be achieved if receptors are characterized as real molecules rather than as figures on a graph.

The first receptor molecule to be purified was the nicotinic acetylcholine receptor isolated by Robertis and colleagues in the late 1960s from membranes isolated from detergent-treated tissues (reviewed by Halliwell 2007). The nicotinic receptor is now known to be one of some 300 ion channels, many of which are of major pharmaceutical interest (see Chap. 6). However, a significant number of current medicines (including those based on adrenaline or histamine) react through a different system, the G-protein-coupled receptors (GPCRs). The prototypic GPCR is the visual transducer rhodopsin, first sequenced in the 1980s (Costanzi et al. 2009). The protein sequence of bovine rhodopsin revealed a serpentine structure which traversed the cell membrane seven times. Work on rhodopsin signaling revealed the action of an enzyme GTPase, thus the term G-protein coupled, or 7TM (transmembrane) receptor entered the pharmaceutical lexicon.

4.1.5.1 Radioactive Ligands

It may seem surprising that up until the 1960s, no one had really succeeded in doing the obvious experiment of directly measuring the binding of ligands to their receptors. The problem lies with having to measure the physically minute amounts of ligand that binds to equally minute amounts of receptor. The numbers involved are so small that there is no question of being able to see the material and weigh it on a balance. The weight of the adrenaline receptor (adrenoceptor) in one gram of lung tissue is approximately 1 nanogram (one thousand millionth of a gram); therefore in order to obtain enough pure receptor to see by eye (as a dried deposit on a tube), at least 5 kilograms of lung tissue would be required. The situation with the ligand is even worse, as it weighs roughly 100 times less than the receptor. This scenario is almost universal in drug discovery science; the main players, be they chemical or biological, operate at levels that are far below the limits of human vision. Nevertheless, they can be detected and measured with great accuracy using technology developed for physics, chemistry, and biology.

One of the most common ways of measuring binding is to physically attach a label to the ligand to make it easily detectable, even at extremely low concentrations. Radioactive atoms fulfill this role perfectly, as the radiation they emit can be measured accurately using suitable instruments. The ligand is then used as a radioactive probe for receptors. Mention the word "radioactivity" to most people and they will think of atomic weapons or nuclear reactors producing radioisotopes such as uranium and plutonium. These emit high levels of harmful radiation that must be contained in highly specialized laboratories under conditions of tight security. Medical isotopes by contrast, although still dangerous if mishandled, are routinely used in hospitals, universities, and biopharmaceutical companies. Some isotopes, like ^{60}Co (60 is the total number of protons and neutrons in these atoms: 27 plus 33, respectively), are used to irradiate cancer cells to drive them into a cell suicide program. It is also used to irradiate white blood cells for immunology research; I recall standing outside the huge concrete doors of the irradiation room to control the system that opened the ^{60}Co source. A brief sound of moving machinery and that was it; the radiation is silent and undetectable, except with a Geiger counter. Radioisotopes like ^{60}Co emit powerful gamma rays and are not suitable for measuring ligand binding to receptors. Historically a range of isotopes were discovered that only emitted weak radiation, so they could even be used in human subjects. The Nobel Prize winning Hungarian scientist George de Hevesy was a pioneer in this field of biological tracers. He started early, because while living in a boarding house as a young man, he placed a small amount of radioactivity in the remains of his dinner to prove that the landlady was serving the leftovers in the following evening's meal.

Since their introduction to laboratories in the 1940s, the following radioactive elements have been used to label a vast range of compounds for both basic and applied research. These are ^3H (tritium), ^{14}C (carbon 14), ^{35}S (sulfur 35), ^{32}P (phosphorus 32), and ^{125}I (iodine 125). All these elements are present in organic molecules in a stable non-radioactive form, so one or more of the natural atoms can be substituted with radioactive ones produced in small reactors. Specialized radiochemical companies have been set up to supply the research community with customized or "off the shelf" radiochemicals, including several ligands for pharmacological receptors. It is also possible to produce radiolabeled proteins and nucleic acids in the laboratory by purchasing prelabeled amino acids and nucleotides which are then incorporated into the larger molecules.

The isotopes mentioned above emit different types of radiation, namely, beta (^3H, ^{14}C, ^{35}S, and ^{32}P) and gamma (^{32}P and ^{125}I). The weaker beta radiation can be measured using a scintillation counter which detects and measures the tiny flashes of light produced when radiation hits a specially formulated liquid, known as scintillation fluid. Samples of radioactive material are added to this strong-smelling organic liquid in glass or plastic vials which are then capped and placed in racks, before being read in a machine. Although this technique has made a fundamental contribution to biomedical research, the filling and capping hundreds of vials, plus the hours of waiting before seeing all the results, must be one of the most tedious procedures undertaken in any laboratory.

4.1.5.2 Binding Assays

Once a ligand has been produced in radioactive form, it is possible to use it in binding assays that measure how much of it binds to its receptor. The word assay originates from the French *essayer*, to try, and originally referred to the determination of the purity of metals. The word is frequently used in biological and pharmaceutical research in the form of the binding assays mentioned already and others such as screening assays, inhibition assays, and competition assays, all of which are described in this book at some point.

Early ligand-receptor binding assays involved overlaying thin sections of whole tissues with radioactive ligand to allow receptor binding. The unbound ligand was then washed away and the tissue section exposed to X-ray film of the type routinely used in hospitals. Because the radiation source was tritium and therefore very weak, the film had to be exposed in the dark for many days or even weeks, to generate a visible darkening where the ligand had bound in the tissue. This autoradiography procedure is still used to locate ligand binding in tissue sections, particularly for visualizing the distribution of experimental drugs in the tissues of whole animals. Unfortunately, the procedure is time-consuming and low throughput; in other words, only a small number of samples can be analyzed at any one time. Autoradiography may help to locate the site of binding, but it does not give a figure for how much ligand is bound. This is where "grind and bind" comes in. In this procedure, whole tissues are ground up in a blender (like the type used in the kitchen) to produce an extract containing fragments derived from damaged cells. Included in these fragments are the cell membranes that bear receptor proteins, so it is normal practice to purify these membranes prior to mixing them together with radioactive ligands to allow binding. This incubation process is carried out in a filtration device that permits the removal of unbound ligand in a simple washing step. Ligand that is specifically bound to receptor on the cell membrane is radioactive, so the amount bound can be determined by simply measuring the radioactivity of the membrane preparation in a scintillation counter. The results are expressed as either disintegrations per minute (dpm) or counts per minute (cpm). This powerful technique is used in competition assays to compare the binding strengths of different drug candidates to a given receptor. They are called competition assays because unlabeled compounds are used to displace the radioactive ligand by competing for sites on the receptor. If these compounds bind to the receptor with a higher affinity than that of the

a

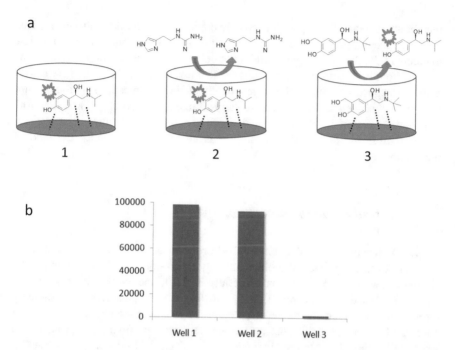

b

Fig. 4.9 Schematic diagram of membrane binding competition assay. (**a**) Receptor protein bound to plastic filter at bottom of microtiter plate well (gray oval). Well 1: radiolabeled ligand molecule is added and allowed to bind (dashed lines). Well 2: effect of adding compound that does not bind to receptor – label remains bound to receptor. Well 3: Unlabeled compound with greater affinity for receptor than ligand displaces radioactive ligand. (**b**) The filter is washed and counted in a scintillation counter to determine the amount of bound ligand and gives a quantitative measure of drug binding

radioactive ligand, the amount of bound radioactivity will be reduced in proportion to the amount of unlabeled compound added to the assay (Fig. 4.9).

This is the information used to construct the IC_{50} graph shown in Fig. 4.7, and it is also used in other calculations to work out the affinity of a drug for its receptor.

These experiments to test the binding of compounds to membrane preparations are now undertaken as a matter of routine. However, problems with non-specific binding can arise where minute amounts of compound bind to equally minute amounts of receptor in the presence of much larger amounts of cell membrane and other biological material. Small ligands can stick to this extra material to produce a high background signal on tissues that do not express the receptor. This problem of non-specific binding is almost universal in biomedical research and gives rise to its own terminology, including the phrase "promiscuous binders," which should be self-explanatory. One way round this problem is to purify receptor proteins away from other cellular material or to produce them in recombinant form by using genetic engineering (see later chapters). Unfortunately, membrane proteins are difficult to produce in a "native form," i.e., in a form that is folded correctly to allow it to bind a ligand (in contrast to the ninety percent of cellular proteins that are not membrane-associated). In practice, genetic engineering is used to introduce (trans-fect) the receptor protein into a cell where it is normally absent; it is then possible to compare the degree of ligand binding to cells bearing the receptor and the same cells without it. Pharmacology has, in a sense, moved from a classical phase (tissue baths) to a "post-modernist" phase (genetically engineered receptors) phase by join-ing forces with cell and molecular biology to understand receptor (and therefore drug target) action in unprecedented detail. To understand how this transition occurred from the late 1980s onwards, it is necessary to introduce the field of biotechnology.

4.1.6   Introduction to Biotechnology

While pharmacology was delivering novel drugs based on receptor-binding small molecules, the early seeds of a biotechnology revolution were being sown in labs around the world. The revolution had really begun in the 1940s with research on antibiotics, but this would move on to transform the biopharmaceutical industry through the development of genetic engineering. Biotechnology is not new, of course; it has been used for thousands of years to create useful products (like baked bread, or alcohol) from living organisms such as yeast[10]. In the twentieth century, the principles of biochemistry and microbiology were applied to the industrial-scale production of vitamins and other important chemicals from living organisms. These products were not only small molecules but also bacterial proteins like the enzymes used in "biological" washing powders. Since microbiology is such an important part of biotechnology, the next section gives an overview of this important scientific discipline.

4.1.6.1   Microbiology

This is the study of the single-celled organisms that include bacteria, fungi, and protozoa, but excludes, for example, mammals and higher plants. Viruses are absent from this list as they really come under the discipline of virology. This is because viruses are unable to survive independently but must infect cells to grow and repro-duce. It is this property that makes viruses such useful tools for modern biotechno-logical processes such as genetic engineering and gene therapy (Chaps. 2 and 8).

Microbiology is important for the biopharmaceutical industry for many reasons: firstly, microbes (or microorganisms) are responsible for infectious diseases, so they are the targets for drugs (sometimes called anti-infectives). More specifically, bac-teria are targeted by antibiotics (or antibacterials), viruses by antivirals, fungi by antifungals, and protozoa by antiprotozoals. Secondly, microbes use the process of

---

[10] The importance of yeast in the early study of biochemistry is illustrated by the fact that the word enzyme literally means "in yeast" since proteins extracted from this fungus had the ability to fer-ment sugar into alcohol in the absence of cells.

fermentation to produce small molecule drugs or nutritional supplements that could not otherwise be synthesized in the laboratory. These molecules may be produced completely from scratch or else from an existing precursor that is fed to the microbe and converted to the final product in a process called a biotransformation. Finally, microbiology is at the heart of genetic engineering and the production of recombinant protein drugs.

Some Background and Terminology

Microorganisms (excluding viruses) are classified as belonging one of the following three evolutionary domains: archaea, prokaryotes (bacteria), and eukaryotes (fungi, protozoa). Archaea and prokaryotes have DNA but no cell nucleus, a structure present in eukaryotes that contains DNA packaged into chromosomes. Plants and animals are eukaryotes but are made up of many cells (multicellular) opposed to yeast, for example, which is single-celled (unicellular).

The systematic names of microbes are written as two italicized words, for example *Escherichia coli*, for the common bacterium, and *Saccharomyces cerevisiae*, for baker's yeast (it is not hard to guess the use to which *Saccharomyces carlsbergensis* is put). Sometimes the first word is shortened to the first letter, so the bacterium becomes *E. coli*.

Microbes/microorganisms are grown in culture or are cultured. A culture medium is a watery (aqueous) solution containing sugars, salts, amino acids, and other nutrients that support microbial growth. Very often it is convenient to grow microbes on a solid medium in culture dishes known as agar plates (the round plastic plates called Petri dishes are often used for this). Agar is a gelatinous material derived from seaweed that behaves like gelatin in cookery. It is dissolved in boiling water, along with the nutrients normally present in liquid cultures, and then allowed to set by cooling. Samples of microbes are literally spread onto the surface of the agar with a glass rod and the plates incubated in a temperature-controlled box for a number of hours, after which colonies of bacteria or fungi appear dotted over the agar. These are picked off the plate using a toothpick and used to inoculate a liquid culture that generates large numbers of organisms within a few hours. Bacteria like *E.coli* will double their numbers every 20 min and yeast every 45 min. This means that the numbers can reach astronomical proportions after a relatively small number of generations (doublings). I have always been impressed by the (untestable) story about placing a grain of rice onto the first square of a chessboard then adding double the number for each subsequent square. By the time the 64th square is reached, the amount of rice used would supposedly be enough to cover the whole of India to a depth of 9 feet!

4.1.6.2 Antibiotics

Life Hinders Life

So wrote the great French scientist Louis Pasteur in 1877, based on his observations that one type of microbe could inhibit the growth of another when cultured together in the same vessel. There was a surprising amount of insight into this process of "antibiosis" well before Alexander Fleming's discovery of penicillin in 1928 and tantalizing accounts of the antibacterial activity of *Penicillium* molds were written as early as the nineteenth century. However, it was not until Fleming directly observed the killing of Staphylococci bacteria on an agar plate by *Penicillium notatum* fungus, that antibiotic research really began to take off. This plate was later preserved in formaldehyde and is now in the British Museum in London. The agent produced by the *Penicillium* fungus then had to be isolated as a pure compound and tested against bacteria in animals and eventually patients. Howard Florey and Ernst Chain took on this problem in Oxford, where they determined the structure of penicillin and showed that it had antibacterial activity in human subjects. Soon the pharmaceutical industry in the USA showed an interest, and a consortium of companies was established to produce large amounts of the antibiotic in deep fermentation tanks. This level of cooperation and intensity of effort was possible because of the military need for antibiotics in the closing years of World War II.

The story of the discovery and development of penicillin is far more detailed than outlined in this summary and in some ways, given the series of coincidences that led Fleming to his original observation, are barely credible. The technical challenges of producing penicillin in large quantities required new experimental techniques. For example, mutant strains of the *Penicillium* fungus were generated by irradiation with X-rays and selected for their enhanced synthetic capacity. The outstanding success of penicillin as a medicinal product, and the clear recognition that microbes were a source of anti-infective compounds, meant that biotechnology rapidly became a mainstream pharmaceutical activity in the post-war years. Despite this, pharmaceutical companies remained traditional providers of small molecule drugs, whether they came from chemistry laboratories or from fermentation tanks. The research organization of these companies included a biotechnology department whose remit was to produce anti-infectives or other natural products from microbes. However, it was not until the first independent biotechnology companies came onto the scene that the modern biopharmaceutical industry really began to take shape.

4.1.6.3 Loosening Up

The social revolution of the 1960s and 1970s had an impact upon the management styles and the formalities of dress and behavior that existed in large pharmaceutical companies. At that time, industry scientists were very conservative by comparison with their counterparts in academia. This all changed with the emergence of the stand-alone biotechnology company that first arose in the State of California[11]. This state produced more than the Beach Boys, as it was (and is) home to some of the greatest research universities in the world. It was here in 1972 that Herbert Boyer (UCSF) and Stanley Cohen (Stanford) developed the recombinant DNA technology that allowed, for the first time, the recombining in the laboratory of DNA from one organism with that of another (actually it happens naturally all the time with viruses). The DNA to be recombined might code for a particular protein in a human cell, for example, so if it is transferred to a recipient organism such as a fast-growing bacterium like *E. coli*, it is then possible to produce far larger amounts of (recombinant) protein than would be possible with normal human tissue. Protein drugs like insulin can therefore be made in quantity by bacterial fermentation rather than through extraction of the natural protein from the pancreatic tissue of pigs. Furthermore, it is far preferable to treat diabetes with human, and not pig (porcine), insulin, because this reduces the risk of provoking an immune response to the protein by the patient.

The details of recombinant DNA technology are covered in a later chapter, but ironically the tools needed for this new biotechnology revolution were provided by traditional microbiology. These tools include the antibiotics that are added to agar plates to ensure that only bacteria containing recombinant DNA can grow; the vast majority of bacteria added to the plate are killed off because they lack a resistance gene that is transferred along with the human DNA.

The world's first venture capital-funded biotechnology company was founded in 1976 by Herbert Boyer and Robert Swanson[12]. The company, Genentech Inc., still thrives, although it is now a subsidiary of Roche and is based at 1 DNA Way, South San Francisco. This company, and similar organizations, created a new model for a drug discovery company, one that did not carry the organizational and historical baggage of a large pharmaceutical enterprise. Equally, however, it did not (at least in the early days) have the latter's financial resources, so its first drug, recombinant human insulin, was marketed through Eli Lilly and Company in 1982. Many biotechnology companies have been formed (and dissolved) since the mid-1970s; some of these have been "spun out" from university departments, so in some cases, the academic ethos has been preserved in an industrial setting. The emergence of the biotechnology company as a separate part of the drug discovery industry means that large pharmaceutical companies are now exposed to competition from "leaner and meaner" research organizations; on the other side of the fence, the universities have been exposed to the realities of drug discovery, sometimes the hard way. Modern biotechnology is now fully integrated into mainstream drug discovery within the large pharmaceutical companies; this has been achieved by internal investment in the technology as well as through the acquisition of small biotech companies.

---

[11] I always enjoy the New Yorker cartoon of a road sign in California that reads "you are now leaving California, please resume normal behavior."

[12] The first stand-alone biotech company was Cetus, founded in 1971.

Table 4.2 Examples of drugs developed during the twentieth century for a variety of medical conditions. The list is far from exhaustive, but does give an idea of the drug development projects considered worthwhile by the biopharmaceutical industry at that time

Condition	Medicine
Bacterial infection	Antibiotics
Inflammation	Corticosteroids, NSAIDs
Hypertension	Beta-blockers, ACE inhibitors, diuretics
Blocked arteries	Statins
Thrombosis	Anticoagulants, e.g., warfarin
Cancer	Cisplatin, fluorouracil, vinblastine
Allergies	Antihistamines
Asthma	Beta-agonists
Ulcers	H2 blockers, proton pump inhibitors
Diabetes	Insulin, metformin
Psychosis	Chlorpromazine
Depression	Fluoxetine
Anxiety	Diazepams
Transplant rejection	Immunosuppressants – cyclosporine, FK506
HIV infection	Reverse transcriptase and protease inhibitors
Pregnancy	Oral contraceptives
Anesthesia	Halothane, benzocaine
Parkinson's	L-DOPA

4.1.7   The Present

Research and development in the biopharmaceutical industry have come a long way since the early days of drug discovery. This is exemplified by the list of medicines produced during what could reasonably be called a "golden age" of drug discovery in the twentieth century (Table 4.2). The targets for these drugs vary from receptors for small molecules to enzymes and (in the case of cancer) DNA itself.

The list above just includes small molecules, but of course biological drugs based on proteins, nucleic acids, and cells have become mainstream and the focus of intense clinical and commercial interest. Biotechnology in the form of genetic engineering is no longer restricted to producing human proteins as drugs since the ability to manipulate genes in patients, as well as cells derived from them, offers new opportunities in the form of gene and cell therapy. All these topics will be covered in the rest of this book along with the key processes used to take a candidate drug from laboratory finding to marketed medicine.

4.1.7.1 Challenges Ahead

The historical timeline of drug discovery covered in this chapter starts with the uncharacterized medicines of antiquity and ends with the synthetic compounds and biotechnology drugs of the present day. Despite the undoubted successes achieved by the modern drug discovery industry, the returns on a significant investment in research and development are not as great as they should be; the number of novel medicines entering the clinic each year is low compared with the past (see Chap. 17). Biopharmaceutical companies are trying to solve this problem by adapting older scientific disciplines to make them compatible with present-day thinking, for example, the transformation of experimental pharmacology into systems biology. Pharmacology has a proven track record in drug discovery, as has been described several times in this chapter. It does, however, suffer from the limitation of using organs or intact tissues to measure drug responses; unfortunately, organs and tissues are like a "black box," an engineering concept where it is possible to measure inputs and outputs (ligand binding and muscle contraction for example) without having any idea of what happens inside the box (tissue). Systems biology (a combination of cell and molecular biology with mathematics and engineering) is an attempt to shine some light into the black box by studying the individual cells and molecules in the tissue and by determining in a holistic way how they interact with each other. One objective of systems biology is to provide a detailed description of how a given drug target operates within a complex biological system; this information may then be used to guide the selection of a drug which has very high selectivity for that target. Alternatively, this information may be used to produce a drug that affects more than one target simultaneously. This multiple activity may enhance the drug's activity against a disease or else produce unwanted side effects. In fact, both situations occur in many of the drugs on the market; the trick is to be able to predict the outcome at an early stage before the drug enters the clinic.

4.1.7.2 Chance and Design

This historical overview concludes with a brief commentary about the role of serendipity in drug discovery. It can be a dispiriting for a drug discovery scientist to reflect that many ground-breaking medicines were discovered by accident rather than through a purely logical process. In his informative book on the chemistry of drug molecules (Sneader 2005), Walter Sneader describes several examples of accidental discoveries (his book also provides in-depth coverage of the history of drug discovery). One of the most famous accidental discoveries was that of penicillin, which has already been mentioned here. Less famous is the story of the platinum-based anticancer compound cisplatin; this life-saving drug was discovered under what can only be described as slightly bizarre circumstances. Barnett Rosenberg and colleagues at Michigan State University were examining the effects of an electrical field on the growth of E. coli bacteria. The experiment consisted of two platinum electrodes inserted into a nutrient solution containing the bacteria. They found to their surprise that when a voltage was applied to the electrodes, the bacteria did not divide by normal cell division but grew to over 300 times their normal length (Rosenberg et al. 1965). It appeared, at first sight, that the electrical current was responsible for the effect; however, after further investigation it became clear that the platinum in the electrode had reacted with the salt solution to produce compounds that arrested cell division. The active platinum compound was eventually identified as cisplatin and shown to be a potent inhibitor of cell division in both bacteria and animal cells. Since uncontrolled cell division is a hallmark of cancer, cisplatin was tested on tumors and found to arrest their growth (Rosenberg et al. 1969). The platinum compound and later derivatives are now used routinely to treat solid tumors and are partly responsible for the extremely high cure rate of testicular cancer.

This story adds further support to Pasteur's dictum that "chance favors only the prepared mind." There is no doubt that more drug discoveries will arise from unexpected directions in the future; this supports the case for preserving the "curiosity driven research" undertaken by individuals who are not necessarily interested in applications, but who simply want to know how the world works.

Summary of Key Points
The use of naturally occurring medicines in plants and animals goes back to antiquity.

The modern era of pharmaceutical research began in the nineteenth century with the development of organic chemistry and the receptor theory of drug action.

Pharmacology is the study of drugs and their effects on the body. Experimental pharmacology uses organs and tissues to identify small molecule drugs that act on a variety of receptor subtypes.

Biotechnology grew out of research into antibiotics and other important small molecules but became synonymous with recombinant DNA technology from the 1970s onwards.

The mid-twentieth century onwards saw a "golden age" of drug discovery that used pharmacology and molecular approaches to develop many of the drugs that are still in use today.

Many drugs have been discovered by accident.

References

Costanzi S et al (2009) Rhodopsin and the others: a historical perspective on structural studies of G protein-coupled receptors. Curr Pharm Des 15:3994–4002

Halliwell RF (2007) A short history of the rise of the molecular pharmacology of ionotropic drug receptors. Trends Pharmacol Sci 28:214–219

ICH M5 EWG Units and Measurements Controlled Vocabulary (2005) https://www.ema.europa.eu/en/ich-m5-ewg-units-measurements-controlled-vocabulary. Accessed 27 Jan 2020

Jespersen B et al (2015) Measurement of smooth muscle function in the isolated tissue bath-applications to pharmacology research. J Vis Exp. https://doi.org/10.3791/52324

Konturek SJ et al (1980) Comparison of ranitidine and cimetidine in the inhibition of histamine, sham-feeding, and meal induced gastric secretion in duodenal ulcer patients. Gut 21:81–186

Prüll C-R, Maehle A-H, Halliwell RF (2009) A short history of the drug receptor concept. Palgrave Macmillan, Basingstoke

Rosenberg B, Van Camp L, Krigas T (1965) Inhibition of cell division in Escherichia coli by electrolysis products from a platinum electrode. Nature 205:698–699

Rosenberg B et al (1969) Platinum compounds: A new class of potent antitumour agents. Nature 222:385–386

Sagan CE (1980) Episode 2: one voice in the cosmic fugue [TV series episode]. In: Malone A (Producer) Cosmos: a personal voyage. Public Broadcasting Service, Arlington

Sneader S (2005) Drug discovery: a history. Wiley, Chichester

Part II
The Drug Development Pipeline: Discovery to Testing in Humans

Chapter 5
Drug Discovery Pipeline Overview

Abstract The highly complex process of discovering and developing new medicines requires the application of many different skill sets over a time frame of years. It therefore helps to visualize the whole process as if it were a pipeline, with an input at one end and an exit at the other. This short chapter provides a brief overview of the drug development pipeline, enabling the reader to get a "helicopter view" before obtaining more detailed information in subsequent chapters.

5.1   Introduction

Figure 5.1 shows a diagram of the drug discovery pipeline, starting at the discovery phase on the left and finishing at phase IV, i.e., when the drug is on the market.

5.1.1   Discovery

- Selection of the disease to be treated
- Identification of a drug target
- Creation of small molecules and/or biologicals in the laboratory
- Screening against the target to identify leads for further development
- (Leads are compounds/antibodies, etc. that are selective for the target and have high levels of potency.)
- Testing leads in disease models (test tube experiments in vitro or animals in vivo)

5.1.1.1   Outcome

The selection of a small molecule or large biological molecules for detailed assessment as a potential drug candidate

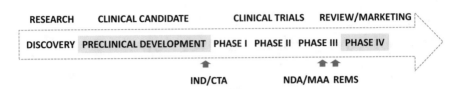

Fig. 5.1 The drug discovery pipeline. The preclinical phase covers the identification of suitable drug targets and the compounds/biologicals that interact with them. Once they have been checked for safety and efficacy, they are tested on human volunteers (phase I) and then on patients (phases II–III and phase IV (post-marketing)). The IND/CTA applications are made to regulators to authorize clinical trials in humans. The NDA/MAA applications and REMS are required for marketing authorization. Further details are laid out in the main text and subsequent chapters

5.1.2   Preclinical Development

• Pharmacokinetics (DMPK or ADME) and pharmacodynamics

Drug metabolism pharmacokinetics or adsorption distribution metabolism and excretion. Studies in animals to determine how rapidly the drug is absorbed, where it goes in the body, how it is broken down by metabolism, and how rapidly it is cleared. Pharmacodynamics studies the effect of the drug on the body.

• Safety pharmacology and toxicology

Determination of maximum safe levels of drug in animals and any side effects that may prevent further development (e.g., causing birth defects or cancer).

• Process development

Experiments to determine a cost-effective and environmentally sensitive way of manufacturing the drug.

• Formulation development

Determining the best way of formulating drugs for dosing in tablets, capsules, etc. This is designed to maximize bioavailability and ease of manufacture.

• Application to conduct trials in humans

If the drug candidate is judged to be potentially safe and effective in humans, an application is made to the regulators to authorize testing in clinical trials. The main transnational regulators are the American Food and Drug Administration (FDA); the European Medicines Agency (EMA) plus the Japanese Ministry of Health, Labour and Welfare (MHLW); and the Chinese National Medical Products Administration (NMPA). The application to the FDA is an IND (Investigational New Drug) and to the EMA, a CTA (Clinical Trial Authorization).

5.1.2.1   Outcome

The decision (and permission) to progress a drug to clinical trials in humans, using a dose which is determined by the preclinical studies in animals.

5.1.3   Clinical Development

- Phase I clinical trials

   The drug is administered to human volunteers to assess the maximum dose that can be tolerated without compromising safety. This information will be used to select doses to test later whether the drug works against the disease. Sometimes a phase I trial will be conducted with patients (particularly with cancer) to assess toxicity at an early stage and speed up development.

- Phase II clinical trials

   The stage at which the drug is tested for efficacy on patients with the disease that the drug is designed to treat. This information is used to plan a series of phase III trials of longer duration, with more patients.

- Phase III clinical trials

   These trials can run for several years and are used to identify any safety issues that may arise over this period. At least two pivotal phase three trials are required to support an application to market the drug.

5.1.3.1   Outcome

A drug that has been shown to work according to the clinical trial design and that has the potential to make a return on investment for the manufacturer.

5.1.4   Product Approval and Launch

- Marketing authorization

   A vast amount of documentation is put together during phase III trials to form a New Drug Application (FDA) or Marketing Authorization Application (EMA). The NDA or MAA includes all the relevant preclinical and clinical data that have been gathered over the years since the drug candidate was first identified.

- Phase IV and post-marketing surveillance

Once the drug is on the market, companies may wish (or be obliged) to conduct further trials to compare its efficacy or safety profile with other medications or to test it in different patient groups (e.g., children or pregnant women). Even though the drug may have been approved for sale, the manufacturer is required to submit a strategy for monitoring and dealing with side effects; this is a Risk Evaluation and Mitigation Strategy (REMS (FDA)) or Risk Management Plan (RMP (EMA)).

5.1.4.1   Outcome

Ideally a medicine that improves the lives of many patients with no safety problems arising in the long term. A commercially viable product that generates sufficient revenues before the patent on the medicine expires allowing generic copies to be sold by other companies

Chapter 6
Target Discovery

Abstract This chapter describes how drug targets are identified in the modern era of drug discovery. It starts with a general description of human disease processes and then leads through a series of investigations that progressively work towards the identification of a drug target protein. In doing so, it covers the key biological disciplines required to make this happen, such as cell biology, biochemistry, molecular biology, genomics, and immunology. Finally, different classes of drug target are described, including receptors, enzymes, and ion channels.

6.1 Introduction

The impressive list of medicines developed in the twentieth century and beyond is testimony to the skill and dedication of the many thousands of drug discovery scientists who work in industry and academia. Despite these successes, the problem remains that many drugs are only effective in relieving the symptoms of disease, and few offer a complete cure. It is true that antibiotics and vaccination have proved highly effective in treating or preventing infectious diseases, but there are very few effective drugs for viral and fungal infections, let alone parasitic diseases like malaria. Furthermore, there is still the likelihood of new infectious diseases appearing from nowhere to take the world by surprise, just as happened in the 1980s with the AIDS epidemic.[1] Those working in the biopharmaceuticals industry are fully aware of these issues and are trying to address them by increasing the efficiency of drug discovery at all levels. Sometimes the term "low hanging fruit" is used to describe the medicines that have already been discovered using traditional pharmacology and medicinal chemistry; the problem now is to reach the "fruit" higher up the tree that may cure serious diseases like cancer and diabetes. This chapter presents a summary of how drug target discovery is currently undertaken in the biopharmaceutical industry and introduces some of the technical specialties, like genomics, which are transforming the search for new medicines. However, in the mathematical sense, human biology is complex in that many variables or interacting factors,

---

[1] At the time of writing (2020), a new coronavirus originating in China has caused many thousands of deaths plus considerable social and economic disruption.

E. D. Zanders, *The Science and Business of Drug Discovery*,
https://doi.org/10.1007/978-3-030-57814-5_6

operate together in ways that cannot yet be precisely addressed with machine learning and supercomputers (see later). The field of complexity can be summarized by the phrase: "the whole is greater than the sum of its parts." To understand biology, and therefore to design drugs that interfere with its processes in a predictable and precise way, it is necessary to have a detailed understanding of how the human body functions in health and disease.[2] This is not a trivial undertaking, since each of us is genetically unique, with different likelihoods of succumbing to disease and having differing responses to the same medicine. Despite the challenges for drug discovery that arise from these complexities, a start along the road to fully understanding human biology has been made by scientists who are drawing up a "parts list" of the body as highlighted in this chapter.

6.1.1   Which Diseases to Treat?

The medical profession and the biopharmaceutical industry exist to provide treatments for the diseases that afflict human beings. Unfortunately, we are spoilt for choice when it comes to the number of different illnesses requiring new medicines (not counting surgical intervention, radiotherapy, or changes in the patient's lifestyle). According to the World Health Organization's International Classification of Diseases (ICD-10 2016),[3] the number of classified conditions runs into the tens of thousands. Clearly this is an impossibly large number for the biopharmaceutical industry to take on, however desirable this may be on humanitarian grounds, and so tough choices are made based on both commercial and scientific grounds. This then generates controversy about industry's lack of enthusiasm for discovering drugs that will not be profitable; this is the case with diseases that affect large numbers of people in the developing world or small numbers of people in the advanced economies. Changes to this attitude are beginning to occur, as will be discussed later in Chap. 17, particularly in the context of rare/orphan diseases. The biopharmaceutical industry is interested in the diseases listed in Table 6.1. These diseases are generally chronic degenerative conditions like arthritis and cancer, and infectious diseases, like influenza and hepatitis, all of which affect large numbers of people in the Western world. These disease areas are grouped into commercial franchises as listed in the table. Franchises are often associated with specific companies, for example, respiratory medicines from GlaxoSmithKline and cancer (oncology) medicines from AstraZeneca. As the marketplace is very fluid, these associations are not fixed forever, as some franchises are lost and others gained (Glaxo, for example, once held a huge gastrointestinal disease franchise because of its Zantac® drug, but once this patent protection was lost, the franchise all but disappeared within a few years).

---

[2] Sometimes, serendipity takes over, and novel targets or mechanisms are discovered by accident (e.g., see Chap 4.).

[3] ICD-10 was implemented in 1990; ICD-11 is due to be released in 2022 and includes more recently described conditions such as gaming disorders.

Table 6.1 Some of the diseases of interest to mainstream biopharmaceutical companies selling into western markets

Disease	Franchise
HIV/AIDS	Infectious diseases
Hepatitis	–
Influenza	–
Septic shock	–
Tuberculosis	–
Hypertension	Cardiovascular
Atherosclerosis	–
Stroke	–
Solid tumors	Oncology
Leukemia	–
Lymphoma	–
Rheumatoid arthritis	Arthritic diseases
Osteoarthritis	–
Osteoporosis	Metabolic diseases
Type II diabetes	–
Asthma	Respiratory diseases
Allergic rhinitis	–
COPD	–
Depression	Central nervous system (CNS)
Anxiety	–
Bipolar disorder	–
Epilepsy	–
Pain	–
Alzheimer's disease	CNS (neurodegeneration)
Parkinson's disease	–
Multiple sclerosis	–
Ulcerative colitis	Gastroenterology
Crohn's disease	–
Macular degeneration	Opthalmology

Although the list is not exhaustive, it does include the main illnesses suffered by an ageing population. The common endings in the name of the disease give a clue about the disease itself. Names ending in "itis," for example, mean inflammatory diseases, "osis" a morbid process, "oma" a tumor, and "emia" relating to blood. Interestingly the medical profession considers pregnancy to be a condition, if not exactly a disease, but this is not a high priority for companies as existing preventative drug treatments are generally successful

6.1.1.1 Medical Terminology

The medical profession is intimately associated with the drug discovery industry by both using their products and by contributing to drug development through clinical research. Medical terminology is therefore employed a great deal in drug discovery and development, so some disease-related terms are highlighted as follows:

- Signs and Symptoms

  Observations made by the doctor and the patient, respectively.

- Etiology

  The agent or mechanism that causes a disease. For infectious diseases, the etiological agent is obvious; influenza is caused by infection with influenza virus. The agents that cause diseases like rheumatoid arthritis or diabetes are much more difficult to pin down. In fact, viruses and bacteria have been implicated in the etiology of these diseases, but by the time the patient presents to the doctor, there is almost no trace of the infectious agent left.

- Pathogenesis

  The pathogenesis of the disease is the biological mechanism through which disease occurs. The pathogenic mechanism for type 2 diabetes, for example, is the development of resistance to insulin in the tissues.

- Morbidity and Mortality

  A feeling of illness and death through illness.

- Comorbidity

  The occurrence of two or more diseases in the same patient.

6.1.1.2 Translational Medicine

A vast amount of public and private money has been invested in biomedical research over the past 60–70 years, and while the return on investment has been difficult to quantify in economic terms, most people would feel that it has produced a significant improvement in healthcare. The factors that limit the rate of progress in discovering new medicines are twofold, namely, the rate at which new biological phenomena are discovered and the rate at which these findings can be translated into clinical practice. Those who undertake biomedical research are either scientists with degrees in subjects like biochemistry or microbiology or else clinicians with medical degrees. Many clinicians have qualifications in both basic science and medicine and have made the transition between the clinic and the laboratory. There is a perceived cultural difference between the basic scientist and the clinician which is reflected in harmless stereotyping by both sides. Most bench scientists are expected to keep the lab tidy by doing their own clearing up. Clinicians, on the other hand, take some time to get used to this as they are used to having others do this for them. Of course, they have advantage of not being squeamish, so the sight of a medical colleague taking his own blood through a large syringe for his experiments was disturbing, but not surprising.

The interaction between basic and clinical scientists has been formalized in recent years by the term translational medicine. The Canadian physician Sir William Osler recognized that medicine is both a science and an art. His teachings to medical students in the early years of the twentieth century focused on the importance of listening to patients and carefully observing their signs and symptoms. This, in a sense, is the art of medicine, which recognizes that each patient has a combination of different physical and psychological factors that are individual to them. The scientific aspect of medicine comes in the classification of disease and the technology used to diagnose and treat it. Eventually it may be possible to call medicine a purely scientific enterprise, particularly when we fully understand the genetic and other factors that make each person unique, but this may not happen for many years. The point about this preamble is that drug discovery research can advance when clinicians, who see real patients, talk to scientists who work in the lab with more abstract models of biology and disease. If successful, this should accelerate the transfer of basic research findings into tangible products that can be tested in the clinic. For this reason, many biopharmaceutical companies have set up translational medicine departments or have at least made sure that a clinician is associated with the drug discovery teams at an early stage.

6.1.2   Understanding Disease Mechanisms

The identification of suitable drug targets for a given condition requires a detailed understanding of the molecular mechanisms of disease. Where the disease is caused by infection, the drug target is most often part of the invading microorganism (although this is not always the case). For complex diseases like cancer or diabetes, things get much more complicated because there are so many possible targets to explore. Where do you start?

Firstly, the features of the disease that are obvious to the patient and doctor may be used as a starting point for a more detailed exploration. Rheumatoid arthritis (RA) is a representative example of a chronic disease that is high on the priority list of many biopharmaceutical companies. Patients with RA have swollen joints and suffer pain and immobility. This immediately focuses attention on the nature of the damage to the joints and how this compares with other joint diseases (arthritides) like osteoarthritis. In other words, is there a specific pathogenesis for RA? From this point onwards, the clinician and scientist must examine a sample of tissue removed from the joint (synovial tissue) and analyze it using a range of sophisticated laboratory tools. Tissue removed from a living patient is called a biopsy or biopsy material; if the person has died, it is autopsy (or necropsy) material. It is always better to use biopsy material, if ethically possible, because cells (and the molecules within them) begin to degrade after death. If nucleic acids like RNA are being analyzed, they must be extracted from tissue as soon as possible, as it is highly labile (unstable). In the past, I have gone to the extreme lengths of attending hip replacement operations so the surgeon could pass me some synovial tissue to freeze within seconds of it leaving the patient. This ensured that the RNA in the tissue was kept intact for later experiments while, at the same time, introducing me to the more spectacular aspects of orthopedic surgery.

Having obtained synovial biopsies from the patient, the next stage is to examine the material under a microscope to identify the different cells that make up the tissue. This introduces the discipline of cell biology, which is introduced below.

6.1.2.1   Cell Biology

Drug targets are proteins that are expressed inside, and on the surface of, cells, or are dissolved in biological fluids, such as blood or lymph. Cells are the fundamental units of living organisms and were first named by Robert Hooke in 1663 after examining cork under the microscope where he compared its structure to the small rooms occupied by monks (hence the derivation of the word from the Latin, *cellula* or small room). The cell theory was developed by Theodor Schwann and Matthias Schleiden in the nineteenth century and is stated as follows:

- The cell is the fundamental unit of structure and function in living things.
- All organisms are made up of one or more cells.
- All cells are derived from preexisting cells through cellular division
- Cells carry genetic material passed on to daughter cells during cellular division.

Cells can bind together in clumps or sheets to form tissues, or, if blood cells, to move freely within the circulation. The human body is comprised of approximately 200 different cell types, all of which arise from a single fertilized egg after conception. The way in which this happens relates to stem cells, which are covered in Chap. 8. The reason for so many different cell types is because each has a specialized function in the body, for example, red blood cells transport oxygen to the tissues and muscle cells create movement. Each cell type has the basic machinery for survival in common but can differ significantly in appearance and function. This is rather like a series of factories that make different products; they all have the same types of offices and layout but are each set up to produce specific items such as cars or washing machines.

To study cells in more detail, they can be examined undisturbed in the tissue environment they occupy (examined in situ) or be purified from the tissue as a unique cell population. Tissue culture is then used to grow larger numbers of these cells in the laboratory for further study (see later). Some examples of different cell types are shown in Fig. 6.1:

a   b

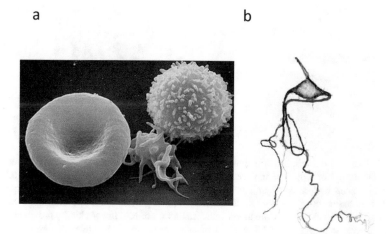

Fig. 6.1 Examples of different cell structures. (**a**) Scanning electron microscope image of blood cells, left to right: red blood cell (erythrocyte), platelet, and lymphocyte. Image taken by Electron Microscopy Facility at The National Cancer Institute US. (**b**) Nerve cell showing long extensions that form a network which distributes signals throughout the nervous system

6.1.2.2 Imaging

Although science deals with abstractions and models, there is nothing more satisfying than directly visualizing the object under study, rather than inferring its existence by other means. Cells are normally visualized under a microscope since they are well below naked eye visibility (a human cell is approximately 10 microns across, i.e., 10 µm or 10 millionths of a meter).

Modern imaging techniques allow the cell biologist to identify a cell by its shape (morphology) and by the types of molecules it expresses. Microscopes vary in sophistication, from the simple apparatus familiar to high school students which uses a white light to illuminate the specimen, to electron microscopes that can resolve features as small as DNA strands. The last few decades have seen the introduction of complex instruments that use lasers and computers to create three-dimensional images at resolutions that were once thought to defy the laws of physics. These latter tools are well suited to studying the organization of cells in complex tissues such as the brain. More routinely, cells and tissues are examined under the light microscopes in the process known as histology. This usually involves freezing lumps of tissue or embedding them in paraffin wax prior to being cut into thin sections using a fine blade in a microtome. The sections may be stained with various dyes so that specific cell types are revealed according to the dye used (histochemistry) or with antibodies targeted at cellular proteins of interest (immunohistochemistry, see Chap. 14.) Both techniques are routinely used in diagnostic laboratories to examine patient biopsies for abnormal cell types (as in cancer). It is also used extensively in the biopharmaceutical industry for target discovery and the analysis of animal and human tissues during the later stages of clinical development.

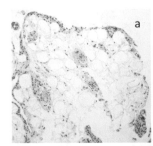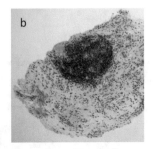

Fig. 6.2 Image of thin section of (synovial) joint tissue taken through a microscope. Tissue samples taken after joint biopsy or replacement surgery were sliced into sections less than 0.1 mm thick and treated with a dye (hematoxylin and eosin (H&E)) that stains cells a dark color. (**a**) Un-inflamed tissue showing a few cells as black dots. (**b**) Highly inflamed synovium showing massive infiltration of tissue with white blood cells, a hallmark of RA. Images kindly provided by Drs Andrew Filer and Dagmar Scheel-Toellner, University of Birmingham, UK with permission

Returning to the RA example, histochemistry on synovial tissue taken from the joints of RA patients reveals the presence of large numbers of white blood cells (leukocytes) that are absent in uninflamed tissues (Fig. 6.2). Leukocytes consist of several different cell types including T and B lymphocytes, monocytes, and macrophages, and all of these are present in the inflamed synovium. Not shown in this figure is the visible damage to cartilage and bone in the RA joint, which again, is absent in unaffected joints. This means that a hallmark of RA is the presence of these cells and the characteristic damage to joint tissue and bone. This points the search for drug targets in the direction of the leukocytes, although at this stage it would be too early to jump to conclusions. Their presence may not be the cause of the problem but simply the consequence of unseen damage or insult to the tissue. Despite this caveat, further examination of these cells would provide a useful starting point for more detailed cellular and molecular analysis of RA.

6.1.2.3 Cell (Tissue) Culture

As highlighted above, much useful information can be gained about the nature of disease by examining the cell types present in diseased tissues. Blood is also a tissue, and its component cells can be readily separated into red and white cell populations (erythrocytes and leukocytes). Histochemistry is a valuable tool for investigating tissues, whatever their origin, but it does not provide information about the functions of each cell type and the molecules associated with them. This is best achieved by physically removing the cells from tissues and maintaining them in culture (tissue culture), using processes that are similar to those used for growing bacteria and yeast (see Chap. 4). To illustrate the jargon, tissue culture produces cultured cells from cell cultures. In practice, cells are grown in sterilized plastic dishes or flasks containing a tissue culture medium consisting of sugars, salts, amino acids, and other nutrients. All manipulations are kept sterile to avoid contamination by airborne yeast or bacteria which grow much faster than the human (or other mammalian) cells in the cultures. For this reason, experiments are conducted in large stainless steel cabinets with special filters to keep contaminated air at bay. These laminar flow cabinets are often seen in TV clips with their white-coated operators when some new medical discovery is being described. Despite the noise of the fan, the cabinets (or "hoods") provide a surprisingly meditative environment in which to escape from the distractions of the outside world.

Once the desired number of cultured cells has been obtained for an experiment, they can be used in a variety of ways, depending on the type of cell and the scientific questions to be addressed. For drug discovery, mammalian cell lines can be used for target discovery or screening or else for producing genetically engineered protein drugs. The cell lines can have the advantage of being essentially immortal so long as there is sufficient growth medium available to keep them alive; this means in practice that cell cultures are divided into more and more flasks or bottles until there are enough cells for the experiment. It is also possible to store cells in liquid nitrogen (at $-196\,°C$) and thaw them for re-culturing when needed. Many cell lines are derived from human or animal cells that have been transformed by viruses to grow indefinitely, so they are often described as transformed cells. A famous example is the HeLa cell derived from a tumor removed from an American woman, Henrietta Lacks; for reasons that are not entirely clear, this line is so robust that it has been grown for decades in laboratories all over the world (for some background to the line's origin, see Johns Hopkins Medicine 2020). Transformed cells are useful because they can be grown to large numbers or be used to study how cell transformation might lead to cancer. Primary cells, on the other hand, are cells that have been taken directly from tissues (like blood) and should in principle retain the properties that they had in the intact body. Transformed cells do not retain these properties and are unlikely therefore to provide an accurate representation of a disease, with the possible exception of cancer. Primary cells are therefore desirable for simulating diseases in as realistic a way as possible, but they are often difficult to obtain in quantity. Stem cells offer a way around this problem by providing a renewable source of different cell types that can be produced in vitro rather than from human donors. More recently, stem cells have been isolated from patients with genetic diseases, thus making it possible to create a range of primary cells that may be altered because of the disease. This is an exciting development and is one reason why biopharmaceutical companies are so interested in stem cells, apart from their potential use in tissue repair.

6.1.2.4 In Vitro Models of Disease

Many diseases are the result of complex interactions that occur in the living person (in vivo) between cells and molecules. If these agents are removed from the body and examined in the test tube[4] (in vitro), some of this complexity will be lost.[5] The in vitro model is therefore designed to be as near to the in vivo situation as possible while recognizing that any targets identified using this approach will ultimately have to be evaluated (in vivo) in living animals and eventually, patients. Many in vitro models use cells that have been removed from tissues and then separated into their individual types before being placed in tissue culture. It is then possible to mix defined cell populations together in a precisely controlled manner and to observe their behavior. One type of cell may cause another to grow, change shape, or produce molecules with effects on other cells or tissues. The tissue culture medium in which the cells have been grown may contain molecules released by the cells. This medium, called the culture supernatant, contains many hundreds of different molecules, the majority of which are not relevant to drug discovery. However, a small minority (mostly proteins) have extremely significant biological activities. For example, they may stimulate cell division when added to nondividing cells in culture, in which case they are called growth factors. These molecules are members of the cytokine family and are of fundamental importance to many aspects of disease.

Some of these points are now illustrated using an in vitro model of rheumatoid arthritis as an example. Since the objective of the drug discovery scientist in this case is to discover drug targets for RA, the model focuses on the cells and molecules that are most likely to produce the hallmark of this disease, namely, destruction of the bone and cartilage in the joint. Histological investigations have already revealed the presence of fibroblasts close to the cartilage, which suggests that these cells may be responsible for the damage to this tissue. Fibroblasts help to form connective tissue in the body and are important components of skin. When these cells are taken from the RA patient and placed in tissue culture, it can be shown that the culture supernatant contains molecules that directly break down cartilage (when added to the in vitro experiment). This implies that, in some way, fibroblasts are activated by the local environment in the RA joint and then release factors that are responsible for the joint damage. Since leukocytes had already been shown to infiltrate the RA joints, it would be reasonable to assume that they may be responsible for activating the fibroblasts. As with the previous experiments with fibroblasts, the leukocytes from RA tissue are placed in tissue culture to observe the effect of the resulting culture medium on other components of the joint. The results of these in vitro experiments show that leukocytes in patients with RA produce factors that activate the

---

[4] The glass test tubes of traditional chemistry are almost never used in a bioscience setting; the modern laboratory has boxes of disposable plastic vessels in various configurations including flasks, plates, dishes, and tubes.

[5] However, there is much research into modeling intact tissues by culturing cells on scaffolds in three dimensions and creating organoids from stem cells.

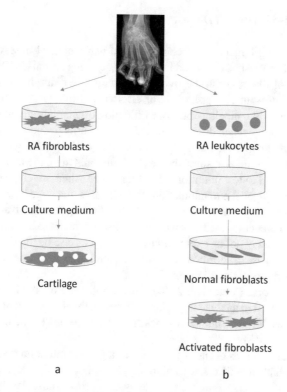

Fig. 6.3 An experiment to unravel the mechanism of rheumatoid arthritis in vitro and identify potential drug targets. A sample of the inflamed tissue from a patient with RA is placed in tissue culture, and the RA fibroblasts allowed to grow. (**a**) The cells are removed, and the culture supernatant containing factors that damage the cartilage is placed in a fresh tube along with a piece of normal cartilage. The agents produced by the fibroblasts (metallopeptidases) eat away at the cartilage as in the patient's joints. (**b**) Leukocytes from the joints of a patient with RA are cultured to see if they produce a factor in the medium that will stimulate normal fibroblasts to become activated as in RA. In real-life experiments, these factors have been shown to be cytokines. The drug target is the receptor to which the cytokine binds, and the drug itself is an inhibitor of this binding

fibroblasts in the joint, which in turn release other factors that cause the destruction of bone and cartilage. The experiment is summarized in Fig. 6.3.

Having identified molecules that have may be responsible for the pathogenesis of a disease, the drug discovery scientists can then home in on the nature of these factors and demonstrate that inhibiting their production or action will also inhibit joint destruction. This logical cause and effect analysis is central to drug discovery. This real-life example with RA did result in the identification of several drug targets, one of the most promising being the cytokine tumor necrosis factor (TNF-α) and its receptor.

6.1.2.5   In Vivo Models of Disease

Once possible drug targets have been identified using clinical observation and in vitro models, they must be evaluated in vivo using a living organism. This is because the properties of the target in the living body may differ significantly from those observed in the test tube. A great deal of effort is put into target validation at the discovery phase of the pipeline, but true validation of a target only really occurs at the clinical trial stage when drugs are tested for their effectiveness in real patients.

Many different experimental designs are used to evaluate a drug target in whole organisms, but there are also some themes in common. Often, the first step is to create a disease model to test inhibitors of a drug target. Alternatively, the target itself can be removed by genetic manipulation. In both cases, the target is worth pursuing further if disease symptoms are ameliorated when the target is inactivated. The main concern for researchers is the relevance of the disease model to human beings since the protein target may not be expressed in the model or else be too dissimilar to the human version. To use the title of a movie of the same name, this "Lost in Translation" between disease models and humans is a major cause of drug failure in the clinic and sometimes worse (Mak et al. 2014). This is another area in which AI/machine learning can be used to analyze large datasets from animal experiments and identify features that are relevant to human responses, for example, in "Found in Translation," Normand et al. (2018).

Some human diseases occur as the result of a genetic mutation that alters a specific protein. Many genetic diseases of this type are monogenic, i.e., they are caused by a single gene defect. Examples include cystic fibrosis, muscular dystrophy, and hemophilia. These result from "loss of function" mutations, where the protein is defective. Unfortunately, these conditions require the replacement of a defective gene (possibly with gene therapy) rather than the inhibition of an overactive one, which is something that is difficult to achieve with a medicine. In the latter case, proteins encoded by these "gain of function" mutations are much more useful as targets as they can be inhibited by drugs. These mutations occur in polygenic diseases like cancer, autoimmune disease, and many other serious chronic conditions, which is why these are of interest to the biopharmaceutical industry. While polygenic mutations have led investigators to new drug targets, most make a small individual contribution to the overall disease and are therefore of little interest, except possibly for biomarkers and diagnostics (see Chap. 14).

Model Organisms

Rodents

Model organisms are used to recreate aspects of human disease in species ranging from single-celled yeast all the way to mammals such as mice and rats. Larger animals are closer to humans in evolutionary terms and therefore more likely to reflect human biology. Rodents, for example, are the mainstay of the in vivo lab because they have a short life cycle and have similar organ systems and physiology to humans. They can also be bred into genetically pure lines with defined mutations induced with chemicals (chemical mutagenesis) or by using recombinant DNA technology. The latter process results in the creation of transgenic animals in which specific genes can be introduced (e.g., a human receptor), or else removed, (a gene knockout). This is enormously powerful technology because it makes it possible to observe the effects of selectively removing every gene in the mouse (or rat) on their biology. The disadvantages of using animals lie in their expense and the practical limitations of how many can be used when there may be many conditions to test. Of course, there is also the ethical consideration of breeding animals for this type of research, something that will be raised again Chap. 11. To overcome the limitations of working with larger animals, a few small nonmammalian organisms are used in target discovery, the most popular of which are briefly described below. Given the rapid pace of research into this area, it would not be surprising if several new model organisms were introduced in the years to come.

Yeast

Yeast, being a single-celled fungus, is one of the simplest model organisms. It is popular with investigators because it grows rapidly and is highly amenable to genetic manipulation. Although it cannot mimic all the facets of a complex disease, it has been used to identify targets for a class of anticancer drugs known as the cell cycle inhibitors. The cell division cycle is the sequence of events that leads to the creation of two cells from one; this process is dysregulated in cancer, so cell division becomes uncontrolled and leads to the growth of tumors. The mechanisms of yeast cell division are not dissimilar to those found in humans, so it has been possible to identify key protein targets in the simpler organism by using genetic approaches and then use that information to produce drugs that interact with their human equivalents. Yeast is a single-celled eukaryote (see Microbiology, Chap. 4), but humans are eukaryotes made up of many different types of cells. This has prompted the search for model organisms that have the same versatility as yeast and yet also reflect the complexity of human biology. Some of the models used in basic and pharmaceutical research are based on flies, worms, and fish as outlined below.

Flies

The fruit fly *Drosophila melanogaster* has been used in pioneering genetic studies for over 100 years because of its ease of handling and short (2-week) life cycle. Many natural mutations that lead to changes in their body layout and behavior have been catalogued and used for studies in basic and applied biology. The complete DNA sequence of fruit flies reveals a close similarity between the genes of this humble fly with those of far more complex human beings. It is estimated that more than 75% of human disease genes have a counterpart in *Drosophila*. Surprisingly, it is possible to observe a considerable range of behaviors in flies and relate them to defects in their nervous systems, something which is clearly relevant to target discovery for neurological diseases. Even drunken behavior can be modeled in fruit flies, since they normally ingest large amounts of alcohol from the fermenting fruit that is their normal diet. They display similar features to human alcoholics (including alcohol tolerance), although it is not known if they can tell the difference between a Bordeaux and a Burgundy.

Worms

Another invertebrate model organism is the nematode worm *Caenorhabditis elegans (C.elegans)*. This small (1 mm) worm grows with a 3-day life cycle on agar plates covered with bacteria and can be frozen down for future use. This last feature is unique for a multicellular organism with a nervous system and other features of higher animals. Since *C.elegans* is transparent, each one of its 959 cells can be tracked in vivo throughout its development cycle. One high-profile area of pharmaceutical research is the mechanism of human ageing; *C.elegans* has provided some startling results that link metabolism to longevity and offers further evidence for a genetic component to ageing. Insulin and related proteins are important regulators of metabolism and energy balance, so the observation that worms with a defect in a member of the insulin receptor family live twice as long as normal, is consistent with the fact that these animals live longer on a restricted diet. Unraveling the connections between these biological phenomena is a complex task, even with relatively simple organism like *C.elegans*, but the power of modern cell and molecular biology makes it possible to make real progress in this area of research.

Fish

Although *C.elegans* worms are useful models, they are still a long distance from human beings in evolutionary terms. They do not have a heart, for example, which is why the tiny zebrafish (*Danio rerio*) has proved so popular in recent years. In addition to sharing many genes in common with humans, the fish is also a vertebrate animal and therefore has similar organs. Like *C.elegans*, zebrafish have small transparent embryos that are produced in large numbers; this combination of size and numbers means that zebrafish embryos can be used to screen compounds for activity against intact animals. The fish are easily manipulated using genetics, so panels of mutant fish with defects that include malformation of the heart and other organs have been used for several investigations into, for example, heart regeneration and leukemia. Zebrafish are also being evaluated as model organisms for toxicology screening in preclinical development (see Chap. 11).

Genetic engineering can be used for just about any application where DNA is transferred from one organism to another; if you can clone it, you can engineer it, hence the appearance of some seemingly bizarre modifications to experimental organisms. One prominent example is the introduction of a jellyfish gene into specific zebrafish cells that can be followed as the fish develops to maturity. The gene encodes a fluorescent protein that emits a vivid color when exposed to ultraviolet light, so the engineered cells literally light up. Green fluorescent protein (GFP) was

Fig. 6.4 Three model organisms used in pharmaceutical research. Clockwise from top left: worm *Caenorhabditis elegans*, fruitfly *Drosophila melanogaster*, zebrafish *Danio rerio*. All are small multicellular organisms that reproduce rapidly in the laboratory and whose genetic and cellular composition have been mapped out in detail. Worm image is from the National Human Genome Research Institute in the USA

the first of the jellyfish genes to be used in this way and has been so successful in advancing cell biology that it earned its discoverers a Nobel Prize for chemistry in 2008. Other proteins that glow red, green, or orange have been expressed in transgenic zebrafish that are sold to hobbyists and science educators by a US company (GloFish 2020).

Three of the model organisms described above are illustrated in Fig. 6.4

6.1.2.6 Target Inhibition

Once a disease model has been established and a possible drug target identified, there is still some way to go before drugs are produced against that target. There must be some confidence that changing the activity of the target (by activating or inhibiting it) will reduce the features of the disease displayed by the model. In the case of arthritis, this might be reduction of joint swelling. Three approaches are often used to achieve this change of activity, namely, genetic engineering of transgenic animals, nucleic acid inhibition, and antibody production.[6]

Transgenic Animals

These have already been introduced in the section on in vivo models of disease. Genetic engineering technology allows the investigator to add or remove genes from a model organism in a controlled way. If a drug gene is removed from the

---

[6] Of course, at this early stage of target selection, a drug may not be available to test in the model.

organism (a knockout), then the effect of this loss on the in vivo disease model can be observed directly. If the drug target is relevant to a disease, its removal should reduce the symptoms. Sometimes the genetic modification inhibits a vital biological function, in which case the embryo cannot develop into adulthood, so it is impossible to perform the desired experiments. Even if the animals do mature in the usual way, the genetic defect may modify other genes and proteins as a compensatory mechanism, so the model may not be a true reflection of the normal biological state. This has led to the development of conditional knockouts, in which the target gene remains silent throughout development into adulthood (e.g., the CreLox system). Once the animal is mature, the gene is knocked out in the live animal using a genetic engineering process that is triggered by a small molecule (e.g., the antibiotic tetracycline) which can be added to the animal's drinking water. DNA technology can also be used to express or inhibit the expression of a gene within a defined cell type, rather than just randomly throughout the animal; this has proved to be particularly useful for studying the cells of the immune system.

Nucleic Acid Inhibitors

Transgenic model organisms are powerful tools for research into drug targets, but they are time-consuming to generate and not always easy to produce. A simpler alternative is to directly administer some agent into the animal that could selectively remove or inhibit the drug target gene. These agents, already been introduced in Chap. 2, are antisense RNA, small interfering RNA (siRNA), micro RNAs, ribozymes, aptamers, zinc finger nucleases (ZFNs), and CRISPR Cas9 editing. siRNA and CRISPR Cas9 technology are currently the tools used most often for blocking target the genomes of even the simplest organisms.

Antibodies

The DNA sequences required for genetic modification of drug target genes are readily available from genome databases, and the experiments themselves are relatively straightforward to perform. Unfortunately, altering a protein target by manipulating the expression of the gene that encodes it is not a reflection of what happens with most small molecule and biological drugs in the patient. Drugs generally interact directly with the drug target itself (a protein) rather than with the gene that encodes it. The ideal solution would be to generate molecules that interact specifically with any protein target of interest. Luckily, this can be achieved by harnessing the power of the immune system to generate antibodies that display exquisite specificity for almost any protein target of interest. Antibodies are large proteins (MW 150,000) that are produced naturally during infection or artificially by immunizing animals with proteins or cells. Antibodies can be added to tissue culture experiments in vitro or injected directly into animals to inhibit their target protein in vivo. This inhibition can occur through the direct blockade of a ligand binding to its receptor or through literally soaking up a protein ligand to remove it from the site of disease. Alternatively, if the target is on the surface of a cell, the antibody can recruit the immune system to destroy the cell in the same way as it would during a viral infection or during surveillance for tumors. It is important to note, however, that antibodies are unable to enter cells, so they are only useful when binding to targets that are present on the cell surface or dissolved in biological fluids like blood.

The rheumatoid arthritis example mentioned earlier in the chapter is now used to illustrate some of these points: the results of in vitro experiments (Fig. 6.3) indicated that leukocytes in the inflamed RA joint produce molecules which cause nearby fibroblasts to destroy bone and cartilage. The obvious next step was to identify the actual molecules responsible for these effects and to see whether they could form the basis of a drug development program. To cut a long story short, the agents produced by fibroblasts which destroy cartilage were identified as enzymes (catalytic proteins) that break down connective tissue proteins like collagen. Since these enzymes (technically, metallopeptidases) can be inhibited with small molecules, several drug candidates were created over the years and tested in arthritis. Unfortunately, none of these are suitable for clinical use because of side effect issues.

Another possibility was to identify the molecules produced by the leukocytes and inhibit those instead. Research on inflammation has been ongoing for years, so there is a long list of possible drug targets that might be involved in RA. One of these molecules is the cytokine tumor necrosis factor alpha (TNF-α) that, as its name suggests, was originally identified as a cytokine that killed tumor cells. Further research revealed that TNF-α has a role in inflammatory reactions, and the results of numerous investigations supported the idea that it might be activating the RA fibroblasts. Antibodies were therefore produced against this cytokine and added to the in vitro cultures of leukocytes and fibroblasts described earlier. This had the desired effect of reducing fibroblast activation, so the antibody was then injected into mouse or rat models of arthritis. These models produce swelling in the paw and bone erosion, both of which were reduced with the antibody to TNF-α but not a control antibody (which is made to an irrelevant protein).

The sequence of events that led to a promising drug target for RA started with the observation of patients, continued with the analysis of cells and molecules in joint tissues and finished with an animal model of disease. The disease pathogenesis was shown to be caused by the overproduction of the cytokine TNF-α, so the drug target was the receptor to which it binds.[7] Antibodies to the cytokine reduced arthritis in vivo by soaking up the excess protein and preventing binding to the receptor. Another approach involved using receptor decoy proteins to act as receptor antagonists, just like the small molecules described in Chap. 3. In fact, both antibodies and decoys have been developed into biological drugs and have become major success stories in the treatment of arthritis (Monaco et al. 2015). One of my former postdoctoral supervisors was responsible for some of this work and has been duly rewarded

---

[7]This is far from being the whole story, as there is much more to learn about the causes and progression of RA; there are also many more target opportunities and drugs in development or in the clinic.

with a share in the Lasker Prize (the medical equivalent of a Nobel) and a knighthood from the British government.

6.1.3 Prospecting for Drug Targets

This chapter has so far described how drug targets are identified through analyzing the disease to be treated. The investigator who uses this approach may have no idea what type of target will eventually emerge from laboratory investigations. The situation could be reversed, however, so that different classes of drug target proteins are defined at the outset and then assigned to different disease states. These two different approaches can be described as "lumping" or "splitting"; research departments in biopharmaceutical companies have been organized along either lumping or splitting principles, depending on the fashion of the day.

The remainder of this chapter covers the analysis of different types of drug target, or the "splitting." It is impossible to cover all the technical background to this analysis, but an attempt will be made to explain the key features of the scientific disciplines that are relevant to this aspect of drug discovery. Since drug targets are mostly associated with cells, attention must now return to these fundamental units of life.

6.1.3.1 Looking Inside Cells: Subcellular Structures and Functions

Until the late 1940s, biologists were limited by the amount of detail they could discern in cells using conventional light microscopy. The invention of the transmission electron microscope meant that it was possible to use the high resolution of electron beams to image subcellular structures, otherwise known as organelles, at high magnification. These organelles have specific functions that contribute to the overall life of the cell including its reproduction through division. The nucleus, for example, contains the genetic material DNA, while the mitochondria are the powerhouses that produce energy. From the drug discovery perspective, the cell membrane is of vital importance, as it presents the outside face of the cell to its environment. This means that signaling molecules, like hormones, neurotransmitters, and cytokines, bind to the outside (extracellular) portions of their receptors which are embedded in the cell membrane. After binding has occurred, a signal is transmitted through the membrane to the inside of the cell to instruct it to perform a specific task, like making the cell divide. Figure 6.5 shows a model of a human cell cut away to reveal the nucleus, mitochondria, and other organelles.

In addition to energy metabolism and cell division, organelles and other cellular components contribute to cell viability and responses to internal and external stress. Some of these functions are highlighted below because of their relevance to disease and the search for useful drug targets.

Fig. 6.5 Model of human cell with outer cell membrane cut away to reveal the organelles. The nucleus, containing chromosomes and DNA, is also cut away and is the large artichoke looking structure in the middle. The small cylinders represent the mitochondria, and the membranous structure at the bottom is the Golgi apparatus involved in transporting molecules through the cell

Apoptosis

Named from the Greek, "the falling of leaves," apoptosis is also known as "programmed cell death" in which mitochondria and other cell systems are involved in a highly regulated destruction of cells when they are no longer required. For example, the number of blood cells produced by cell division must be kept to a certain level; otherwise, the body would be overwhelmed very quickly. This number is maintained by their apoptosis and removal by scavenger cells of the immune system. Chemotherapy and radiotherapy of cancer drive tumor cells into apoptosis, but resistance can develop because of upregulated anti-apoptotic or downregulated pro-apoptotic mechanisms. Such mechanisms involve a range of potential targets for drugs designed to overcome treatment resistance.

Non-apoptotic Cell Death: Necrosis, Pyroptosis, and Ferroptosis

Other cell death mechanisms have been discovered, both uncontrolled (necrosis) and those resulting from oxidative (ferroptosis) and infections (pyroptosis). Drug targeting opportunities exist here for reducing inflammation or, as with apoptosis, enhancing cell death in cancer therapy. For a review of these and further "subroutines" of cell death, see Tang et al. (2019).

Autophagy

Literally meaning "self-eating," autophagy in cells is a recycling process in which cellular components are digested and reused in response to stresses like starvation. As well as being important in cancer, drug discovery efforts in neurodegeneration (e.g., Alzheimer's and Parkinson's diseases) are focused on encouraging autophagy in damaged neurons to restore brain function (see Cheng et al. 2013 for a review).

Proteostasis

Once proteins are synthesized on ribosomes in the cell, they are subject to competing pathways involving folding, trafficking, and breakdown. Under normal circumstances, proteins are held in their correct three-dimensional conformation by chaperones such as the heat shock protein hsp90, but loss of this control can lead to diseases such as Alzheimer's disease which are characterized toxic protein aggregates (for a general review of drug discovery efforts, see Shrestha et al. 2016). Misfolded proteins are removed by endoplasmic reticulum-associated degradation (ERAD) using a waste disposal system based on the small peptide ubiquitin (and a family of related peptides) plus a large cylindrical protein complex called the proteasome. Proteins marked for degradation are tagged with ubiquitin chains and broken down in the proteasome, thereby releasing ubiquitin for reuse. The above mechanisms are useful for drug discovery for several reasons. For example, inhibitors of hsp90 and related chaperones could unfold and disable proteins that are of interest as drug targets, such as kinases (see later). Inhibition of some of the many components of the ubiquitin system is the subject of intense research activity for a range of diseases, and two proteasome inhibitors have been clinically approved for multiple myeloma (Huang and Dixit 2016). Finally, the ubiquitin/proteasome system is being exploited to selectively degrade drug target proteins using proteolysis-targeting chimeras (PROTACs) (Pei et al. 2019). Based on this approach, a small molecule drug that selectively degrades the androgen receptor is undergoing clinical trials for treating prostate cancer.

6.1.3.2   Biochemistry

Biochemistry is the chemistry of life processes that examines all the large and small molecules that make up living organisms. Although biochemistry and organic chemistry are both based on the chemistry of carbon compounds, they differ because of the conditions under which these compounds react together. Organic chemists synthesize compounds in the laboratory by using extremes of temperature and pressure or highly reactive chemicals. This would obviously be impossible in living cells since they would be destroyed. Instead, a large variety of chemical reactions occur at a benign 37 °C without the need for harsh conditions. This is achieved through the action of enzymes, a large class of proteins that orchestrate nearly all the chemical reactions in cells. Enzymes are biological catalysts which accelerate chemical reactions by lowering the amount of energy required for chemical substances to react together; the key point is that catalysts are not consumed in chemical reactions and can be reused many times. Industrial catalysts for the manufacture of plastics or removal of car exhaust emissions are quite simple molecules, often based on the precious metals, platinum, or palladium, although enzymes may work on the same principle, they have an important extra feature. Being proteins, they can selectively bind to many different types of molecule, so they have evolved to catalyze highly specific chemical reactions. Enzymes of the protease class, for example, catalyze reactions with proteins, while nucleases catalyze reactions with nucleic acids. It will be noticed that enzymes have a common suffix "ase" associated with the type of molecule with which it reacts. Some common names are still used, for example, pepsin and trypsin, which are proteases found in the digestive system. Because of the large number of different chemical reactions catalyzed by enzymes, the International Union of Biochemistry and Molecular Biology (IUBMB 2020) has devised the EC number system to create a logical classification; so, pepsin becomes EC 3.4.23.1 and trypsin EC 3.4.21.4.

Enzymes are central to biochemistry, which was renamed from the earlier "physiological chemistry" by Carl Neuberg in 1903. Many of the chemical reactions occurring in cells, such as the conversion of sugars to energy and the creation of waste products like urea, were elaborated in the first half of the twentieth century. These reactions are collectively termed the metabolism of the cell and occur in metabolic pathways, each step of which is driven and controlled by enzymes. As more and more reactions were discovered in different organisms, these pathways became extremely complicated and even quite esthetic when printed on a color poster. At the time I was a biochemistry student, these posters were sometimes displayed alongside more conventional images of Led Zeppelin and other rock stars of the era. These charts have now been superseded by computer databases of metabolic pathways, such as the ones provided by the Kyoto Encyclopedia of Genes and Genomes (KEGG 2020). A small part of a typical metabolic pathway is shown in Fig. 6.6. The lines in the figure represent a series of chemical transformations, each catalyzed by an enzyme, which form a pathway from one molecule (represented by a dot) to another. The pathway can be linear, branched, or even circular, as in the Krebs (or TCA) cycle illustrated in the figure.

Metabolic pathways were traced out by blocking the activity of individual enzymes in a pathway and then analyzing the small molecules that accumulated in the cell because of the pathway being blocked. Alternatively, compounds labeled with radioactive atoms were added to cells and converted into new molecules by metabolism. In this procedure, a simple radioactive compound (e.g., ss glucose) was added to cells, which were then analyzed at different points in time for the presence of new radioactive molecules. The radiolabel was "chased" into different compounds, so it was then possible to work out the flow of metabolism over time.

Compounds that are metabolized by enzymes in cells are called metabolites and account for about ten percent of the mass of a cell. The rest of the cell, in comparison, is made up of approximately 60% protein, 16% lipid, 6% carbohydrates, 5% nucleic acids, and 3% inorganic compounds like metals. Metabolism and metabolites will be revisited in Chap. 11 when discussing drug metabolism in the liver.

The elucidation of metabolic pathways has been one of the great triumphs of biochemistry leading, among other things, to improvements in nutrition through the discovery of vitamins and cofactors. Biochemistry has also revealed much about the

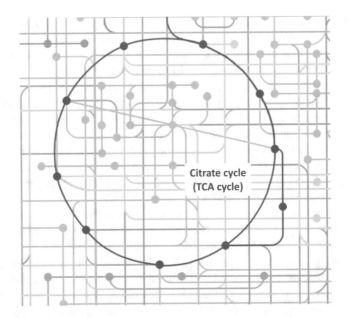

Fig. 6.6 Part of metabolic pathway. Each enzyme that catalyzes a biochemical reaction is represented as a dot linked in a specific path. The circular pathway is the Krebs cycle, a component of the system that converts glucose to energy. Original image courtesy of Kyoto Encyclopedia of Genes and Genomes (KEGG)

structure of large molecules, such as proteins, nucleic acids, carbohydrates, and lipids. Of these, proteins are central to many discovery programs as they are the targets of most drugs. Another aspect of biochemistry relates to the properties of nucleic acids, something that is now described in the sections below.

6.1.3.3 Molecular Biology

In the immediate years following World War II, many physicists turned their attention to biological problems, particularly the nature of the genetic material and the 3D structures of large molecules. Although these activities were, strictly speaking, the pursuit of biochemists, they became subsumed within the term molecular biology, whose greatest triumph was the discovery of the double helical structure of DNA by Watson and Crick in 1953, made possible by X-ray crystallography data from Rosalind Franklin and insights into the basic chemistry of nucleotides by Chargraff and others. X-ray crystallography was later used by Perutz and Kendrew to determine the 3D structures of proteins which provided insights into how the shape of a large molecule determines its function. Macromolecular (i.e., large) structures of enzymes, structural proteins, and nucleic acids are now determined routinely at the level of individual atoms using X-ray diffraction, NMR, and other techniques. It is now even possible to determine the three-dimensional structures of subcellular structures comprised of dozens of individual proteins using cryo-electron microscopy

In addition to providing structural information, molecular biology also reveals how genetic information, encoded in the DNA molecule, is turned into protein. Francis Crick[8] coined the term "central dogma of molecular biology"; this states that information flows from DNA through an RNA intermediate (messenger RNA) to specify the type and order of amino acids added to the final protein. Like most dogmas, this does not always hold true since information can also flow backwards from RNA to DNA. Retroviruses, such as HIV, integrate their RNA genomes into the DNA of infected cells by a process of reverse transcription, i.e., they produce a DNA copy from an RNA template. This is catalyzed by the enzyme reverse transcriptase, a prime target for small molecule drugs that inhibit the replication of HIV.

The actual mechanisms by which proteins are produced in the cell under the instructions of messenger RNA are well understood and described in many books and online resources. The following description of the way DNA sequences specify a genetic code is relevant to later sections on genomics and DNA sequencing. The basic principles are illustrated in Fig. 6.7 below:

The genetic code is made up of three letters (i.e., DNA nucleotides), which specify a single amino acid. In the example in Fig. 6.7, AAA on one strand (the coding strand) of DNA is copied into a messenger RNA intermediate. The AAA code then instructs the protein synthesis machinery to select the amino acid lysine (Lys) for inclusion into the growing protein chain. All 20 amino acids are specified by a

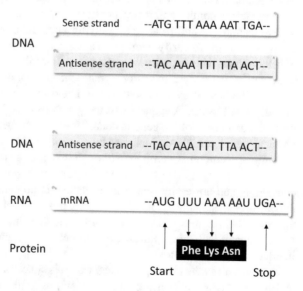

Fig. 6.7 Conversion of DNA code into protein via an RNA intermediate. Double-stranded DNA (top) with sense and antisense strands (see Chap. 2). Messenger RNA (mRNA) is transcribed (copied) from the antisense strand after the double helix has unwound. The nucleotides in the mRNA copy are read as groups of three letters, i.e., a triplet, each triplet specifying an amino acid or a stop/start signal

[8] Crick had a rather jaundiced view of life, for example: "I myself was forced to call myself a molecular biologist because when inquiring clergymen asked me what I did, I got tired of explaining that I was a mixture of crystallographer, biophysicist, biochemist, and geneticist, an explanation which in any case they found too hard to grasp."

triplet code (codon); in addition, there are start and stop codons that tell the cell when to start the protein chain and when to stop it. This genetic dictionary is extremely important for molecular biologists, as it allows them to predict the length of a protein and its amino acid sequence by just knowing the DNA sequence. This will be discussed later in connection with genomics and bioinformatics.

6.1.3.4 Genomics

The word genome was originally coined by the botanist Hans Winkler in 1920 as follows:

> I propose the expression Genom for the haploid chromosome set, which, together with the pertinent protoplasm, specifies the material foundations of the species ….

The study of genomes, i.e., genomics was introduced as journal of the same name in 1987 (The Scientist 2001). Since then it has broadened out to define a scientific discipline that studies the genetic complement of an organism, in other words the DNA sequences, their protein coding capacity, regulation, and changes in disease. Genomics is distinct from genetics, the study of inheritance, whose basic principles were understood many years before DNA was shown to be the genetic material. The suffix "omics" has now been used to the point where there is an epidemic of "omics" technologies. Proteomics was one of the first, dealing, unsurprisingly, with the protein complement of an organism (see later). The meaning of others, like glycomics and metabolomics, can be deduced from the prefix in the name.

In the pre-genomics era, it was possible for a biochemist or molecular biologist to devote almost an entire career to studying a handful of genes and proteins. Heroic efforts were made to purify microgram amounts of a protein of interest from tissues or cell cultures. This quantity was often far too small to be of much further use, so a strategy was employed to vastly increase the amount of protein that could be produced in the laboratory. Firstly, the amino acid sequence of the protein was determined, so that the DNA sequence of the gene that encoded it could be worked out using the dictionary of the genetic code. This DNA sequence could be turned into a physical sample of DNA which was then introduced into bacteria to produce large amounts of recombinant protein. Although this brief description makes this appear to be very straightforward, the library screening procedures were very time-consuming and not easily scaled up to handle large numbers of proteins. Despite these limitations, many cytokines, and other proteins that were only present in minute amounts in tissues and cells, could, for the first time be produced in milligram to gram quantities. Some of these proteins became therapeutic proteins, notably the interleukin family of cytokines.

Now we are in the era of genomics, which is concerned with the DNA sequence and protein-coding capacity of an organism, as well as the way in which these sequences are regulated. Rather than studying a few genes at a time, as in the pregenomic era, it is now possible to study the complement of an entire genome. Genomics makes it possible to calculate the number of proteins that an organism can make, since DNA sequences can be used to infer protein sequences. In principle, this knowledge should provide an upper limit to the number of (protein) drug targets in an organism and therefore the size of the "playing field" for drug discovery. Furthermore, the deduced amino acid sequences can be used to group proteins into families of known targets for drugs, such as receptors and enzymes. Surprisingly, the protein coding regions only take up about 2–3% of the human DNA sequence of three billion base pairs. The rest, initially referred to as junk DNA, encodes a series of RNA molecules that are involved in gene regulation, including miRNAs (microRNAs) and lncRNAs (long noncoding RNAs). In addition, lncRNAs encode small peptides, thus expanding the protein repertoire of human cells and therefore opportunities for target discovery (Yeasmin et al. 2018).

The regulation of gene expression occurs in part through the action of transcription factor proteins that bind to defined DNA sequences which are scattered around the genome; this binding results in the activation or repression of specific protein-coding genes. Although transcription factor binding is a central mechanism for regulating gene expression, additional processes are involved, including noncoding RNA binding and a series of processes which come under the term epigenomics, or epigenetics. More details will be given later, but there is literally more to life than raw DNA sequence. The regulation that turns a DNA code into a complex organism can be illustrated by comparing a caterpillar and a butterfly. Both have identical DNA sequences and should therefore produce the same proteins, yet they clearly look quite different. This is because the proteins that specifying the butterfly (e.g., wings) are turned on during development, and those related to distinctive caterpillar structures are switched off.

The Genomics Toolbox

The development of genomics was possible because of the development of new technologies by scientists and engineers working in universities and instrumentation companies. Foremost among these technologies are automated chemistry, high-throughput DNA sequencing, bioinformatics, and micro-fabrication, all of which are covered in the following sections.

Automated Chemistry

Manipulating and sequencing the DNA of any genome would not be possible without oligonucleotides. These are normally synthesized as strings of around 20 nucleotides (A, C, G, T) for use in several applications, including recombinant DNA production or for chemical reactions that synthesize DNA or RNA in vitro. One of these so-called priming reactions is part of the DNA sequencing process (see below). The synthesis of oligonucleotides was not a trivial exercise in the early days of molecular biology. In 1972, an entire issue of the prestigious Journal of Molecular Biology was taken up with papers describing the creation of a gene built from chemically synthesized oligonucleotides (e.g., Khorana 1972). Although the gene was a modest size by today's standards, it required a great deal of sophisticated chemistry which was performed using standard laboratory procedures. This was acceptable for pioneering research, but not for routine use at high throughput. To overcome this limitation, engineers and chemists produced a small bench top oligonucleotide synthesizer in which all the necessary chemical reactions were automated. Oligonucleotides can now be ordered on the Internet from supply companies and delivered by post in freeze dried form, ready to mix with water and use in sequencing or other experiments.

DNA Sequencing

The dideoxy sequencing system for revealing the order of A, C, G, and T nucleotides in a piece of DNA was developed by Fred Sanger in Cambridge, UK.[9] The "Sanger Method" originally employed radioactive nucleotides since these could be detected using medical X-ray film. The characteristic "DNA ladders" seen in sequencing images are the darkened areas of a photographic emulsion that correspond to a DNA fragment. The position of each fragment corresponded to a DNA letter, so by reading the film from the bottom upwards, a scientist could obtain around 100 nucleotides of sequence. However, this throughput was far too low to make sequencing the genomes of even the simplest organisms remotely feasible. This only became possible with the invention of automated sequencing machines which used fluorescent dyes instead of radioactivity. Fluorescence can be readily detected with an electronic detector linked up to a computer and a graphical output to display the sequence on a screen. Now that many sequencing reactions could be run in parallel, i.e., multiplexed, the number of bases that could be sequenced in a day rose to thousands, culminating in the Human Genome Project, completed in 2003. The original project to sequence this genome took 13 years and cost the US taxpayer approximately 2.7 billion dollars. Although it provided a "reference genome" from which to identify possible protein coding sequences and other features from human subjects, it did leave many questions unanswered about how genomes differ between individuals in health and disease. The Sanger technology was used to generate more sequences across the kingdom of life but was ultimately superseded by faster and cheaper methods under the heading "next generation sequencing" with the stated goal of bringing the costs of sequencing a human genome down to $1000. For a review of sequencing technologies and their history in genome assembly, see Giani et al. (2020)

---

[9] This achievement won him his second Nobel Prize, the first being awarded for his sequencing a protein (insulin) for the first time.

Next-Generation Sequencing (NGS)

Advances in microfluidics, electronics, and solid-state chemistry allowed instrument makers to design machines that can rapidly sequence millions of bases. Several different companies like 454, Solexa, Illumina, Applied Biosystems, and Ion Torrent used their own technology to rapidly generate long sequences from many smaller "reads" but sometimes at the expense of accuracy. Since single base pair changes in a genome can be highly significant (see SNPs later), a high degree of accuracy is essential.

Third-Generation Sequencing (TGS)

Third-generation technologies are designed to sequence single DNA molecules without prior amplification and can produce far longer reads than NGS. Two systems are in current use, sequencing by synthesis and nanopore sequencing, providing a vast amount of reliable sequence data in a cost-effective manner.

An example of a sequence trace is given in Fig. 6.8 below:

Bioinformatics

Strings of letters in DNA, RNA, or proteins are strings of letters and nothing else unless meaning can be extracted from them. One of the triumphs of molecular biology was the deciphering of the genetic code, so that DNA sequence could be used to predict the amino acid sequences of proteins. Sequence information was also used to understand gene regulation through identifying the stretches of DNA that are targets of transcription factors. Most importantly, for medicine and drug discovery, sequencing could uncover genetic mutations. These are areas in DNA that differ between individuals, possibly at a single nucleotide, or else in larger stretches (see

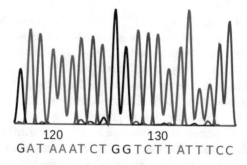

Fig. 6.8 DNA sequencing trace showing part of a sequence starting 117 nucleotides into a piece of DNA and stopping at 137 nucleotides. Each peak corresponds to A, C, G, or T depending on the color of the trace. A sequencing machine will generate traces several hundred nucleotides long. The information is stored in a computer for later analysis

Chap. 14). Although mutations are important drivers of evolution, they can have serious consequences. Sickle cell anemia, for example, occurs in patients who have a single nucleotide mutation in the hemoglobin gene. A single nucleotide change will create a single amino acid change in the hemoglobin protein, which now aggregates (i.e., sticks together) and forces the red blood cells into a sickle shape. This is an example of a mutation in the coding region of a gene, but there are also mutations in noncoding regions that may affect gene regulation. These mutations can contribute to disease pathology by altering protein expression.

A comprehensive analysis of DNA and protein sequences is only possible with the assistance of computers. The advent of powerful mainframe and personal computers in the 1990s helped to bring about the field of bioinformatics and of course continues into the era of machine learning and artificial intelligence. Centralized computer databases of nucleic acid and protein sequences, such as GenBank and Ensembl, are freely available to the research community. Computer programs have been written to perform specific tasks like, for example, the comparison of sequences between individuals or between different species. One of the most widely used comparison tools is the Basic Local Alignment Search Tool, developed in 1990 by the US National Institutes of Health (NIH). Commonly known as BLAST, it aligns sequences together to create the maximum overlap, so that if they are identical, they share 100% homology. BLAST also reveals mismatches that occur because the genes (or proteins) have altered through evolution (i.e., they come from different species) or they are altered between individuals because of mutation.

Bioinformatics opened the field of comparative genomics, which is important for drug discovery; this is because it allows comparisons to be made between drug target sequences in humans with the same sequences in a model organism. For example, if the target sequences differ significantly between species, it is unlikely that drugs directed against the human target will work in that model. Genome comparisons are also used to identify proteins that are only present in infectious microbes, as these could be new targets for antibiotics, antivirals, or antifungals. Finally, sequence comparisons can be made within a species to classify proteins into families that share features in common, an important tool for the selection of drug targets.

A typical BLAST result is shown in Fig. 6.9. In this example, the sequence of the human CD4 protein is compared with the same molecule present in the mouse. The degree of homology (i.e., similarity) between the CD4 proteins from these two species is important when choosing animal models of HIV infection. The HIV virus uses the CD4 molecule as a receptor to bind to the human T lymphocyte prior to entering the cell. However, since the protein sequence of the murine (mouse) CD4 is too dissimilar to the human homolog (equivalent), the virus cannot bind to mouse cells. The differences in amino acid sequence are clearly shown in the figure (which could have also been constructed using nucleotide sequences). Genome databases are now so comprehensive that it is possible to compare sequences of a gene or protein of interest across the entire kingdom of life.

Nucleotide

```
Query   805   CTATAAGAGTGAGGGAGAGTCA-GCGGAGTTCTCCTTCCCACTCAACTTTGCAG-AGGAA   862
              |||||||| |||||  ||  ||| ||||||||||  |||  |||| |||    || |  ||
Sbjct   835   CTATAAGAAAGAGGGGGA-ACAGGTGGAGTTCTCCTTCCCACTCGCCTTTACAGTTGAAA   893

Query   863   A----ACGGG-TGGGGAGAGCTGATGTGGAAGGCAGAGAAGGATTCTTTCTTCCAGCCCT   917
              |    ||||| || |||| ||||||||  |||| |||| |||| || ||| | || | ||
Sbjct   894   AGCTGACGGGCAGTGGCGAGCTGTGGTGGCAGGCGGAGAGGGCTTCCTCCTCCAAGTCTT   953

Query   918   GGATCTCCTTCTCCATAAAGAACAAAGAGGTGTCCGTACAAAAGTCCACCAAAGACCTCA   977
              ||||| ||||  ||   |||| ||| ||| |||| |||| |   |||  | |||| |||| |
Sbjct   954   GGATCACCTTTGACCTGAAGAACAAGGAAGTGTCTGTAAAACGGGTTACCCAGGACCCTA   1013

Query   978   AGCTCCAGCTGAAGGAA-ACGCTCCCACTCACCCTCAAGATACCCCAGGTCTCGCTTCAG   1036
              ||||||||| || | || |  |||  ||| || |||||  | |||||| ||| || |||
Sbjct   1014  AGCTCCAGATG-GGCAAGAAGCTCCCGCTCCACCTCACCCTGCCCCAGGCCTTGCCTCAG   1072

Query   1037  TTTGCTGGTTCTGGCAACCTGACTCTGACTCT-GGA-C--AAA-GGGACA-CTGCATCAG   1090
              | |||||| ||||| ||||| || ||| || | |  |   ||| ||| || | ||||||||
Sbjct   1073  TATGCTGGCTCTGGAAAACCTCACCCTGGCCCTTGAAGCGAAAACAGGAAAGTTGCATCAG   1132

Query   1091  GAAGTGAACCTGGTGGTGATGAAAGTGGCTCAGCTCAACAATACTTTGACCTGTGAGGTG   1150
              ||||||||||||||||||||||| ||||| |||||| | | |  |||||||||||||||||
Sbjct   1133  GAAGTGAACCTGGTGGTGATGAGAGCCACTCAGCTCCAGAAAAATTTGACCTGTGAGGTG   1192

Query   1151  ATGGGACCTACCTCTCCCAAGATGAGACTGACCCTGAAGCAGGAGAACCAGGAGGCCAGG   1210
              || ||| |||| ||| |  ||| ||||     ||||  | |||||||| |||||||| | |
Sbjct   1193  TGGGGACCCACCTCCCCTAAGCTGATGCTGAGTTTGAAACTGGAGAACAAGGAGGCAAAG   1252
```

Protein

```
Query   24    TQGKTLVLGKEGESAELPCESSQKKITVFTWKFSDQRKILGQHGKGVLIRGGSPSQF-DR   82
              TQGK +VLGK+G++ EL C +SQKK    F WK S+Q KILG  G   L +G PS+  DR
Sbjct   23    TQGKKVVLGKGDTVELTCTASQKKSIQFHWKNSNQIKILGNQG-SFLTKG--PSKLNDR   79

Query   83    FDSKKGAWEKGSFPLIINKLKMEDSQTYICELENRKEEVELWVFKVTFSPGTSLLQGQSL   142
              DS++  W++G+FPLII   LK+EDS TYICE+E++KEEV+L VF +T +  T LLQGQSL
Sbjct   80    ADSRRSLWDQGNFPLIIKNLKIEDSDTYICEVEDQKEEVQLLVFGLTANSDTHLLQGQSL   139

Query   143   TLTLDSNSKVSNPLTECKHKKGKVVSGSKVLSMSNLRVQDSDFWNCTVTLDQKKNWFGMT   202
              TLTL+S   S+P  +C+  +GK + G K LS+S L +QDS  W CTV  +QKK  F +
Sbjct   140   TLTLESPPG-SSPSVQCRSPRGKNIQGGKTLSVSQLELQDSGTWTCTVLQNQKKVEFKID   198

Query   203   LSVLGFQSTAITAYKSEGESAEFSFPLNFAEE--NGWGELMWKAEKDSFFQPWISFSIKN   260
              + VL FQ +    YK EGE  EFSFPL F  E    G GEL W+AE+ S  + WI+F +KN
Sbjct   199   IVVLAFQKASSIVYKKEGEQVEFSFPLAFTVEKLTGSGELWWQAERASSSKSWITFDLKN   258

Query   261   KEVSVQKSTKDLKLQLKETLPLTLKIPQVSLQFAGSGNLTLTLDK--GTLHQEVNLVVMK   318
              KEVSV+   T+D KLQ+ + LPL L +PQ    Q+AGSGNLTL L+   G LHQEVNLVVM+
Sbjct   259   KEVSVKWVTQDPKLQMGKKLPLHLTLPQALPQYAGSGNLTLALEAKTGKLHQEVNLVVMR   318

Query   319   VAQLNNTLTCEVMGPTSPKMRLTLKQENQEARVSEEQKVVQVVAPETGLWQCLLSEGDKV   378
               QL   LTCEV GPTSPK+ L+LK EN+EA+VS+ +K V V+ PE G+WQCLLS+  +V
Sbjct   319   ATQLQKNLTCEVWGPTSPKLMLSLKLENKEAKVSKREKAVWVLNPEAGMWQCLLSDSGQV   378

Query   379   KMDSRIQVLSRGVNQTVFLA-CVLGGSFGFLGFLGLCILCCVRCRHQQRQAARMSQIKRL   437
              ++S I+VL         +A  VLGG  G L F+GL I  CVRCRH++RQA RMSQIKRL
Sbjct   379   LLESNIKVLPTWSTPVQPMALIVLGGVAGLLLFIGLGIFFCVRCRHRRRQAERMSQIKRL   438

Query   438   LSEKKTCQCPHRMQKS     453
              LSEKKTCQCPHR QK+
Sbjct   439   LSEKKTCQCPHRFQKT     454
```

Fig. 6.9 A BLAST result taken from the NCBI online database (NCBI BLAST tool 2020). Top: part of the nucleotide sequence of the human CD4 protein (query) is aligned with the mouse equivalent (sbjct). Bottom: protein alignment using the single letter code for amino acids (Chap. 3). The numbers refer to the position of this stretch of sequence within the full-length protein. Areas of identity and difference between the two species are shown in the middle line; a + indicates that while the amino acids may differ, they have similar chemical properties which may not significantly affect the protein structure. These are called conservative substitutions

Transcriptomics: Measuring Gene Expression

So far, this overview of genomics has focused on DNA sequencing as a tool for identifying drug targets, by using the genetic code to reveal protein sequences. Once the gene sequence for a potential drug target has been identified, it is important to know where in the body it is expressed, to anticipate any side effects that might arise with a drug. For example, if a drug is required to affect the lung, but the target is also expressed in the heart, there could be problems with cardiotoxicity during clinical development. Alternatively, it may be desirable to analyze the effect of existing drugs on gene (and protein) expression in the cell or animals; identifying drug-induced changes in specific genes or proteins can give clues about how the drugs are working in the body. Whichever question is asked, the answer depends on having technology that reliably measure the expression of many thousands of different proteins or messenger RNAs. The total mRNA complement of a cell is called the transcriptome, and the analysis is called transcriptomics. Several well-established techniques for detecting mRNA and proteins are used routinely by biochemists in academia and industry. Genomics, however, is characterized by its high-throughput analysis of genes and genomes, so the low- throughput methods in routine use are unsatisfactory. What follows is a description of two methods used to measure mRNA expression in tissues down to the level of a single cell (for a review, see Lowe et al. 2017).

Microarrays and Gene Chips

Microarrays allow the simultaneous measurement of thousands of different molecules using a highly compact format. Microarrays for detecting mRNAs consist of DNA which has been arrayed in precise rows onto glass slides at high density. For example, one million DNA samples can be supported on a slide within an area about the size of a fingernail. This is achieved through either the inkjet technology used to spray ink onto printer paper or else through the process of photolithography used to manufacture silicon chips. This latter procedure, developed by Affymetrix in the USA, gives rise to the "gene chip," a term that is sometimes used to describe DNA microarrays. CCD-based imaging devices and computer software make it possible to detect and measure a fluorescent signal from each mRNA that binds to its correct target on the chip. In practice the mRNA is converted *en masse* to a more stable DNA copy, called complementary DNA (cDNA) which is labeled with red or green fluorescent dyes. The use of different colors means that mRNAs from different sources can be compared on the same slide. For example, mRNA could be extracted from a cancer cell, and its cDNA labeled red; at the same time, a cDNA copy of the mRNA present in a normal cell could be labeled green. When mixed and applied to the same DNA microarray, genes that are present in the cancer cell will produce red dots, and those expressed in the normal cell will produce green dots; genes expressed at the same level in both will produce a mixture of the two (yellow). Figure 6.10 is

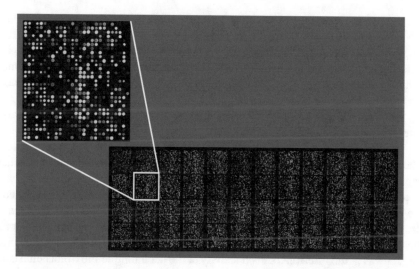

Fig. 6.10 DNA microarray showing the fluorescent signal created by the binding (hybridization) of cDNA copies of mRNA to DNA spotted onto a glass slide. Red dots, mRNA upregulated in one sample, green dots, mRNA downregulated, yellow dots, no difference between mRNA levels in the two samples

an example of a DNA microarray showing the fluorescent signals from thousands of different DNA binding events on a glass slide.

In addition to detecting mRNA (*via* a cDNA copy), DNA microarrays are used to identify DNA mutations. The DNA can be extracted from a patient's blood, labeled with fluorescent dye and then applied to a microarray that contains pieces of DNA with specific mutations. If the patient has a mutation in a corresponding piece of DNA this will show up as a hybridization signal. This type of DNA microarray is an important tool for large-scale genomics surveys (see later) as well as clinical development and personalized medicine covered in Chap. 14.

Finally, even though DNA microarrays are easier to fabricate, many investigations will be concerned with proteins, rather than the genes that encode them. Proteins can be detected, in clinical and other samples, on microarrays containing immobilized antibodies. If a fluorescently labeled protein is recognized by an antibody, it will bind and, as with DNA arrays, give a colored signal at a specific location on the slide. This large-scale detection of proteins is one of the characteristic features of proteomics.

RNA-Seq

RNA-Seq employs high-throughput DNA sequencing of transcript cDNAs derived from mRNAs the number of counts from each transcript giving an indication of relative abundance. In recent years, ten^9 transcript sequences can be recorded which is now sufficient for accurate quantitation of entire human transcriptomes. When this sequencing technology is coupled with the polymerase chain reaction (PCR) for DNA amplification (see Chap. 8), it is possible to measure mRNA transcripts from single cells. Understanding how genes are expressed at this level of sensitivity provides insights into key biological processes; for example, mRNA levels in single cells can vary significantly in what appears to be a homogeneous cell population.

Interpreting and Using DNA Sequences

The biopharmaceutical industry invests a vast amount of financial and intellectual capital in new technologies and procedures in the hope that drug discovery will become more efficient. Since the transition period between "traditional" pharmacology and biotechnology and the new molecular disciplines in the 1990s, the face of drug discovery has changed significantly, with the introduction of industrial scale laboratory research underpinned by powerful computers. The Human Genome Project was at the forefront of such activity at the beginning of the twenty-first century. It was established by a consortium of institutions worldwide to determine the nucleotide sequence of all three billion bases of human DNA thereby creating a catalog of all human genes, and by implication, all proteins expressed in the body. From a drug discovery perspective, knowing the full protein complement in human cells gives an upper limit on the number of possible drug targets (since most drug targets are proteins). Furthermore, different types of RNA molecules were discovered in the process of the sequencing, providing new insights into biological processes.

Once the human genome sequence was completed, many scientists were surprised to find that there are only about 20,000 protein coding regions, since previous work suggested that there could be as many as 140,000. This comparatively small number of genes seems strange for such a complex organism as *Homo sapiens*; the answer lies, in part, with the complex regulation of cell function that occurs independently of genome sequence. This area of epigenomics or epigenetics deals with the functions of DNA and RNA that are independent of the gene sequences inherited from both parents (see later).

Since 2003, thousands of important genomes have been sequenced, and the data deposited in online databases. The organisms studied include pathogens such as malaria parasites, as well as ones more likely to please the gourmet, such as the truffle genome. A new field of metagenomics has arisen, which uses sequencing technology to identify the genomes of individual microbes from within DNA that has been extracted from a mixture of organisms. Microbial DNA has been isolated from a variety of sources, both far away in the oceans and close to home in the human gut.[10] Identifying microbes, through their DNA sequence, is important from a general biology perspective because only a tiny proportion of this hidden world

---

[10]The adult human body is made up of about ten trillion cells, but at least the same number of bacteria live inside and on the skin (the microbiota).

has ever been cultured; it is possible to identify many new species without even having to grow them on agar plates. Sequence information may also be used to identify novel metabolic pathways that could be activated by specialized growth media, thus in a sense bringing these microbes back to life. This is potentially useful for drug discovery because microbes are sources of novel compounds. Furthermore, from the perspective of digestive diseases like ulcerative colitis and Crohn's disease, these studies help to identify the microbes that reside in the human intestines, some of which may trigger the immune system in the gut to become overactive. Interestingly, sequencing has been used to identify bacteria associated with a particular type of seaweed used in sushi; these bacteria were found in the intestinal bacteria of Japanese but not North American subjects, so the effect of diet on human populations can be followed in great detail (Nature 2010).

Despite the widespread interest in sequencing the whole of the living world, the human genome is still the focus of attention from a medical, pharmaceutical, and anthropological point of view. The first published human sequence was obtained from the pooled DNA of different individuals, but the first individual sequence was obtained in 2007 from Craig Venter's[11] own DNA, followed by others including James Watson and Archbishop Desmond Tutu. Since then, the number of full genome sequences is approaching one million, providing insights into human origins (e.g., ancient interbreeding with Neanderthals and Denisovans) as well as medically important findings, including potential drug targets. In the latter case, the low cost of sequencing and data processing has enabled researchers to explore the genetic basis of complex diseases like cancer in unprecedented detail as illustrated by the work of the International Cancer Genome Consortium (ICGC) described in the following section.

A large-scale analysis of 2,658 whole genome sequences from 38 tumor types was published in early 2020 by the Pan-Cancer Analysis of Whole Genomes (PCAWG) consortium (ICGC/TCGA consortium 2020). The multinational scale of the enterprise and the need for cloud computing to deal with the data pipelines are a good illustration of the challenges that can be overcome with this type of industrial-scale research. The results presented so far could lead to useful drug targets or early diagnostics (see Chap. 14) by providing novel insights into the number and types of mutations present in different cancers.

6.1.3.5 Epigenetics/Epigenomics

During the life of an organism from development to adult, specific parts of their genome can be turned on or off by a variety of chemical modifications, collectively known as the epigenome. The consequences of these modifications may be passed down through subsequent generations using mechanisms that are independent of the

---

[11] Sequencing pioneer who led the commercial effort to sequence the human genome in parallel with the publicly funded enterprise.

individual's genome sequence, i.e., they are epigenetic. The field is attracting a great deal of interest from researchers because of its relevance to diseases such as cancer and to the biology of stem cells (Kaiser 2010). Epigenetic mechanisms are varied and complex but involve proteins that are potential drug targets. The genetic material in a human cell is packaged into 23 pairs of chromosomes inherited from both parents. The chromosomes have the task of replicating the DNA each time a cell divides, which is quite an achievement since the length of DNA in a cell would stretch out to about a meter if pulled apart. To compress this length of DNA into chromosomes, it has to be supercoiled, as shown in Fig. 6.11.

As well as providing structure, the chromosomes also control the exposure of stretches of DNA to the enzymes that produce messenger RNA. These enzymes must physically interact with the DNA double helix, but they do so in conjunction with a class of chromosomal proteins called histones. The histone proteins are modified by the addition of specific small molecules to create a chemical code based on whether the small molecule is present or absent. Three classes of proteins are required to interpret the "histone code" depending on the small molecule modification, i.e., acetylation or methylation. These are writers: (histone acetyl transferases, histone methyltransferases), readers: (e.g., BET domain adaptor and PHD proteins), and erasers: histone deacetylases and histone demethylases. These enzymes and adapter proteins are under intense scrutiny as targets for small molecule drugs that can alter the expression of specific genes through epigenetic mechanisms.

Fig. 6.11 (**a**) Twenty-three pairs of human chromosomes imaged under a microscope (original image courtesy of NIH). Each pair has a number, and the sex chromosomes are labeled X. Females have two X chromosomes and males an XY pair. (**b**) Each chromosome contains genes in the form of double-stranded DNA that is supercoiled in combination with histone and other proteins to package a long strand into a small volume. Original image courtesy of National Human Genome Research Institute

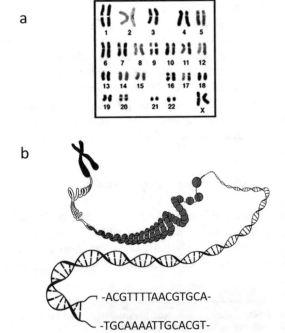

In addition to the histone proteins in chromatin, DNA itself can also be methylated, in this case at cytosine residues in "CpG islands." Methylation serves to repress gene transcription and can be involved in carcinogenesis through repression of the tumor suppressor genes that normally serve to keep tumor cells in check.

Finally, RNA itself is epigenetically modified by methylation at position 6 on the adenine base (m^{6}A) with consequences for gene expression, including a possible role in leukemia.

6.1.3.6 Proteomics

Proteins are present in each of the 200 different cell types in humans but can differ significantly in their level of expression. Some of them, the so-called housekeeping proteins, are expressed in all cell types, as they are involved in common processes such as replication and energy production. Other proteins are expressed in individual cell types, for example, hemoglobin in red blood cells and myosin in muscle.

Since proteins are central to drug discovery, scientists realized that the genomics approach to DNA research could also be applied to proteins. The (inevitable) name proteomics was coined in the early 1990s and is still very much in use today.

The Proteomics Toolbox

Proteomics is more challenging than genomics for several reasons, including the fact that the abundance of individual proteins in human tissues can vary from picogram to milligram levels, i.e., by a factor of a billion. Blood, for example, contains hundreds of proteins that are swamped by comparatively enormous levels of albumen and immunoglobulins (antibodies). This can be a real problem when attempting to detect low levels of blood proteins for a diagnosis. Despite these challenges, a range of highly sophisticated proteomics techniques has been developed to separate proteins from cell extracts or biological fluids and identify them. These techniques will now be described in the following sections.

Gel Electrophoresis and Mass Spectrometry

Mixtures of proteins can be separated based on their electrical charge, in a process called electrophoresis. The most common format is SDS polyacrylamide gel electrophoresis (SDS-PAGE) where polyacrylamide is a synthetic jelly-like matrix and SDS (sodium dodecyl sulfate) is a powerful detergent that is (rather alarmingly) used in shampoos. Proteins are dissolved by the SDS, which then confers on them a negative charge. This means that the proteins migrate through the polyacrylamide towards the positive electrode (anode) in the electrophoresis apparatus if an electric current is applied to both ends. The smaller the protein, the faster it migrates, so proteins are distributed through the gel at a distance that is proportional to their molecular weights. Once separated, the proteins can be stained with special dyes to reveal their positions. When standard proteins of defined molecular weight are run alongside the samples, the molecular weights of unknown proteins can be determined by simply comparing their migration relative to the standards. The identities of the separated proteins can be determined in several ways, one of which is through using specific antibodies and Western blotting (see Chap.14). This of course can only work if such antibodies are available. Another technique, protein mass spectrometry, can be used to identify any protein and is, in fact, a hallmark of proteomics. Mass spectrometry uses an electron beam to break molecules into charged fragments; these are accelerated through an electric field and separated according to their mass in a magnetic field. This provides highly accurate values for the molecular weights of the fragments that can be used to identify a protein through its amino acid sequence. The method of peptide mass fingerprinting (PMF) is now routinely used to identify proteins that have been literally cut out of the gel used for SDS-PAGE. The proteins are digested by the enzyme trypsin, which cuts the chain into peptides of defined size only where certain amino acids are present. The resulting peptides are passed through a mass spectrometer and their exact molecular weights determined. Protein sequences in the genome databases can be virtually "digested" by trypsin in the computer to create a set of peptides corresponding to all the proteins that the genome can encode. Each of these "virtual" peptides has a defined molecular weight which can be compared with that found in the real protein under investigation. If the peptides from the separated protein correspond with those obtained in the computer, the protein is identified. This powerful technique has been used to analyze complex mixtures of proteins from many different cell types and provides information about how these proteins interact together to perform their biological functions. The technology has advanced to the stage where it is now feasible to employ proteomics at the single-cell level using a "barcoding" system to identify minute amounts of protein using antibodies or mass spectrometry (Slavov 2020). The core technologies for proteomics described above are summarized in Fig. 6.12.

Proteins link together in specific complexes in the cell to form molecular machines, such as signaling pathways that transmit information from outside the cell to the inside. The association between proteins is a dynamic process that can occur very rapidly. This is important for drug discovery, since drugs that interact with a target that is part of a complex network, may perturb other pathways in the cell. These off-target effects may, or may not, be desirable. In an ideal world, it would be possible to predict all off target effects in advance of clinical trials, but our current understanding of complex biological interactions is inadequate. Therefore, a great deal of basic and applied research is being directed towards systems biology (see also Chap. 4). This approach to biological research is heavily reliant upon data generated from among other things, genomics, and proteomics experiments. The objective is to build models of how the cell operates, independently and within the intact body, to make drug discovery more predictive than is currently the case.

Another approach is to study the interactions between small molecules and protein targets in intact cells and tissues. If this were achieved on a proteome-wide scale, it would help to identify on-target and off-target effects with high precision. Thermal shift profiling based on quantitative mass spectrometry to measure changes in the thermal stability of proteins bound with small molecule ligands. This can be achieved using tissues and whole blood and therefore represents another approach to drug target discovery (Perrin et al. 2020).

Structural Proteomics

A protein's function is related to the three-dimensional shape that is dictated by its primary amino acid sequence. The importance of protein shape for drug binding has been outlined already in Chap. 3 and highlights the importance of structure as well as sequence. Structural proteomics is a term for the high-throughput generation and analysis of protein structures using tools such as X-ray crystallography and nuclear magnetic resonance (NMR). The former method requires, as the name implies, that the protein can be crystallized before being bombarded with X-rays to determine the exact position of each atom within the molecule. Crystals can be difficult to grow in the laboratory, particularly when the protein is normally found in cell membranes. In addition, the crystal "freezes" the structure in a fixed position, so an X-ray structure is a snapshot of one moment in time. Proteins have flexible portions whose movement can be visualized using NMR; the preferred method of structure determination, however, is still X-ray crystallography because of its high degree of atomic resolution. However, the technique of cryo-electron microscopy is beginning to compete with X-ray analysis, particularly for membrane proteins and large protein assemblies such as ribosomes. For this structure determination, proteins are flash frozen then bombarded with electrons to give a 3D image of the molecule. Advances in hardware and software are making this an increasingly popular method for protein structure determination despite the high cost of the equipment and the need for specialized laboratories.

Analyses of large numbers of structures (nearly 161,000, Protein Data Bank 2020) have revealed specific protein folds or domains that are used like building blocks in different protein families.[12] This "mixing and matching" of domains creates a great variety of protein structures that can be grouped together in families by virtue of their similarity. As will be seen later, these protein families are important for drug target selection.

6.1.4 How Many Drug Targets Are There?

It is obviously impossible to give a precise figure, not least because many potential drug targets are associated with microorganisms, rather than human cells. It is possible, however, to give an approximate figure for the number of targets for current drugs. An often-quoted paper was published by Overington et al. (2006), who arrived at a figure of 248 unique targets for small molecules and 76 for biologicals.

---

[12] Domain names often have quite esoteric origins, for example, Tudor domains are so named because the *Drosophila* Tudor protein causes offspring lethality when mutated, an outcome like with England's Tudor King, Henry VIII.

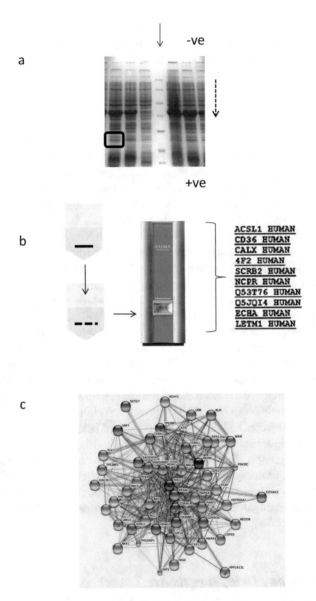

Fig. 6.12 Some of the tools of proteomics. (**a**) SDS-PAGE used to separate proteins through a gel by applying an electric field from top to bottom; -ve indicates the negative connection and +ve the positive. The proteins are applied in separate lanes to the top and, being negatively charged, are attracted to the +ve end and migrate to the bottom. Each protein appears as a thin band that is revealed by staining the whole gel with a dye. The arrow at the top indicates where seven molecular weight marker proteins have been separated within the gel. Since their precise molecular weights are known already, their relative migration can be used to determine the molecular weights of all the other proteins on the gel. (**b**) Determining the identity of proteins in each band. The box indicates where a protein band has been cut out of the gel. The piece of gel is then placed in a tube and the protein digested to smaller peptides using the enzyme trypsin. The peptides are introduced into

These figures cover existing drugs, but if drug discovery is to be made more systematic, as its practitioners hope, then it would be useful to know the maximum number that could be discovered. The tools of genomics and proteomics now make it possible to make at least a rough estimate of this figure. Analysis of the human genome sequence using bioinformatics has identified just over 19,000 protein coding genes and nearly 8000 noncoding RNAs (HGNC 2020). On the face of it, there should be a maximum of about 19,000 drug targets if every protein in the cell were involved in some disease. Unfortunately, biology is never this straightforward, there is always some complication. In this case the number of possible proteins far exceeds the number of protein-coding genes. There are two reasons for this: firstly, the phenomenon of alternative splicing means that each gene can be processed into mRNA with different lengths, so some the resulting proteins are derived from only a part of the original DNA sequence. It has been estimated that 95% of human mRNA molecules are spliced in some way. A second reason for the mismatch between the number of genes and proteins is the process of posttranslational modification. This is where proteins are chemically modified by the addition of small molecules. One example of this is the addition of carbohydrate molecules (glycosylation) to proteins that are destined to protrude from the cell surface. Other important chemical modifications, such as phosphorylation, will be highlighted later. It is now clear then, that the human proteome consists of hundreds of thousands of chemically unique proteins that are originally derived from only about 19,000 genes. Estimates of how many of these might be drug targets vary between one and ten thousand, based on existing and speculative targets, the latter based on observations in the biomedical literature (Zanders 2016). Whatever the true number, there are quite enough targets to keep the drug discovery industry occupied for a few years yet.

6.1.4.1 Target Classes

Estimates of the number of possible human targets for drugs are based on a mixture of guesswork and prior knowledge of drug target families. This prior knowledge is summed up by the rather inelegant term "druggable targets"; these are proteins with structural features which allow strong binding by small molecules. Targets that are affected by marketed small molecule drugs are, by definition, druggable, so one way of identifying novel targets is to search for related protein family members and produce small molecules against them to test in the laboratory. Some of the main protein families that contain known drug targets are listed in Table 6.2.

Fig. 6.12 (continued)   a mass spectrometer instrument that precisely identifies the size of the peptides; a computer then searches the protein databases to identify the best matches. These are shown as a list of names to the right of the instrument. (c) A protein interaction map drawn using the STRING database (STRING 2020). Proteins that strongly associate together as a complex within cell have been isolated and then analyzed using SDS-PAGE and mass spectrometry. In this way, it is possible to draw a network of interactions between each component protein in the complex. In this example, the p53 protein that is involved in cancer is at the center of the network

Table 6.2 Main protein families containing druggable targets. Cytokine receptors are an exception because they are not inhibited with small molecules, but they are an important target class for biologicals

Target class	Representative example	Role in disease	Associated drug
Receptors			
GPCRs	Histamine H2	Stomach ulcers	Ranitidine
Nuclear	Estrogen receptor	Breast cancer	Tamoxifen
Cytokine	TNF-α receptor	Rheumatoid arthritis	Enbrel (biological)
Enzymes			
Proteases	ACE	Hypertension	Lisinopril
Kinases	Abelson tyrosine kinase	Leukemia	Gleevec
Phosphatases	Calcineurin	Transplant rejection	Cyclosporine A
Ion channels			
Ligand gated	GABA$_A$ receptor	Anxiety	Benzodiazepines
Voltage gated	Calcium channel	Angina/hypertension	Norvasc
Transporters	Monoamine transporters	Low serotonin- depression	Fluoxetine

The above list represents a small fraction of possible drug targets, but it does highlight those families that are still the subject of intense interest by the biopharmaceutical industry. This chapter will conclude with a brief description of each of these classes in turn.

- G-Protein-Coupled Receptors (GPCRs)

Over 20% of marketed drugs bind to this class of receptor, which is expressed on the surface of many different cell types. The protein is anchored to the cell membrane by threading through it seven times, hence the alternative name seven transmembrane (or 7TM) receptor. G proteins are molecules that couple to this type of receptor to transmit to the inside of the cell signals generated after ligand binding. The human genome contains over 1000 GPCRs, many of which are involved in sensing tastes and smells (i.e., gustatory and olfactory). However, around 400 GPCRs have potential as drug targets although there is still much to learn about their normal roles in the body. All have the potential to bind ligands, but those proteins without a known ligand are called orphan receptors, until someone finds a home for them, which is happening with increasing frequency. GPCRs are coupled to two types of signaling protein (arrestins and G proteins) such that drug binding can result in multiple actions on the cell. As a result, it may be possible to design small molecules that exhibit biased signaling towards a specified pathway to create a desirable outcome. One prominent example is the possibility of creating safer opioid drugs as painkillers (Ehrlich et al. 2019) although there is some controversy over the safety benefits of drug candidates in this area (Servick 2020).

- Nuclear Receptors (NRs)

These small molecule receptors bind ligands inside the cell rather than on the surface. After ligand has bound, a complex formed between the ligand and the receptor moves into the cell nucleus to switch target genes on or off. Nuclear receptors bind small lipophilic molecules like steroids and retinoids (e.g., the sex hormones and vitamins A) which have profound effects on human physiology. The human genome codes for about 50 NRs of which as many as half are orphan receptors without known ligands.

- Cytokine Receptors

A varied family, consisting of more than one protein chain anchored into the cell membrane. The cytokine ligands are proteins such as interleukins, interferons, and growth factors. As small molecules are not able to disrupt cytokine binding to its receptor, larger protein drugs must be used. These are either antibodies or else recombinant forms of the receptor that act as decoys to prevent ligand binding.

- Proteases

Enzymes that clip proteins at defined points by targeting specific amino acid sequences. Proteases have many roles in the human body, such as blood clotting, processing of peptide hormones, and the life cycle of viruses like HIV. Although inhibitors of proteases are challenging to produce as drugs, there are several successful medicines on the market that target this enzyme class.

- Protein Kinases

A large family of enzymes that attach a phosphate group (PO_3) to a protein molecule to change its activity in the cell. This process of protein phosphorylation is used by the cell to transmit signals from external receptors through the membrane and into the cell nucleus where they switch on gene expression. Kinases are of importance in cancer; this is because cells grow independently of external growth factors due to the permanent activation of specific kinase enzymes. Protein kinase inhibition is now a major area of small molecule drug discovery.

- Protein Phosphatases

Protein phosphatases are enzymes that perform the reverse role of protein kinases, that is, they remove phosphate groups from proteins instead of adding them. This is an important mechanism for cell signaling, as it acts as an "off switch" to keep control over cellular processes. Although phosphatase inhibitors would find use in several diseases, they have been difficult to turn into medicines. Despite this, there is still plenty of research activity aimed at producing drugs based on this target class, including, possibly, a small molecule substitute for insulin.

- Other Enzymes

The enzyme families listed above (except for the phosphatases) make up the majority of current drug targets, but not all of them. Prominent examples are the HMGCoA reductase inhibitors (otherwise known as statins) that lower cholesterol and the phosphodiesterase (PDE) inhibitors, of which Viagra® is the most famous (infamous?) example.

- Ion Channels

Ions are atoms or molecules that bear a positive or negative charge. Metal ions, such as calcium (Ca^{2+}), sodium (Na^+), and potassium (K^+), play vital roles in many physiological processes. Ions pass through pores in cell membranes called ion channels. The two main types are known as ligand-gated and voltage-gated ion channels. Each channel is selective for an ion that moves from the outside of the cell inwards or vice versa. Ion movements are involved in the function of nerve cells and are implicated in conditions such as epilepsy, pain, and heart arrhythmias. There is considerable interest in ion channels as drug targets, with a few important drugs already on the market.

- Transporters

Transporter proteins carry molecules within the body, specifically through the cell membranes which would normally block their passage into cells. For example, nutrient molecules are carried across the stomach by transporters. Transporters are molecular targets in several diseases, the best known being depression, which is associated with low levels of serotonin in the brain. The monoamine transporter removes serotonin from nerve cells, but this neurotransmitter can be maintained at higher levels by inhibiting the transporter with drugs like fluoxetine (Prozac®), thereby alleviating the symptoms of depression. Other transporter drug targets of interest include glucose transporters, whose inhibition is useful for treating diabetes.

6.1.5   Closing Remarks

The search for drug targets that makes up the beginning of the drug discovery pipeline is complex and technically challenging as has been made clear in this long chapter. This is only the beginning, however, since enormous challenges lie ahead in finding drug molecules with all the properties required to turn them into safe and effective medicines. It is worth making this point, because however exciting a new target opportunity seems to its discoverer, the clinical reality can be disappointing. This could of course either be due to the target being unsuitable or the drug itself being problematical; however, there is certainly no shortage of drug target ideas coming from academia and industry (Zanders 2016). The description of how researchers approach the selection of drug targets is now concluded. The next stage in the drug development pipeline involves the discovery of small or large molecules that selectively interact with the targets.

Summary of Key Points

Observations of the signs and symptoms of disease give clues that allow the researcher to identify specific tissues in the patient which can be further investigated at the level of cells and molecules. This process is formalized as translational medicine.

In vitro experiments are undertaken to tests the hypothesis that a specific drug target protein is relevant to a disease mechanism.

In vivo experiments are undertaken in animals and other model organisms, such as zebrafish, to show that modifying the target with an antibody, or other agent, will ameliorate a disease symptom.

Modern biological disciplines, based on cell biology, biochemistry, and molecular biology, provide the tools to understand diseases and drug targets at the level of genes, proteins, and other molecules. More recent offshoots of these disciplines include genomics, epigenomics, transcriptomics, proteomics, systems biology, and bioinformatics.

The Human Genome Project identified about 19,000 genes that could encode human proteins, but the number of proteins produced by cells is much higher.

There are approximately 400 targets for current medicines, but there could be as many as 10,000 targets in human and other cells to be exploited in the future, based on estimates of protein numbers.

Many drug targets are distributed in protein families, such as GPCRs and nuclear receptors.

References

Cheng Y et al (2013) Therapeutic targeting of autophagy in disease: biology and pharmacology. Pharmacol Rev 65:1162–1197

Ehrlich AT et al (2019) Biased signaling of the Mu opioid receptor revealed in native neurons. iScience 14:47–57

Giani AM et al (2020) Long walk to genomics: history and current approaches to genome sequencing and assembly. Comput Struct Biotechnol J 18:9–19

GloFish (2020) https://www.glofish.com/. Accessed 25th Feb 2020

HGNC (2020) https://www.genenames.org/download/statistics-and-files/. Accessed 19th Feb 2020

Huang X, Dixit VM (2016) Drugging the undruggables: exploring the ubiquitin system for drug development. Cell Res 26:484–498

ICD-10 (2016) https://icd.who.int/browse10/2016/en/. Accessed 5th Feb 2020

ICGC/TCGA consortium (2020) https://doi.org/10.1038/s41586-020-1969-6. Accessed 18th Feb 2020

IUBMB. https://www.qmul.ac.uk/sbcs/iubmb/iubmb.html. Accessed 13th Feb 2020

Johns Hopkins Medicine. https://www.hopkinsmedicine.org/henriettalacks/. Accessed 8th Feb 2020

Kaiser J (2010) Epigenetic drugs take on cancer. Science 330:576–578

KEGG. http://www.genome.jp/kegg/. Accessed 13th Feb 2020

Khorana HG (1972) Studies on polynucleotides. 103. Total synthesis of the structural gene for an alanine transfer ribonucleic acid from yeast. J Mol Biol 72:209–217

Lowe R et al (2017) Transcriptomics technologies. PLoS Comput Biol 13:e1005457

Mak IWY et al (2014) Lost in translation: animal models and clinical trials in cancer treatment. Am J Transl Res 6:114–118

Monaco C et al (2015) Anti-TNF therapy: past, present and future. Int Immunol 27:55–62

Nature (2010) https://www.nature.com/news/2010/100407/full/news.2010.169.html. Accessed 25th Feb 2020

NCBI BLAST tool (2020) https://blast.ncbi.nlm.nih.gov/Blast.cgi?CMD=Web&PAGE_TYPE=BlastHome. Accessed 17th Feb 2020

Normand R et al (2018) Found In Translation: a machine learning model for mouse-to-human inference. Nat Methods 15:1067–1073

Overington JP et al (2006) How many drug targets are there? Nat Rev Drug Discov 5:993–996

Pei H et al (2019) Small molecule PROTACs: an emerging technology for targeted therapy in drug discovery. RSC Adv 30:16919–17520

Perrin J et al (2020) Identifying drug targets in tissues and whole blood with thermal-shift profiling. Nat Biotechnol 38:303–308

Protein Data Bank (2020) http://www.rcsb.org/pdb/results/results.do?tabtoshow=Current&qrid=280DEA66. Accessed 19th Feb 2020

Servick K (2020) Safety benefits of 'biased' opioids scrutinized. Science 367:966

Shrestha L et al (2016) Heat shock protein (HSP) drug discovery and development: targeting heat shock proteins in disease. Curr Top Med Chem 16:2753–2764

Slavov N (2020) Unpicking the proteome in single cells. Science 367:512–513

STRING (2020) https://string-db.org/. Accessed 25th Feb 2020

Tang D et al (2019) The molecular machinery of regulated cell death. Cell Res 29:347–364

The Scientist (2001) https://www.the-scientist.com/commentary/ome-sweet-omics%2D%2D-a-genealogical-treasury-of-words-54889. Accessed 13th Feb 2020

Yeasmin F et al (2018) Micropeptides encoded in transcripts previously identified as long non-coding RNAs: a new chapter in transcriptomics and proteomics. Front Genet. https://doi.org/10.3389/fgene.2018.00144

Zanders ED (2016) Human drug targets a compendium for pharmaceutical discovery. Wiley, Chichester

Chapter 7
Medicinal Chemistry

Abstract Medicinal chemistry is concerned with the synthesis of small molecules which are passed to biologists for testing against drug targets. The vastness of chemical space means that the number of different molecules that could theoretically be made is almost infinite; in practice, however, there are constraints based on the limitations of synthetic chemistry and the effects of certain functional groups on the human body. This chapter describes the goals of medicinal chemistry and introduces the range of technologies used to produce active drug molecules, including computer-aided drug design and natural product chemistry.

7.1 Introduction

The target discovery process at the beginning of the drug discovery pipeline was covered in some detail in the previous chapter. By this stage it can be assumed that a suitable protein target has been identified, so the next challenge is to discover molecules that interact with the target; these molecules will then be turned into candidates for full clinical development. This chapter covers the discovery of the small molecules that still represent most drugs, despite the rapid rise of biotherapeutics (see Chap. 8). The primary aim of small molecule discovery is to provide compounds (leads) that can be further developed into the final medicinal product. It is quite possible however that the final medicine will differ significantly from the first identified compounds (hits) because of the modifications needed to improve their pharmacokinetic properties (such as adsorption and metabolism) that will be discussed in later chapters. Almost all small molecule drugs are produced using the tools of synthetic organic chemistry which have been developed over approximately 200 years. Organic chemistry, like other sciences, has its own subdivisions, which in the biopharmaceutical industry are medicinal, analytical, and process chemistry. Medicinal chemists (sometimes referred to as pharmaceutical chemists) undertake the compound development work from the beginning, right up to the point where the compound is ready for clinical testing. Once a compound is likely to be progressed further, process chemists start to develop a safe, environmentally sensitive, and cost-effective strategy for producing it in large amounts for clinical testing and ultimately manufacture. Analytical chemists are important at all stages of drug development, as they must devise ways of testing the purity of compounds and measuring their concentrations in blood and other biological fluids.

7.1.1 The Vastness of Chemical Space

In Chap. 6, biological information was used to pose the question: "how many drug targets are there?" A related question could be asked of small molecules, this time based on chemical information, namely, how many organic molecules would it be theoretically possible to make? The term chemical space is used to describe all these possible combinations. Theoretical calculations of the number of possible structures of molecules with molecular weights of less than 700 have arrived at the staggeringly large number of 10^{60} combinations (Bohacek et al. 1996). In fact, just adding 150 common functional groups in combination to the simple n-hexane molecule (containing six carbon atoms in a chain) would alone produce 10^{29} combinations (Lipinski and Hopkins 2004). Chemists have produced roughly 100 million (10×10^7) compounds, a figure that is clearly a minute fraction of the total number that could theoretically be made. In practice, organic synthesis is a far from trivial exercise; this means that a realistic estimate of the number of small molecule drugs that are likely to made, although numbered in the millions, are a minute proportion of the theoretical maximum. Table 7.1 is adapted from a review of chemical space by Mullard (2017) and puts these numbers in context.

7.1.2 The Goals of Medicinal Chemistry

The aim of medicinal chemistry is to produce compounds with the following properties:

Table 7.1 The vastness of chemical space

	Number
Compounds with drug-like characteristics	10^{60}
Atoms in the Solar System	10^{54}
Stars in the universe	10^{24}
Bytes in Facebook's data warehouse	10^{18}
GDB-17 virtual library	1.7×10^{11}
Neurons in human brain	10^{11}
ZINC virtual library	10^8
PubChem cataloged small molecules	10^8
Approved drugs	6×10^3

Figures taken from Mullard (2017)

- Potency and selectivity against the target in vitro and in vivo
- Lack of toxicity
- Good pharmacokinetics (stability in body and other factors to be detailed later)

It is rarely possible to produce a molecule that immediately fulfils all the above criteria. A useful analogy to compound optimization is the notorious Rubik's cube game devised by the Hungarian Ernő Rubik. The cube consists of six faces made up of nine stickers displaying one of a total number of six colors. The colors can be mixed up through a series of independent turns, and the objective is to produce a single color on each face. An inevitable feeling of triumph ensues when one face has been manipulated into the same color, followed by frustration when attempts to create further uniform faces result in the first face disappearing. This exactly parallels what can happen in medicinal chemistry when, for example, a potent compound is identified that is not very selective. Subsequent efforts to improve selectivity by modifying the molecule may only serve to reduce potency and so on.

7.1.2.1 Starting Points

Some medicinal chemistry case histories have already been described in Chap. 4. The histamine H2 antagonists, for example, were discovered by modifying histamine and testing these modified compounds (i.e., analogues) in a biological assay to determine their activity. The starting point was therefore the natural molecule whose structure was already known. The type of starting molecule will vary depending on the drug target class. For receptors like GPCRs and nuclear receptors, these will be ligands such as hormones, while for enzymes they will be the natural substrates. Taking proteases as an example, the enzyme binds to the substrate (i.e., a specific protein) and clips it at a specific amino acid, producing two or more smaller fragments. This chemical reaction occurs via a so-called transition state, and the starting point for medicinal chemistry might be a chemical analogue of this based on a peptide sequence. In many cases, however, it is not possible to use a natural molecule as a starting point because no such molecule exists, or the natural molecule is too difficult to synthesize or modify. There may also be patent issues if other companies have covered the obvious chemical features with their own drugs. The alternative method is to identify completely novel compounds that interact with the target by randomly screening large collections of diverse chemical compounds. This is one of the most common starting points in the industry and is often the first port of call for any small molecule discovery program. Screening will be covered in more detail in Chap. 9. Another starting point is, of course, the target protein itself. If a three-dimensional structure is available, this can provide details about the various binding pockets that could be accessed by small molecules. This is made more straightforward if the protein structure already has a small molecule bound to it, so much effort goes into producing crystals with both molecules bound together (co-crystallized).

Finally, it is possible to design molecules in the computer as starting points (in silico drug design). This is used extensively in medicinal chemistry to optimize the interactions between drugs and their targets and, with advances in machine learning, is promising to revolutionize the search for small molecule drug candidates (see later).

7.1.2.2   The Search for Pharmacophores

The collection of molecular features contributing to the specific action of a drug on a target is known as a pharmacophore. The objective of a medicinal chemistry program is to identify a suitable pharmacophore and modify it to achieve the desired biological outcome (e.g., agonist or antagonist). The histamine H2 receptor case history given in Chap. 4 is a good illustration of this. The pharmacophore was identified as the imidazole ring in histamine which was then retained in cimetidine in a slightly modified form (addition of a methyl group) and replaced completely in ranitidine. This replaced group (a furan) is known as an isostere since it has the same shape as the imidazole and can substitute for it, producing a drug molecule with as good, if not better, activity than the original molecule. This process of isosteric replacement is widely undertaken by medicinal chemists who want to create greater compound diversity for improving their biological properties and establish composition of matter patents.

Statin drugs provide another example where pharmacophores lead to useful drugs, in this case, among the best-selling medicines of all time. Statins lower the levels of cholesterol in the blood by inhibiting its synthesis, the result being a significant reduction in the incidence of coronary heart disease in high-risk groups of patients. Cholesterol is a lipid, built up in a stepwise fashion from small chemical units by a series of enzymes, the key one being HMGCoA reductase which controls the rate of cholesterol synthesis. Inhibition of this enzyme reduces cholesterol levels in the blood and the associated buildup of fatty deposits in the arteries (atherosclerosis). HMGCoA inhibitors have been developed by several companies and marketed as statin drugs. Figure 7.1 shows the molecular structures of the natural molecule hydroxymethylglutaryl coenzyme A (HMGCoA), along with two of the best-selling statin drugs marketed under the names Crestor® and Lipitor® by AstraZeneca and Pfizer, respectively.[1] The pharmacophore in HMGCoA is indicated by the dotted box in the figure and is replicated (with some modification) in the two totally synthetic drugs. The resemblance between them and the natural molecule ends there, as will be obvious even to the untrained eye. In these examples, the addition of a fluorophenyl group (highlighted) has enhanced the binding of the synthetic statins to the enzyme target. The fluorine atom (F) is rarely found in natural molecules but is used extensively in synthetic drugs as it can improve metabolic stability or binding to the protein target.[2]

---

[1] These are now off-patent and sold as cheaper generic products (see Chap. 16).

[2] Other "nonnatural" substitutions are also used to optimise drug-like properties, including deuteration (replacing hydrogen with the deuterium isotope) or using silicon as a substitute for carbon.

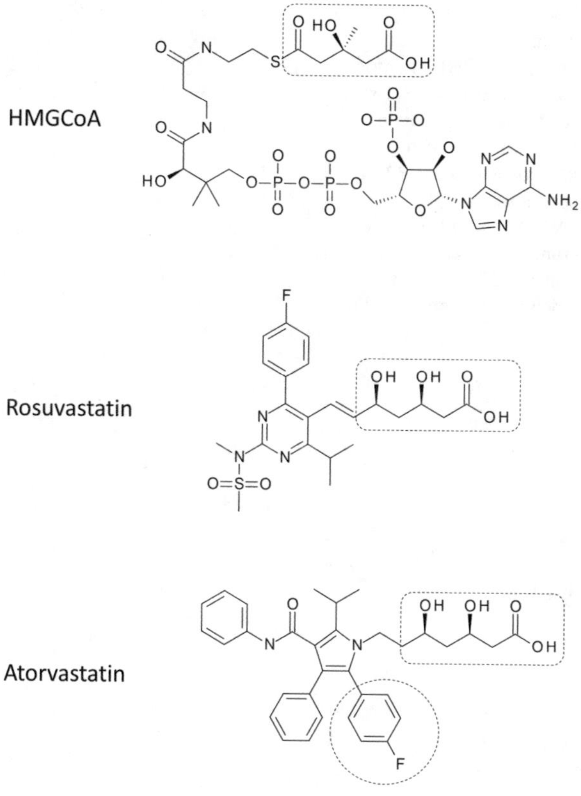

Fig. 7.1 Molecular
structures of cholesterol-
lowering statin drugs and
the pharmacophore present
in the natural substrate for
HMGCoA reductase. The
fluorophenyl group is
highlighted (see text). The
reason why some bonds
are drawn as solid or
dotted wedges is explained
later under stereochemistry

HMGCoA

Rosuvastatin

Atorvastatin

7.1.2.3 Visualizing Chemistry

The "think like a chemist" section in Chap. 3 highlighted the way in which a medicinal chemist can interpret two-dimensional chemical structure diagrams to identify different functional groups and get some idea of three-dimensional structure. This chemists' technique of visualization goes back to the nineteenth century when pioneers such as August Kekulé tried to determine the structures of organic molecules using the ideas about valency that were being put forward at the time. Kekulé is credited with working out the ring structure of benzene, apparently after visualizing it in a series of dreams (one of them on top of a London bus) (Rocke 2010; Robinson 2010). The routine use of computer graphics makes the visualization process much more straightforward, allowing different molecular features such as charge and shape to be added and subtracted at will.

The chemical modifications that gave rise to histamine antagonists and the statin drugs reflect the skill and knowledge acquired by medicinal chemists during their work. This skill is almost like having a natural feel for the system under

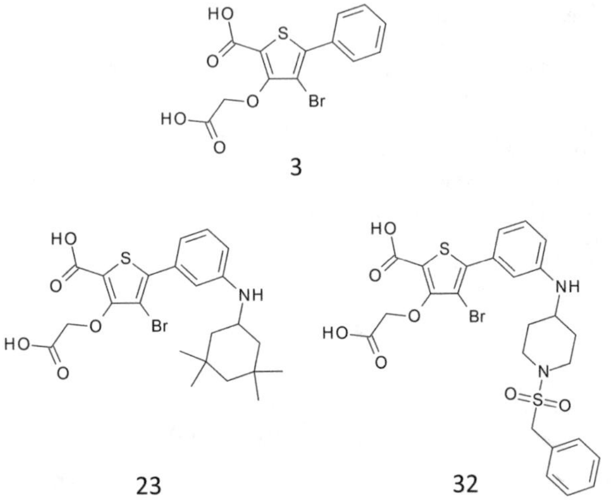

Fig. 7.2 Medicinal
chemistry program to
produce PTP1B inhibitors
based on structure 3. Extra
functional groups added to
basic structure with
compounds 23 and 32
shown as examples.
(Adapted with permission
from Douglas et al. (2007),
copyright 2007 American
Chemical Society)

3

23

32

Table 7.2 Activity of three
inhibitors of PTP1B from
Fig. 7.2

Compound	PTP1B	CD45
3	3.2	280
23	.036	151
32	.004	77

Figures represent the potency
of inhibition measured (in
micromolar units) by a value
called the K_i. The lower the
value, the greater the potency
(see any resource on enzyme
kinetics). The potency of compound 32 is therefore much
greater than compound 3,
because of the difference
between 3.2 micromolar and
0.004 micromolar, (or 4 nanomolar). Adapted with permission from Douglas et al. (2007)
copyright 2007 American
Chemical Society

investigation, rather like a gardener having "green fingers" for growing plants.[3] The
Structure Activity Relationship (SAR) is central to medicinal chemistry; as the
name implies, each structure chosen for biological testing will have an activity that
varies in a defined way according to the type of chemical groups added to the starting molecule. This is illustrated by an example taken from a study of enzyme

---

[3] Drug discovery scientists like James Black and Paul Janssen had this feel and, with their teams,
were responsible for major pharmaceutical innovations.

inhibitors by scientists at Wyeth Research. Full details are available in their publication, a highly technical description of attempts to create potent inhibitors of the protein phosphatase enzyme PTP1B (Douglas et al. 2007). There is considerable interest in this enzyme as a drug target, since its inhibition would enhance the response of cells to insulin signaling; this signaling defect is responsible for the insulin resistance that is central to type 2 diabetes. Compound 3 in Fig. 7.2 was used as a starting point for a series of modifications to the basic structure aimed at making the molecule large enough to bind to different parts of the enzyme and enhance its activity. Compounds 23 and 32 represent two such modifications, 32 being larger and more complex.

The activities of each compound against PTP1B and CD45, another protein phosphatase, are shown in Table 7.2. The smaller the figure, the more potent the inhibitor. By adding an extension to molecule 3, its activity is increased about 100-fold with 23 and about 1000-fold with 32. This brings the activity from 3.2 micromolar to 4 nanomolar. This is the level of potency required for this type of inhibitor if it is to progress further in development. Table 7.2 shows the comparative lack of activity of all the compounds on the phosphatase CD45; this means that the compound is unlikely to cause side effects related to inhibition of this enzyme.

The Quantitative Structure Activity Relationship (QSAR) is a key part of chemoinformatics, which is the analysis of virtual (imaginary) compounds using computer algorithms. QSAR involves some relatively complex statistical analysis, using values such as molecular weight and lipophilicity (solubility in fats) to predict the functional groups which are most likely to enhance biological activity. In addition to identifying groups with positive attributes, QSAR is used to eliminate potentially toxic groups, for example, the thiourea group in the H2 antagonists (Chap. 4). Medicinal chemists have access to computer databases of known toxic groups, so they can pre-screen a collection of compounds before committing any to synthesis and biological testing.

7.1.2.4 What Makes a Good Drug Candidate?

Not all compounds interacting with a selected target will be suitable for administration to human subjects. This has been understood the hard way by generations of drug discovery scientists and nothing has changed since. It is therefore important to define exactly what chemical properties a new compound must possess if it is to have any chance of reaching the marketplace. Medicinal chemists are rather like doctors, in the sense that it is always possible to get a second (or third or fourth) opinion on what they think is promising and what is not worth pursuing. I always used to feel dispirited in project meetings when the chemists would review just about every compound that we had selected with good biological activity only to reject it on the basis that it was "a Michael acceptor," a chemical feature that usually raises problems during clinical development.

This rather ad hoc way of reviewing structures led Pfizer chemist Peter Lipinski to formulate the "Rule of Five" for selecting compounds for further development. This "Lipinski's Rule" states that for compounds to be orally active, they should have:

- A molecular weight of less than 500 daltons
- No more than 5 hydrogen bond donor atoms
- No more than 10 hydrogen bond acceptor atoms
- A logP of less than 5

This last point is a measure of the lipophilicity of a drug that is either determined theoretically or by measuring how much dissolves in an organic solvent (octanol) compared with water. As logP is a logarithmic scale, a compound with a logP of 6 is ten times more lipophilic than one with a logP of 5.

Other rules have been developed by the medicinal chemistry community, and all serve a useful purpose if followed to the spirit, if not the letter, of the law.

At this point, it is important to stress that compounds selected initially for their potency and selectivity in biological assays may not be suitable as drugs because of pharmacokinetic or toxicity issues (see Chap. 11). If such issues do exist, it may be necessary to modify the chemical structure of the initial compound to a point where the structure of the final drug is significantly different. Specific functional groups in the initial structure may be replaced by bioisosteres. Examples include the replacement of phenyl rings by bicyclopentanes (BCPs) conferring increases in solubility, membrane permeability and resistance to metabolism (Zhang et al. 2020), and the "magic methyl effect" of inserting a methyl group to increase binding affinity (Leung et al. 2012).

7.1.2.5 Stereochemistry

The stereochemistry (i.e., shape) of small molecule drugs determines how they will bind to their target, and this must be considered when planning which compounds to synthesize. Many molecules exist in isomeric forms, i.e., they have the same number and type of atoms but are arranged differently (Introduction to Stereochemistry 2019). Some are mirror images of each other; a phenomenon is termed chirality, from the Greek for hand.[4] This is illustrated by looking at the left and right hand: the fingers are the same, but they are ordered as mirror images and cannot be superimposed. This has implications for drug binding, since only one mirror image form (enantiomer) will bind to a protein target, while the other is therefore inactive or will possibly interact with a totally different protein.

The chiral forms of the amino acid valine are shown in Fig. 7.3. The L and D enantiomers have identical atoms but cannot be superimposed on each other. The drawing uses the Fischer projection, in which bonds projecting out towards the

---

[4] Other types of stereochemistry are important, including geometrical isomerism, all of which can be found in educational resources for organic chemistry.

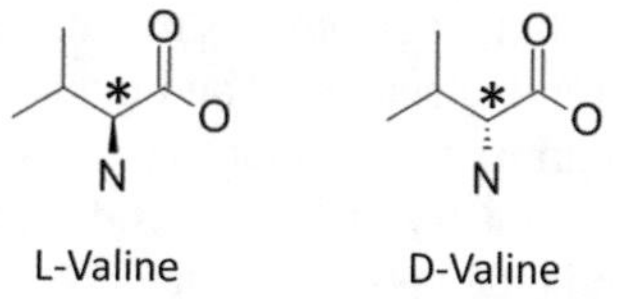

L-Valine   D-Valine

Fig. 7.3 Example of chiral forms of the amino acid valine with Hs omitted. The two molecules are mirror images of each other. Only the L-form of this and other amino acids are used in proteins, suggesting that this mirror image form was favored during the early stages of life on earth

viewer are indicated with a solid wedge and those with a dotted wedge project backwards. The asterisk indicates the carbon atom positioned at the point of symmetry of the chiral molecule.

Many complex drug molecules have more than one chiral center; mathematically if there are N chiral centers, there will be 2^N enantiomers, which becomes worryingly high for chemists planning the synthesis of a single chiral form. Examples of chiral drugs and their importance for toxicology and patent extension will be given later in the book.

7.1.3   How Are Compounds Synthesized?

This is where the author must concede defeat when trying to explain the ramifications of this vast and highly technical subject. However, some attempt must be made to convey at least a feeling for how synthesis is performed in the modern chemistry laboratory.

The early literature on organic chemistry was written in the nineteenth century and is still referred to directly or indirectly by chemists wanting to convert one set of molecules into another. Many chemical procedures are used "off the shelf" to produce specific functional groups as required and are often named after the people who developed them. Examples include the Williamson ether synthesis and the Diels-Alder reaction. An illustration of how the Diels-Alder reaction can be used to form complex rings from simpler molecules is shown in Fig. 7.4 below:

Synthetic methods are still being developed to solve chemical problems which previously may have seemed intractable; one example is the selective breaking of

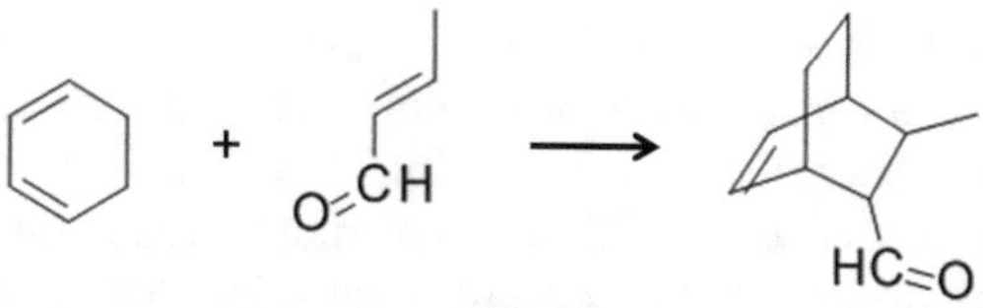

Fig. 7.4 Example of complex ring synthesis from simple alicyclic (nonaromatic ring) compound and a linear compound (an aldehyde). This uses the Diels-Alder reaction invented by two German chemists which earned them a Nobel Prize for chemistry

C-H bonds to introduce new functional groups into molecules. Synthetic chemistry is one area in which innovation in academic and industrial laboratories can make a real impact upon the biopharmaceutical industry. New tools are appearing all the time, for example, "click chemistry" used to produce complex compounds from simple modular units, often in the presence of metals as catalysts.

It is hard to overestimate the technical challenges of devising a reaction scheme for synthesizing drug compounds, particularly where chiral molecules are involved. The actual synthesis of organic compounds is not totally dissimilar to cooking, although the products of that pleasurable activity are unlikely to blow up in your face. The following example of a typical synthetic procedure is taken (more or less) verbatim from a paper published in the Journal of Medicinal Chemistry by a group investigating receptor agonists as potential treatments for multiple sclerosis (Bolli et al. 2010) adapted with permission Copyright 2010 American Chemical Society.

> To a solution of isopropylamine (1.31 g) in methanol (25 mL), phenyl isothiocyanate (3.00 g) is added portionwise. The mixture which became slightly warm (approximately 30 °C) was stirred at room temperature for 3.5 h before bromoacetic acid methyl ester (3.39 g) followed by pyridine (2.63 g) was added. Stirring of the colorless reaction mixture was continued for 16 h. The resulting fine suspension was diluted with 1N Hydrochloric acid (100 mL) and extracted with diethyl ether (150 mL). The separated organic phase contains crude 4a. The pH of the aqueous phase was adjusted to pH 8 by adding saturated aqueous $NaHCO_3$ solution. The aqueous phase was extracted with diethyl ether (4 × 150 mL). The combined organic extracts containing crude 3a were dried over $MgSO_4$, filtered, and concentrated. The remaining crystalline solid was washed with heptane and dried under high vacuum to give 3a as an off-white crystalline solid.

This example illustrates the long cooking times needed to transform one compound into another. It also highlights the potential hazards and discomforts involved in working in a chemistry laboratory, as ether is explosively flammable, and pyridine has a highly unpleasant smell. Industrial chemists and biologists have traditionally been kept well apart in separate buildings, partly because of sociological differences, and more seriously, because of the nature of the science. This is beginning to change in some institutions, as scientists from different disciplines are becoming more integrated within single working areas to allow cross fertilization of ideas.

The field of organic synthesis is not standing still as there are many challenges to be faced in exploring new chemical space and designing cost-effective and environmentally sensitive reaction schemes. Some examples of the way in which these challenges are being met are shown below

Novel Synthetic Chemistry: Redox Reactions Using Light or Mechanical Pressure

Photoredox catalysis, or the creation of redox (oxidation/reduction) reactions using visible light, is capable of transforming drug-like molecules in previously unachievable ways. Alternatively, it is possible to use mechanoredox reactions by applying mechanical stress to piezoelectric materials in a ball mill, a device used for industrial applications for over 100 years. The material acts as a redox catalyst and has the advantage over photocatalysis in not requiring potentially harmful solvents (Kubota et al. 2019).

Automated Chemistry: The Chemputer

The automated synthesis of oligonucleotides was described in Chap. 6 as part of the genomics toolbox. Along with peptide and oligosaccharide synthesis, this is relatively easy to automate as the chemical reactions required for each step are similar. The world of drug synthesis is far more varied, however, so the challenges of automation are greater. Nevertheless, using computer software in the form of a chemical programming language, along with a modular robotic platform in the lab, it is possible to achieve this automation for a disparate range of drug molecules (Steiner et al. 2019).

Machine Learning for Chemical Synthesis

The ability of computers to extract useful information from datasets in the absence of human intervention would appear to be ideal for designing synthetic reaction schemes. There is of course no shortage of interest in this area but to answer the question posed by Maryasin et al. (2018), "Machine Learning for Organic Synthesis: Are Robots Replacing Chemists?" the answer is, not yet, but the robots are advancing.

7.1.3.1 Analytical Chemistry

Once a compound has been synthesized, it is purified from all the other molecules present in the chemical reaction through a variety of well-established techniques such as filtration, crystallization, or chromatography. Chromatography is widely used in chemistry and biochemistry and will be described in later chapters. Analytical chemistry is a key part of medicinal chemistry as it is used to ensure that a synthesized compound is what the chemist thinks it should be. It would obviously be impossible to obtain reliable data on compounds that have not been properly identified. Standard methods of analysis include the use of spectroscopy, which is based on the interaction of radiation with matter. Different forms of spectroscopy are listed below:

- UV spectroscopy

Absorption of ultraviolet light by compounds to produce characteristic fingerprints relating to presence of specific chemical groups.

- IR spectroscopy

The same principle as UV but using absorption of longer wavelength infrared radiation.

- NMR spectroscopy

Nuclear magnetic resonance spectroscopy is a commonly used technique for identifying chemical structures. It is based on the interaction of certain elements with radio waves produced inside an NMR spectrometer.

- Mass spectrometry

Its use for determining the identities of proteins by peptide mass fingerprinting has already been described in Chap. 6. It is also used to for accurate structure determinations of small molecules.

- X-ray analysis

This is used to determine the three-dimensional positions of atoms in a compound, thereby providing a definitive structural view called the absolute configuration.

- Electron diffraction

A technique being used for determining the structures of large proteins (Chap. 6) but now being applied to small molecules. It has the advantage of not requiring large crystals, which vastly expands the number of structures that can be determined.

7.1.4   Natural Products

The extraction of medicinal compounds from natural products formed the basis of early pharmacy and led to the development of many of the drugs in use today (see Chap. 4). The modern biopharmaceutical industry became strongly committed to natural products during the early days of antibiotic discovery, when thousands of cultured microorganisms were screened to find alternatives to penicillin. Antibiotics, and other natural products, are examples of secondary metabolites, as opposed to the primary metabolites of living cells, such as the amino acids, nucleotides, lipids, and carbohydrates that are essential for basic cellular functions. The purpose of secondary metabolism in bacteria fungi and plants has been debated, but with a consensus view that it is related to communication between living organisms. Microbes use small molecules to communicate with each other; they live closely together in varying degrees of harmony, ranging from peaceful coexistence to outright warfare. Some of their communication molecules influence animals, including humans. As a result, there is some hope among scientists that an understanding of the basic ecology of plants and microbes will guide the search for natural products with medicinal properties.

7.1.4.1   Sources of Natural Products

The range of secondary metabolites that an organism can produce varies according to its environment, so in theory, the more exotic the habitat, the more chance of finding novel compounds. Therefore, there is interest in prospecting for drugs from marine sources, or plants living in tropical rain forests. As an employee of a pharmaceutical company with a major interest in natural product drug discovery, I was encouraged, along with my colleagues, to bring back soil samples from holidays abroad; the purpose of this was to supply the microbiology department with new bacteria and fungi for their screening collections. This was presumably encouraged by the success of Sandoz (now part of Novartis) who, under a similar scheme, discovered the powerful immunosuppressant drug cyclosporine in extracts of a fungus. The fact that cyclosporine was discovered in soil samples brought back from the USA and Norway shows that it is not necessary to go to exotic places to discover useful natural products.[5]

7.1.4.2 Natural Product Chemistry

Secondary metabolites are often complex chemical structures, which in terms of drug discovery, is part of their attraction. The more exotic the functional groups, the more chance of exploring novel chemical space. In the past, the determination of natural product structures was a major challenge, but modern analytical techniques (see Analytical Chemistry previously) make this relatively straightforward. What is not so simple, however, is the total synthesis of complex molecules extracted from nature. These compounds have been created in living organisms through the action of specialized enzyme pathways that transform chemicals in ways that may never have been previously encountered in the laboratory. A number of heroic efforts have been made to synthesize useful compounds, but often the motivation of the chemists has been similar to that of mountaineers: "I did it because it was there," which, of course, is completely impractical for most commercial drug discovery efforts. This is not meant to be dismissive, since these individual challenges have resulted in new synthetic methods that are extremely useful for medicinal and process chemists. Furthermore, there are cases in which a biopharmaceutical company will manufacture a natural product by undertaking a complex multistep chemical synthesis. An example of this is the compound eribulin, derived from a sea sponge and marketed by Eisai Pharmaceuticals for the chemotherapy of breast cancer (Ledford 2010). The drug requires a massive 62 synthetic steps to manufacture, so this may prove to be an isolated case because of the seriousness of the disease.

The molecular structures of three natural products used directly (or as starting points) for drugs are shown in Fig. 7.5 to give an idea of their complexity.

Once an active natural product compound has been identified, it is necessary to find ways to scale up its production. Firstly, it may be possible to extract the intact compound directly from a plant, animal, or fungal source; this assumes that there is enough of this source available to satisfy the demand for the drug by patients. Sometimes there may be large amounts of a naturally occurring precursor molecule, that is, a compound that requires some chemical modification in the laboratory for conversion into the final drug product. In this case, the conversion process is known

---

[5] However, the immunosuppressant rapamycin was isolated from a *Streptomycete* living at the bottom of an extinct volcano in Easter Island (Rapa Nui).

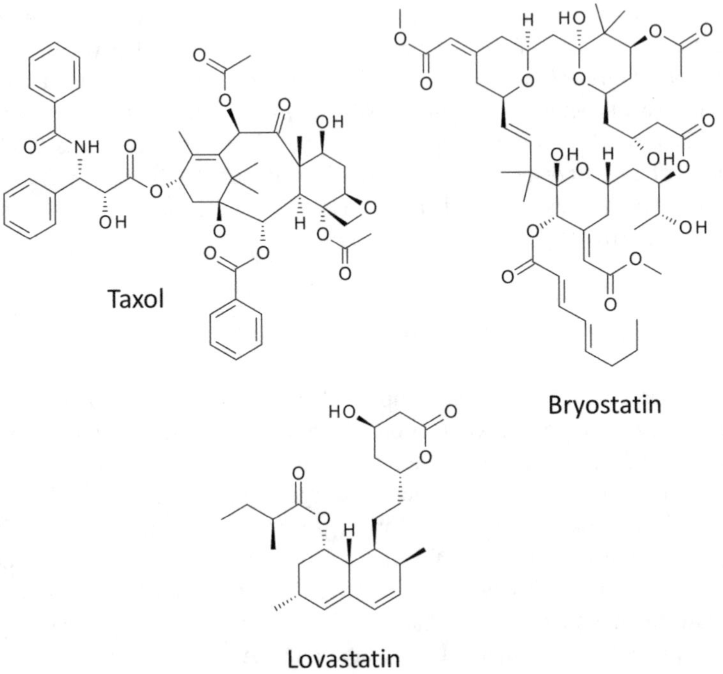

Fig. 7.5 Some compounds extracted from natural sources showing the complexity of their structures. Taxol is derived from yew bark and has powerful antitumor properties. Bryostatin is derived from a marine invertebrate and inhibits a cell signaling enzyme. Lovastatin was isolated from fungi and was the first HMGCoA reductase inhibitor for lowering blood cholesterol

as semisynthesis. Secondly, the enzymes that synthesize the compound in the plant or microbe can be transferred to a simple bacterium or yeast using genetic engineering techniques; in this case, the natural product is produced in large amounts because the growth of the microorganisms can be scaled up considerably using large fermenters. It is, in effect, recombinant DNA technology applied to the production of small molecules, rather than proteins. Although this is technically challenging, there have been some successes, for example, in the production of the antimalarial compound artemisinin extracted from the sweet wormwood plant. A Californian group has engineered yeast to produce artemisinic acid from simple sugars, using a series of enzymes cloned from the *Artemisia annua* plant (area reviewed by Kung et al. 2018). In an example of a semisynthesis, the artemisinic acid is subsequently converted to artemisinin in the laboratory to produce the drug in quantity. This is illustrated in Fig. 7.6.

Since the demand for this antimalarial drug is understandably high, and the cost to patients in the developing world must be kept low, alternative sources are being actively explored, including genetically engineered organisms and cultured plant cells.

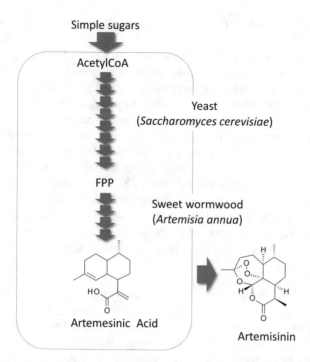

Fig. 7.6 Production of antimalarial drug artemisinin using genetically engineered yeast cells. Each gray arrow in the metabolic pathway represents a conversion of one compound into another that is catalyzed by a specific enzyme. The first steps in the pathway use yeast enzymes to convert simple sugars (i.e., glucose or similar) via acetyl coenzyme A to farnesyl pyrophosphate (FPP), a precursor of cholesterol and other lipids. The remaining enzymes have been introduced from the *A. annua* plant to convert FPP to artemisinic acid which can be purified from the yeast cells after they have been harvested. Some chemical manipulations in the laboratory convert artemisinic acid to artemisinin (Ro et al. 2006)

The above example of artemisinin semisynthesis uses intact yeast cells, but it is also possible to use isolated enzymes, either singly or as a cascade. The modification of enzyme specificity and action using genetic engineering helped in the creation of a cell-free "biocatalytic cascade" for producing the small molecule drug islatravir in a highly efficient process (Huffman et al. 2019).

Finally, the structures of natural products have been used to identify pharmacophores that can then be synthesized in the laboratory and used in a conventional medicinal chemistry program. The synthetic statin drugs were based on a pharmacophore discovered in lovastatin (see Fig. 7.5), produced by the fungus *Aspergillus terreus* and oyster mushrooms (see Fig. 7.1).

7.1.4.3 Natural Product Discovery in the Biopharmaceutical Industry

It should be clear from the foregoing sections that natural product chemistry has played a major part in small molecule discovery by biopharmaceutical companies. This was certainly the case from the 1940s to the 1990s, but the advent of combinatorial chemistry and other sources of synthetic compounds led to the closure of many natural product discovery departments in the major companies. There was frustration with the fact that it was almost impossible to consistently produce measurable amounts of active compounds in crude extracts of plants and microbial cultures. These extracts were also difficult to test in sophisticated biological assays because of the presence of interfering substances, like tannins in plants, which generated false hits. Sometimes researchers could be led in the wrong direction; I recall the excitement of having a good clear "hit" in a screen of fungal media on an immunology target only to find, after exhaustive purification and analysis, that the active compound was a non-specific inhibitor of cellular respiration, something that was totally unexpected.

The situation is beginning to change, however, even though the basic problems outlined above remain still. Firstly, there is recognition that most of the synthetic compounds produced for screening do not have sufficient structural complexity to make useful leads, so there is no escaping the fact that natural products are a rich source of chemical diversity. Secondly, the capabilities of the analytical instruments at the disposal of the natural product chemist have increased out of all recognition. High-throughput chromatography, coupled with mass spectrometry, has made it possible to analyze thousands of components in complex mixtures and to identify novel structures to explore in the laboratory. Finally, the full extent of microbial and plant diversity is nowhere near being fully explored, for natural products or for anything else. Only a tiny fraction of the world's species of microbes have ever been grown and tested, so there is still a lot to play for.

7.1.5 Matching Chemical Structures to Drug Targets

7.1.5.1 Focused Compound Libraries

Certain target classes, like the protein kinases and ion channels (Chap. 6), contain binding sites that will accept a specific small molecule chemical scaffold or pharmacophore; this may bind to every member of a particular protein family and can be made highly selective for individual family members by adding extra functional groups. These scaffolds form the basis of focused compound libraries which are commercially available from specialist medicinal chemistry companies.

Figure 7.7 shows a computer-generated model of about 200 individual compounds overlaid onto each other; all sharing a scaffold ring structure that projects out to the bottom left of the picture. The various side chains that protrude round the

Fig. 7.7 Three-dimensional structures of compounds in focused library overlaid onto each other. The common ring structure faces towards the bottom left hand of the picture and the individual side chains protrude from the scaffold like a brush

edges can explore different binding pockets in the target protein, so that compound with the best fit should, in theory, have the best biological activity.

These libraries have the advantage of being much smaller than the huge collections of random compounds used in high-throughput screening (see Chap. 9), but they lack the diversity required for exploring novel chemical space to any depth. The solution to the balance between practicality and finding novel chemotypes may lie in virtual screening and computer-aided drug design as discussed later.

7.1.5.2   DNA-Encoded Chemical Libraries

In many cases it will be necessary to produce as many randomly diverse compounds as possible rather than the relatively few in the focused libraries described above. Combinatorial chemistry was introduced in the last part of the twentieth century as a means of synthesizing large numbers of compounds by reacting together combinations of simple starting materials like amides and carboxylic acids. The number of compounds produced can exceed millions very rapidly as the growth is exponential. This form of chemistry fell out of favor as the quality of the compounds as drugs did not compare well with more conventional candidates such as natural products. The situation has changed with the adoption of DNA-encoded chemical libraries (DELs) by the biopharmaceutical industry. Here, the starting compounds are tagged with an oligonucleotide "bar code." Millions of tagged compounds can be stored in single sample tube and used to screen against a target of interest (Chap. 9). Once a tagged compound has been selected by binding to the target, it can be identified by sequencing the oligonucleotide tag and developed further (Ottl et al. 2019). The size of DELs is now massive, with the Danish company Nuevolution (possible Amgen acquisition) producing libraries with 40 billion compounds.

7.1.5.3   In Silico Screening and Drug Design

Medicinal chemistry requires a great deal of painstaking compound synthesis, biological testing and optimization to produce leads with the potential to become a drug. Despite genuine advances in the field, the medicinal chemistry process is still, to a large degree, hit and miss. In principle, it would be far better to "synthesize" compounds in the computer and then check them against target proteins for optimal binding prior to doing any laboratory work. This is an important area of research that aims to minimize the number of compounds made and tested but maximize the chance that a compound will be active. Another attraction is the ability to automatically reject those compounds with chemical groups that are known to cause problems with toxicity or other clinical development issues. There are two main aspects to in silico drug design, namely, virtual screening, and de novo drug design. Both techniques require a prior knowledge of the three-dimensional structure of the target protein. Virtual screening involves docking small molecule structures into the binding pocket(s) on the protein and identifying those with the best fit. Unfortunately, there are number of different ways in which a single molecule can bind, so it is not always easy to select the one conformation that should occur in vivo. In this case, it may be necessary to co-crystallize the compound with the protein and determine the structure of the small molecule bound into the protein; this is something that is not undertaken lightly. Despite this problem, virtual screening is routinely used to optimize the binding strength and selectivity of compounds during the early stages of drug discovery. The size of virtual libraries, along with computer processing power, is increasing to the point where it is now possible to screen millions of compounds against a given target. For example, Lyu et al. (2019) have screened 170 million virtual compounds against a bacterial target (beta-lactamase) and neurological target (dopamine D4 receptor) to find novel chemotypes, some of which are highly potent. In the case of the D4 receptor, the processing power of roughly 190 desktop computers was used to the screen the library in just 1.2 days (Gloriam 2019).

De novo drug design, as its name implies, is the process of designing chemical structures from scratch using the target structure as a template. Academic groups and small companies have been involved in de novo design for some years, but there are still significant technical challenges to be overcome. Firstly, protein structures often move during ligand binding and are therefore in a sense moving targets for de novo design. There are also unknown quantities, such as the number of water molecules that are bound, and some features like charge or hydrogen bonding may not be adequately defined. Another key problem is whether it is possible to make the compounds once a design is available.

Although the full potential of in silico drug design has not yet been realized, applications of AI/machine learning are beginning to make an impact on this area. Generative modeling uses deep learning with a training set of existing structures to identify a "latent space" that can be used as a template for generating new molecules. The use of machine learning for antibiotic discovery was published in 2020 (Stokes et al. 2020). Here, >170 million virtual compounds were used as a training set in conjunction with bacterial growth inhibition to identify novel antibiotics with structures that diverge significantly from known classes. This work represents only the tip of the iceberg, and it is not unreasonable to suggest that most small molecule drugs of the future will have their origins in a computer rather than a bottle.

7.1.5.4   Fragment-Based Design

Fragment-based screening is a combination of physical (i.e., laboratory) screening and in silico drug design that is used to create small molecule drug candidates. It works on the principle that very weak binding of small chemical groups to target proteins can be measured using techniques such as NMR or X-ray crystallography. This level of binding makes the compounds unsuitable as drugs, because they lack any potency and selectivity. However, once a number of these groups have been identified for a single target, they can be joined together chemically to create drug-sized molecules with much higher binding affinity. This is illustrated in Fig. 7.8 below:

7.1.6   Medicinal Chemistry Challenges: Protein-Protein Interactions

One of greatest challenges for the medicinal chemist is finding small molecules to inhibit protein-protein interactions (PPIs), in other words, the binding of proteins to other proteins. In the case of drug target proteins, these are said to be "undruggable." There are many such examples, including the binding of cytokines to their receptors or the interaction of cells via adhesion molecules. There are also many proteins that bind to each other to form a series of "on-off" switches inside cells;

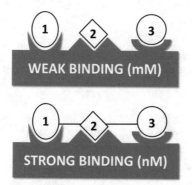

Fig. 7.8 Diagram of fragment-based drug design. Three small chemical fragments 1–3 bind to a protein target with low but detectable affinity. The fragments may be little bigger than the benzene ring (molecular weight 78, compared with 500 for an average drug). Once a group of fragments have been identified, they can be linked together in the laboratory to produce a compound with a higher molecular weight and much stronger binding affinity example here, 1 million-fold (milli-molar to nanomolar)

these are part of the signal transduction systems that regulate cell division and other cellular functions in health and disease. There is no universally applicable solution to the problem of protein-protein interaction inhibitors, despite continuing efforts by medicinal chemists; this is a major reason for the development of protein drugs (see next chapter). All is not lost, however, as success stories are emerging of small molecules that inhibit these types of interactions. A now venerable example is the class of small molecule drugs based on a motif of three amino acids (RGD in single letter code). These drugs inhibit the protein-protein interactions that occur during thrombus (blood clot) formation. Around 2010, a series of small molecule inhibitors of BET reader proteins (see epigenetics, Chap. 6) were described which blocked their binding to acetylated histone proteins via interaction with the so-called bromo-domain. Early clinical development of these inhibitors for cancer has revealed on-target toxicity effects, but a newer generation of BD2 inhibitors shows potential as therapeutics for prostate cancer (Faivre et al. 2020).

Finally, one of the greatest challenges in oncology drug development has been the search for inhibitors of KRAS a member of the RAS family of signaling proteins. Mutations in KRAS are responsible for up to 25% of cancers, but until recently, this target has been intractable to small molecule inhibition due to shallow binding pockets on the protein. The situation has now changed, with several companies realizing the fruits of many years of research in getting KRAS inhibitors into the clinic. As an example, BI-2852, the Boehringer Ingelheim KRAS inhibitor with nanomolar affinity compound, could only be developed using fragment-based screening of weakly binding small molecules. A conventional screen of 1.7 million compounds was unable to provide suitable lead compounds for further development (Kessler et al. 2019). This success story and others like it make it likely that many targets previously considered to be "undruggable" may one day yield to perturbation with potent small molecules.

Summary of Key Points

The diversity of chemical structures of small molecules based on carbon is known as chemical space.

Chemical space is vastly greater for compounds that could theoretically be made compared with the approximately 100 million compounds made in the laboratory.

Medicinal chemistry aims to produce small molecules to interact with a drug target with high potency and selectivity.

Chemical modifications are made to pharmacophores from natural ligands (e.g., histamine), random screening hits, natural products, or molecules designed in silico.

Drug design and machine learning in the computer are in early stages of development but are beginning to identify novel compounds for synthesis and testing against drug targets.

Natural products derived from plants, animals, and microbes are a rich source of drugs, either as starting points for medicinal chemistry or as final drug products.

References

Bohacek RS, McMartin C, Guida WC (1996) The art and practice of structure-based drug design: a molecular modeling perspective. Med Res Rev 16:3–50

Bolli MH et al (2010) 2-imino-thiazolidin-4-one derivatives as potent, orally active S1P1 receptor agonists. J Med Chem 53:4198–4211

Douglas P et al (2007) Structure-based optimization of protein tyrosine phosphatase 1B inhibitors: from the active site to the second phosphotyrosine binding site. J Med Chem 50:4681–4698

Faivre EJ et al (2020) Selective inhibition of the BD2 bromodomain of BET proteins in prostate cancer. Nature 578:306–310

Gloriam DE (2019) Bigger is better in virtual drug screens. Nature 566:193–194

Huffman MA et al (2019) Design of an in vitro biocatalytic cascade for the manufacture of islatravir. Science 366:1255–1259

Introduction to Stereochemistry (2019) https://chem.libretexts.org/Bookshelves/Ancillary_Materials/Worksheets/Worksheets%3A_Inorganic_Chemistry/Structure_and_Reactivity_in_Organic%2C_Biological_and_Inorganic_Chemistry/05%3A_Stereochemistry/5.1%3A_Introduction_to_Stereochemistry. Accessed 6th Mar 2020

Kessler D et al (2019) Drugging an undruggable pocket on KRAS. Proc Natl Acad Sci 116:15823–15829

Kubota K et al (2019) Redox reactions of small organic molecules using ball milling and piezoelectric materials. Science 366:1500–1504

Kung SH et al (2018) Approaches and recent developments for the commercial production of semi-synthetic artemisinin. Front Plant Sci. https://doi.org/10.3389/fpls.2018.00087

Ledford H (2010) Complex synthesis yields breast cancer therapy. Nature 468:608–609

Leung CS et al (2012) Methyl effects on protein-ligand binding. J Med Chem 55:4489–4500

Lipinski C, Hopkins A (2004) Navigating chemical space for biology and medicine. Nature 432:855–861

Lyu J et al (2019) Ultra-large library docking for discovering new chemotypes. Nature 566:224–229

Maryasin B et al (2018) Machine learning for organic synthesis: are robots replacing chemists? Angew Chem Int Ed Eng 57:6978–6980

Mullard A (2017) The drug-maker's guide to the galaxy. Nature 549:445–447

Ottl J et al (2019) Encoded library technologies as integrated lead finding platforms for drug discovery. Molecules. https://doi.org/10.3390/molecules24081629

Ro DK et al (2006) Production of the antimalarial drug precursor artemisinic acid in engineered yeast. Nature 440:940–943

Robinson A (2010) Chemistry's visual origins. Nature 456(36)

Rocke AJ (2010) Image and reality: Kekule, Kopp and the scientific imagination. University of Chicago Press, Chicago

Steiner S et al (2019) Organic synthesis in a modular robotic system driven by a chemical programming language. Science 363:144

Stokes JM et al (2020) Deep Learning Approach to Antibiotic Discovery. Cell 80:688–702

Zhang X et al (2020) Copper-mediated synthesis of drug-like bicyclopentanes. Nature 580:220–226

Chapter 8
Biotherapeutics

Abstract Biotherapeutics are large molecule drugs developed for "undruggable" targets that are difficult to modify using small chemical compounds. This chapter covers these agents in the form of protein drugs, the genetic engineering used to produce them, gene therapy vectors used to modify DNA directly, and therapies based on genetically modified cells.

8.1 Introduction

The synthesis of small chemical molecules for testing against biological targets was covered in the last chapter. While these orally active small molecules are still the preferred offering of the biopharmaceutical industry, the larger molecules of nature are being used to literally access the parts of drug targets that small molecules cannot reach. From the 1960s onwards, drug discovery chemists invested a great deal of time and effort trying to find agonists or antagonists of small peptide hormones; as a result, there was a certain amount of medicinal chemistry expertise in this area of (small) protein ligands. Unfortunately, the technical difficulties experienced by these earlier researchers were forgotten by a newer generation of scientists[1]. There was, in the words of the British writer Samuel Johnson, a "triumph of hope over experience" in trying to find small molecules that could inhibit the binding of larger proteins (e.g., the cytokines) to their receptors[2]. Large pharmaceutical companies and some of the new biotechnology startups screened many thousands of compounds trying to find small molecules that could block the binding of protein ligands to their receptors but to no avail. This was immensely frustrating, because discoveries of new targets for protein ligands were being made on a regular basis. Furthermore, the cell biology and clinical work in industry and academia made it quite clear that drugs affecting these targets could provide enormous medical benefit to patients; the example of tumor necrosis factor (TNF-α) inhibitors for rheumatoid arthritis has already been given in Chap. 6. Something had to be done to get

---

[1] Some successes have been described at the end of Chap. 7, and there will no doubt be many more in the future.

[2] Although the author was referring to second marriages

© The Editor(s) (if applicable) and The Author(s), under exclusive license to
Springer Nature Switzerland AG 2020
E. D. Zanders, *The Science and Business of Drug Discovery*,
https://doi.org/10.1007/978-3-030-57814-5_8

around this problem, and eventually a solution, or at least a partial one, was found that has dramatically changed the traditional chemistry-based pharmaceutical business from the 1990s onwards.

Fig. 8.1 illustrates the nature of the problem by showing small molecule drug (aspirin) interacting strongly with a target protein (phospholipase AII) alongside a protein-protein interaction between tumor necrosis factor (TNF) and its receptor.

8.1.1 Terminology

There are several ways of describing biotechnology-based medicines; some common terms are listed as follows:

- Biologicals.
- Biologics.
- Biotherapeutics.
- Protein therapeutics.
- Antibody therapeutics (antibodies are also proteins, so can be included in all the above categories).
- Cell therapeutics.

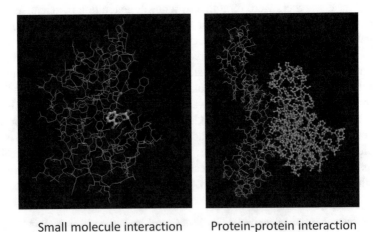

Small molecule interaction   Protein-protein interaction

Fig. 8.1 Comparison of small molecule interaction with tight binding to a drug target protein with an interaction between two large protein surfaces. Left: aspirin bound to PLA2 enzyme. Right: tumor necrosis factor bound to soluble 55Kd receptor. Structures from PDB (1OXR and 1TNR)

8.1.2 The Emergence of Protein Therapeutics

Protein drugs like insulin have been introduced already in Chaps. 2 and 4. In the early years of the new biotechnology industry (1980s), large pharmaceutical companies were wary of becoming involved with protein drugs, since their whole business had been founded upon small molecule chemistry. There were good reasons for their concerns about protein drugs, since these products cannot be administered by mouth and are difficult to consistently manufacture to the purity standards that are demanded of small molecule drugs. Nevertheless, large pharmaceutical companies noticed the commercial success of the first recombinant protein drugs and so began the now established practice of buying small biotech companies to acquire the drugs and molecular technology behind them.

The early biologicals were mostly genetically engineered replacements of natural growth factors and enzymes which could be produced in large quantities in fermentation tanks; this avoided the need to purify them from human or animal cadavers. Several hematopoietic growth factors were developed as medicines designed to stimulate the growth of red or white blood cells in cancer patients undergoing radiotherapy or chemotherapy. One of the most successful of these is erythropoietin, a hormone that stimulates the growth of red blood cells; a recombinant form is used to treat anemia in patients with kidney disease as well as cancer. It also has the dubious honor of being yet another illegal performance-enhancing drug that sports authorities must watch out for. Erythropoietin, brand name Epogen®, was a great commercial success for its developer Amgen Inc.; it allowed the company to grow from a small "pure" biotechnology company to a significant pharmaceutical business with products in the top ten of prescription drug sales. This level of sales means that the consequences of generic competition are now the same with protein drugs as they are already with small molecules. There is one difference, however; when patents on small molecule medicines have expired, competitors can manufacture a generic product that is chemically identical to the original, but this is not so easy with protein drugs. Therefore, the industry is now grappling with biosimilars, a term used to describe generic copies of biotherapeutics (see also Chap. 16). Some examples of recombinant protein therapeutics are shown in Table 8.1:

Table 8.1 Some of the first-generation recombinant therapeutic proteins. All are relatively small cytokine/growth factor-like proteins, except for tPA and streptokinase which are enzymes

Name	Disease
Somatotropin	Growth retardation
Insulin	Diabetes
tPA	Cardiovascular
Streptokinase	Cardiovascular
Erythropoietin	Cancer
Interferon beta	Multiple sclerosis

8.1.3   Technical Aspects of Recombinant Protein Production

Some of the basic elements of molecular biology and protein biochemistry which are relevant to genetic engineering have been touched upon earlier in this book. What follows is a fuller description of the process that begins with the identification of a protein of interest and continues with the genetic engineering needed to produce it in industrial quantities for use as a medicine. In this example, the cells used for protein expression are the *E. coli* bacteria, which were used for the first recombinant protein drugs.

Recombinant protein expression involves the following experimental steps:

1. Isolate human DNA that codes for the protein of interest (e.g., human insulin).
2. Stitch the human DNA into a vector that will "carry" the DNA into bacteria.
3. Grow the bacteria to produce the human protein.
4. Purify the human protein.

8.1.3.1   Step 1 Finding the Needle in the Haystack

The length of DNA encoding an average sized protein is about 1000 bases, which corresponds to about 350 amino acids (because of the three-letter genetic code). Total human DNA contains about 3,000,000,000 bases, so locating a sequence of interest is rather like finding a needle in a haystack. Luckily, techniques have been developed which make it a relatively simple task to locate and amplify a DNA sequence of choice, however low its abundance. The most important method is the polymerase chain reaction, commonly known as PCR; this is based on the hybridization of specific oligonucleotide primers to DNA, followed by the synthesis of new strands using an enzyme called DNA polymerase. At this point, the reader is referred to the many Internet resources which use animations to illustrate the PCR process. The sequences of the oligonucleotide primers correspond to the beginning and end of the desired protein sequence, so the resulting fragment of double stranded DNA contains only the protein-coding information. If 10 milligrams of total human DNA is isolated from a tissue such as blood, it will contain roughly 40 picograms (million millionth grams) of DNA coding for the specific protein sequence. The amount of amplified fragment required for inserting into a vector is in the order of micrograms, so the original gene must be amplified at least a million times. One of the powerful features of PCR is its ability to add any desired sequence to the oligonucleotide primers so that they are incorporated into the final DNA product. This can be particularly useful for attaching linkers which allow the fragment of copied human DNA to insert into vector DNA at exactly the right place. PCR can also be used to introduce mutations into the DNA, making it a central part of the genetic engineering toolbox now used to manipulate the genomes of everything from viruses to humans.

While the principle of amplifying protein-coding fragments of DNA is straightforward enough, there is a slight complication; it is not possible to do this directly on the DNA isolated from human cells. The reason is that protein-coding genes are split into sections called exons which are separated by regions of DNA called introns. The introns are later removed in the messenger RNA (mRNA) intermediate that is used to prime the synthesis of proteins, while the exons are joined to give the correct sequence for amplification. Unfortunately, PCR cannot be performed directly on the RNA, so a DNA copy called complementary DNA (cDNA) is created as a template for amplification. The steps described above are summarized in Fig. 8.2.

8.1.3.2 Step 2 Introduce the Human DNA into an Expression Vector

The human DNA fragment from step 1 contains all the sequence information required to express the human protein of interest, but further manipulation in the

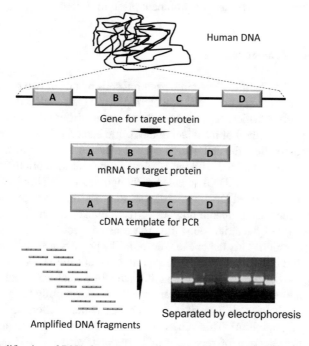

Fig. 8.2 Amplification of DNA fragments containing the gene to be introduced into bacteria to produce human protein. The gene to be amplified is broken up into four exons (A B C D) separated by introns (solid line). During the process of mRNA formation in the human cell, the introns are removed (spliced out) and the exons joined into one piece of RNA with all the genetic code in place to specify a protein. An exact cDNA copy of the mRNA is made in vitro and is amplified millions of times in a PCR reaction to physically create enough DNA for the next step. The DNA fragments are separated using agarose gel electrophoresis and visualized by staining with a fluorescent dye (ethidium bromide or equivalent)

form of genetic engineering is required before it can program the synthesis of protein in other cells. Genetic engineering was built upon the practical application of microbiology, i.e., the study of bacteria, viruses, and fungi, to the problem of moving DNA from one cell to another. Viruses use their own proteins to inject DNA into cells during the process of infection[3]. The viral DNA is then turned into mRNA and translated into viral proteins that are used to build new viruses. Viruses infect many cells from plants and animals, but they also target bacteria, in which case they are called bacteriophages or phages. These have been used for much of the pioneering work on molecular biology and were later employed as vectors that could transfer foreign DNA into bacteria. Although viruses that infect bacteria and other cell types are used extensively in genetic engineering, they are not the only vectors available to the molecular biologist. Bacteria contain a small piece of circular DNA called a plasmid, which is separate from the main chromosomal DNA that contains most of the bacterial genes; actually, bacterial DNA is not packaged into separate chromosomes like in eukaryotes, but the term is used all the same. The circular plasmid DNA can be purified from bacteria, cut into fragments, and then reformed with foreign genes to produce a recombinant plasmid vector.

8.1.3.3 Step 3 Transformation

The vector DNA can be introduced into bacterial cells by one of two processes, both of which may seem rather bizarre at first sight. Bacteria are enclosed in a cell wall, a protective barrier which is impermeable to many different molecules, DNA included. The first method of introducing DNA into bacteria involves treating them with a chemical to allow the cell wall to open at elevated temperatures. The treated *E. coli* cells are simply mixed with the DNA in a tube and then briefly incubated at 42 °C to allow the plasmid DNA to enter. Although this is a highly inefficient process, the number of added DNA molecules is so great, that enough get inside the bacteria to produce a reasonable number of recombinant colonies on agar plates. The other method of bacterial transformation uses an electric current in a process called electroporation. This rather Frankenstein-like process stretches the DNA out into a thread and creates pores in the bacterial cell wall; the DNA is then literally threaded through the cell wall under the influence of the electric current. Whichever transformation method is used, the bacteria that have successfully taken up the plasmid (or bacteriophage) will have also taken up a gene that confers resistance to an antibiotic (e.g., ampicillin). This gene is built into the plasmid or phage vector to allow antibiotic selection of recombinant bacteria. If ampicillin is added to the growing culture, the untransformed cells will die because they are sensitive to the antibiotic and only the transformed ones will survive and grow.

---

[3] Some viruses, including HIV, use RNA as their genetic material

8.1.3.4 Step 4 Production of Recombinant Protein

Transformed bacteria are grown in laboratory-scale cultures of a few hundred milliliters of culture medium before being analyzed for the expression of human protein. The resulting protein will either be associated with the bacterial cells or will have been secreted into the culture medium. In either case, it should be easily detectable using the gel electrophoresis technique described in Chap. 6. The next stage involves the purification of the human recombinant protein from the proteins and other molecules present in the cells and growth medium. This is accomplished by chromatography (discussed in later chapters) and must be conducted to rigorous standards if the protein is to be administered to human subjects. This is because bacteria contain various molecules, such as lipopolysaccharide (LPS), which are powerful stimulators of the immune system and are potentially dangerous, even in trace amounts.

Once a purification scheme for the protein has been established, the production is scaled up inside production plants using large fermenters. The volumes of culture media used to produce protein on the kilogram scale can be as high as 27,000 liters per fermenter. The different stages of recombinant protein production in the laboratory are summarized in Fig. 8.3.

8.1.3.5 The Problem with Sugars

Bacterial expression of recombinant proteins in *E. coli* is routinely performed in the research laboratory and is generally a quite efficient process. Unfortunately, there are many proteins of pharmaceutical interest that have complex sugar molecules (glycans) attached at various points along the chain of amino acids. Such proteins, called glycoproteins, are only present in eukaryotes, because the (glycosylation) machinery required to produce them is absent in bacteria. Figure 8.4 shows the three-dimensional structures of glycosylated interferon beta-1a, highlighting the chains of sugars that extend from the basic protein core.

The therapeutic efficacy of proteins can be significantly improved by the presence of sugars; this is because of improved stability in the blood, resistance to heat, and, in the case of antibodies, higher biological activity. The requirement for fully glycosylated proteins has driven the search for expression systems that add sugars onto recombinant proteins. This can only be achieved in eukaryotic cells, so one of the first systems to be used was based on *S. cerevisiae*, or baker's yeast. This single-celled fungus can be grown in large amounts, so it seemed to be ideal for the large-scale production of recombinant proteins. Unfortunately, *S. cerevisiae* introduces sugars that are significantly larger than the natural forms, so alternative species of host cells have been evaluated, including another type of yeast, *Pichia pastoris*. Although yeast has the desirable property of rapid growth in simple media, the problem with unnatural glycosylation is a real handicap when considering proteins for therapeutic use. This problem does not occur in higher organisms, so cell lines have been produced from both insects and mammals which can produce

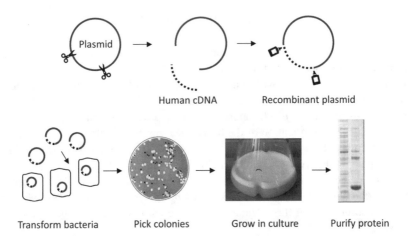

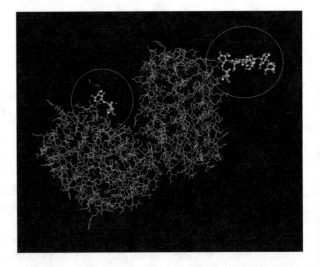

Fig. 8.3 Stages of recombinant protein production in bacteria. Circular plasmid DNA is cut using restriction enzymes and a piece of human (or other foreign) DNA produced according to Fig. 8.2 pasted in using ligase enzyme to reseal the circle. The resulting recombinant DNA with the human gene is introduced into bacteria by transformation and the cells selected by growing on an agar plate. White colonies (i.e., those with plasmids containing the DNA insert) are picked and grown in liquid media in flasks. The bacteria multiply in the growth medium and produce human protein as well as proteins encoded by the bacterium's own DNA. The recombinant human protein is purified away from the other proteins in the bacteria and growth medium and can be analyzed using SDS-gel electrophoresis (see proteomics Chap. 6)

Fig. 8.4 Three-dimensional structure of glycosylated interferon beta-1a showing protruding carbohydrate side chains

recombinant glycoproteins in quantity after transformation by genetically engineered plasmids or viruses. One of the most used cell lines, called CHO, was originally derived from the ovary of a Chinese hamster (a typically exotic) source of biological material. For industrial-scale production of proteins, CHO cells are suspended in large bioreactors; these are regularly replenished with fresh growth medium to produce recombinant proteins in yields of up to 5 grams of protein per liter of culture.

Finally, protein expression in insect cells, while a well-established laboratory tool, is now being evaluated as a system for producing biopharmaceutical proteins. The insect cell expression system is based on the virus, called a baculovirus, which infects the Sf21cell line that is derived from the ovaries (again) of the moth *Spodoptera frugiperda*. Genetic engineering is used to modify baculovirus DNA, allowing the expression of human proteins in large-scale insect cell cultures. There is interest in the baculovirus system as a source of vaccines, since it can produce very pure antigen preparations in a relatively short period of time when compared with conventional production methods, for example, influenza vaccines in chicken eggs.

8.1.4   Antibodies

On the face of it, antibodies do not look very promising as drugs because of their size (150,000 daltons) and complexity. The fact that antibody drugs are now some the best-selling medicines in the world is testament to the fact that they can offer features that are not found in other types of biologicals. The main characteristic of antibodies is their ability to bind with exquisite specificity to a protein which can differ from other proteins by only a few atoms. They are also unique in being designed by nature, figuratively speaking, to attack and destroy cells that have been infected by viruses. Once an antibody binds to the surface of an infected cell, it recruits white blood cells to literally break open the cell and clear up the debris afterwards. Therapeutic antibodies, such as Herceptin®, bind to cells (in this case, from breast tumors) and cause their destruction through the above mechanisms.

Other antibodies, such as infliximab, are designed to "mop up" cytokines, such as tumor necrosis factor (TNF-α), to prevent them binding to their receptors. Infliximab marketed as Remicade®, was among the first wave of therapeutic antibodies to be introduced into the clinic. The antibody was developed by the US biotech company Centocor, now a subsidiary of Johnson and Johnson, and commercialized in partnership with large pharmaceutical companies. This was new territory for the traditional biopharmaceutical industry which, at best, was lukewarm about protein drugs, particularly antibodies. The financial success of Remicade®, and other emerging antibody therapeutics, finally encouraged the large companies to take these products more seriously. The acceptance of proteins as products on an equal footing with small molecules, and the later acquisition of antibody companies to develop the technology, all stemmed from this period in the mid-1990s.

8.1.4.1 Terminology

The word antibody is used to describe the class of proteins called immunoglobulins. These proteins are divided into different subclasses, or isotypes, called immunoglobulins A, D, E, G, and M (shortened to IgA, IgD, IgE, IgG, and IgM). Nearly all therapeutic antibodies are of the IgG isotype, while those responsible for allergic reactions, like hay fever, are IgE antibodies. The isotypes are themselves subdivided into subtypes, e.g., IgG1, IgG2, IgG3, and IgG4 in humans.

8.1.4.2 Generating Antibodies

Immunology is a large and complex subject which is medically important because of the immune system's central role in fighting infection, rejecting organ transplants and causing a wide range of allergic and autoimmune diseases. Antibodies are produced by a class of white blood cells called B lymphocytes and are a central part of the adaptive immune system. There are many sources of information about antibodies and the white blood cells that produce them, most of which are outside the scope of this book. However, to understand how therapeutic antibodies are produced by the biopharmaceutical industry, it is necessary to provide some background information. Antibodies are produced in mammals and birds, either by natural infection or by deliberate immunization. It is generally unethical to deliberately immunize human beings with a protein for research purposes, so animals like mice and rats are normally used instead. There is another reason: if a human protein is injected into a human, the immune system recognizes that protein as "self," or a harmless part of the body. This phenomenon is called immunological tolerance and is the reason why we generally do not mount an immune response to ourselves unless we suffer from an autoimmune disease such a multiple sclerosis.

In practice, antibodies are generated over a period of a few months by injecting animals with antigens that are purified proteins or cells that bear protein targets on their surface[4]. It is certainly possible to create antibodies to proteins which reside on the inside of cells, but these will not be useful therapeutically as they cannot penetrate the cell membrane. Once the immune system has been stimulated with repeated doses of antigen for a sufficiently long time, the amount of antibody that has been generated is measured in blood samples using immunoassays (see Chap. 14). Immunoassays are important for analyzing therapeutic antibodies during the preclinical development phase (Chap. 11) and for diagnostics (Chap. 14). Once animals have been identified that produce large amounts of antibody to the antigen of interest (i.e., they have high titers), they are sacrificed (standard terminology) for their spleen cells. The spleen contains large numbers of antibody-producing B lymphocytes, each one of which produces a single unique antibody; because there are

---

[4] It is possible to produce antibodies to carbohydrates and other molecules, but protein antigens make up by far the greatest number of targets for antibody drugs.

many different B cells in a living spleen, the overall repertoire of antibodies produced against a single antigen is quite varied. This is called a polyclonal response and is the sum of all the antibody-producing activity of single (monoclonal) antibody-producing cells. Since only one specific antibody is required for development as a protein drug, individual B cells must be isolated and grown in culture. This was first achieved by Koehler and Milstein, who developed the technique of producing monoclonal antibodies from spleen cells. This essentially involves immortalizing the clones of antibody-producing B cells isolated from spleen; this is done by fusing them with a tumor cell line to create a hybrid cell, i.e., a hybridoma, which can be grown indefinitely in large numbers. The monoclonal antibody is secreted into the tissue culture medium used to grow the cells; this medium is the starting material from which therapeutic antibodies are purified during the manufacturing process (see Chap. 10).

8.1.4.3   The Importance of Being Human

Since antibodies are usually produced in mice or rats, the immunoglobulin proteins will have amino acids specified by mouse or rat DNA; the proteins will therefore be considered as foreign by the human immune system. This is a highly undesirable property for a therapeutic antibody that will be repeatedly injected into patients; their own immune system will generate antibodies to the antibody, thereby neutralizing its beneficial effect. This problem has forced companies to devise ways of making antibodies that are indistinguishable from the version that would be produced if human beings were injected with the protein antigen. This process of humanization involves the application of genetic engineering technology to monoclonal antibody production and is otherwise known as antibody engineering.

Figure 8.5 is a schematic diagram of the immunoglobulin molecule showing the structure of an IgG antibody as four protein chains tightly linked together by covalent bonds (Chap. 3). The large and small components are called heavy and light chains (H and L chains), respectively. The arms of the Y shape molecule contain stretches of amino acids that make up variable regions (V regions) which, as the name suggests, vary between individual antibodies. It is these regions that make up the antigen-binding site, each having a unique shape and charge in the same way as drug target proteins.

When a therapeutic antibody is produced in mice or rats (hereafter referred to as rodents), these V regions contain amino acids relating to these species; as already discussed, these rodent sequences will eventually cause problems when the antibody is administered to patients. The remainder of the antibody contains constant regions (C regions) that are the same for every antibody of a given subclass. If the rodent constant regions can be changed to human sequences without compromising the ability of the antibody to bind to its target, this will make it more useful as a biological drug. Such molecules are known as chimeric antibodies, one example being Remicade®. They are quite acceptable as therapeutics, particularly if they are not repeatedly injected over a long period, but there is still the risk that the human

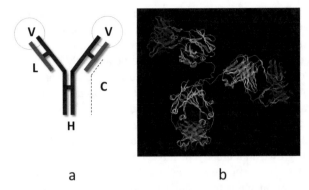

a   b

Fig. 8.5 Structure of an antibody molecule. (**a**) Schematic diagram of immunoglobulin G (IgG) molecule showing two heavy (H) and two light (L) chains linked together. The top parts of the H and L chains have amino acid sequences that vary to create a unique binding site for target antigens (V regions). The rest of the molecule is the same for all immunoglobulins from the same species and with the same isotype. (**b**) Ribbon diagram of IgG molecule with the structure determined by X-ray crystallography. The strands represent the amino acid backbone of each of the four protein chains. Each chain is folded into protein domains with characteristic structures. The whole antibody molecule has a molecular weight of 150,000, which is significantly greater than most cytokines and hormones

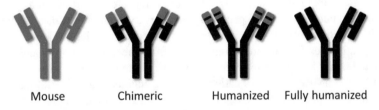

Mouse   Chimeric   Humanized   Fully humanized

Fig. 8.6 Modifications of mouse antibodies to create human therapeutic antibodies. Mouse (or rat) antibodies will provoke an immune response if injected into humans; chimeric antibodies use all the rodent variable (V) regions joined to a human constant region. Humanized antibodies have rodent sequences removed from the V regions to make them even less likely to stimulate the human immune system, thus fully humanizing the antibody

immune system will make antibodies to the rodent variable regions. The solution is to progressively change the rodent variable regions into human sequences, until the antibody is fully humanized, as illustrated in Fig. 8.6.

Antibody Engineering

Chimeric and humanized antibodies are produced using genetic engineering to manipulate the DNA sequences that code for constant and variable regions. The techniques require PCR and cloning in bacteria, both of which have been introduced earlier in the chapter. The outcome, after a complex set of procedures, is the creation of two heavy and two light chains expressed in mammalian cells; the chains are correctly glycosylated and ready to be assembled into antibodies which are almost identical to their natural counterparts.

Bispecific Antibodies and Antibody-Drug Conjugates

Immunoglobulins can be modified by genetic engineering to create bispecific antibodies consisting of two chains linked together, each recognizing different antigens. This is a promising approach to cancer therapy, as it targets both the surface of the white blood cells primed to destroy tumors and the surface of the tumor cell itself; the result is that the two cell types are drawn into close proximity to allow the cell killing to take place. This has been taken even further in the form of trispecific antibodies engineered to recognize three targets simultaneously (Garfall and June 2019).

Lastly, antibody-drug conjugates, of which there are several in the clinic and many under development, bind to tumor cells and kill them with a cytotoxic drug or radioisotope. The drug is chemically attached to a tumor-specific antibody via a small linker that is degraded inside the cell thus releasing the cytotoxic payload. Examples of the latter include highly potent auristatin compounds that inhibit cell division by disrupting microtubule polymerization. For a review of different aspects of therapeutic antibodies see Lu et al. 2020.

8.1.4.4 An Immune System in a Test Tube

Immunizing rodents to produce antibodies has served basic and applied biological research well and will continue to do so for a long time. Millions of lives have been saved, or improved, by therapeutic antibodies that were originally created in these animals (sometimes irreverently called "furry test tubes"). The downside is the time taken to produce antibodies and the fact that not all human molecules will stimulate the rodent immune systems (i.e., they are not immunogenic). An ideal solution would be to avoid the use of animals altogether and to create the antibody in vitro. This is not so farfetched, given the versatility of genetic engineering and the ingenuity of scientists. Phage display was invented in 1985 and exploited as a tool for creating large numbers of antigen-binding molecules (McCafferty 1990). Specifically, the M13 bacteriophage is used as a vector to carry human DNA into *E. coli* bacteria which then express a human protein. For antibody engineering, the M13 phage are constructed, not with a gene like human insulin but with genes coding for the huge number of V regions present in human heavy and light chains. These phage DNA libraries may contain 10^{11} or more individual sequences which are expressed as proteins on the surface of the phage virus itself. The technical details of how these V regions are identified in the first place are quite complex, but the result is a collection of phages that can bind to target antigens in the test tube, just like natural antibodies. In practice, a protein of interest is attached to a solid support and then covered with a solution of phage particles to allow binding. The phage with the appropriate V region sequences will stick to the target, while the rest are washed away. Since the bacteriophages that stick are fully capable of infecting bacteria and growing up to large numbers, they can be amplified in this way and used for further experiments. The binding process is usually repeated several times to select the strongest and most specific phage binding. Throughout this process, the investigator has no idea of the sequence of the V region DNA since its selection has been an entirely random process. Once a candidate phage has been identified, however, its DNA sequence is determined and used to construct synthetic V regions. These can then be inserted into genuine human immunoglobulin protein scaffolds that have been engineered to lack V regions; in this way, a fully humanized antibody can be constructed without having to immunize an animal.

A summary of the phage display process is shown in Fig. 8.7.

The first commercial antibody that incorporated this technology was adalimumab, developed by Cambridge Antibody Technology (now MedImmune/Astra Zeneca) and marketed by AbbVie as Humira®. It is one of several antibodies to TNF-α developed for treating arthritis and other inflammatory diseases and has consistently been one of the top selling branded medicines, making billions of dollars in revenues and a hot target for biosimilar development (Chap. 16).

8.1.4.5 Next-Generation Antibodies

Although antibodies have proven their worth as drugs, they do suffer from several technical problems, including their slow clearance rate after injection into the body. Furthermore, an IgG molecule is over five times larger than the average therapeutic protein and, if used against solid tumors, will not fully penetrate the cancerous tissue to destroy it. This has prompted the search for much smaller proteins that retain both the antigen-binding capability of large antibodies and their ability to remove invading organisms (i.e., their effector functions). A new generation of antibody-like proteins is under development that retains V region sequences within a much smaller protein framework. One of these is the single domain nanobody, developed by Ablynx (now part of Sanofi), which has antigen-binding capability in a molecule of around 12,000 daltons (Ablynx 2020). Nanobodies are derived from the antibody repertoire of camelid species (camels, llamas) and sharks, all of which have strong antigen-binding capacity but lack light chains (Leslie 2018). Caplacizumab has reached the market for use against the blood disorder thrombotic thrombocytopenic purpura (aTTP) and represents the first nanobody drug to be approved at the time of writing.

Affibodies are another type of engineered protein designed with antibody-like properties but reduced size (Fregd and Kim 2017). Figure 8.8 illustrates the relative sizes of a small molecule drug, affibody and full-size IgG antibody.

Finally, phage display technology, as described earlier, is also being used to create so-called bicycle therapeutics. These are genetically engineered peptides that are selected by phage display screening and chemically modified to form bicyclic molecules that are held in rigid conformations for effective binding to targets. The

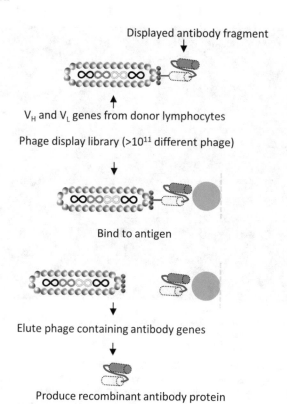

Displayed antibody fragment

V_H and V_L genes from donor lymphocytes

Phage display library (>10^{11} different phage)

Bind to antigen

Elute phage containing antibody genes

Produce recombinant antibody protein

Fig. 8.7 Production of antibodies using phage display. Variable region heavy and light chain genes (V_H and V_L) are obtained from the lymphocytes of human donors by genetic engineering and incorporated into M13 phage DNA. The encoded antibody proteins containing random V regions on both H and L chains are expressed on the surface and are essentially an immune system in a test tube. The phage library can be mixed with target antigen (gray disc) on a solid support (which could also be whole cells), and the single phage that recognizes the antigen is bound to it, while all the others are washed away. The single phage is eluted from the support and grown to larger numbers (amplified) in bacteria. This is like the normal process of recombinant protein production, so the antibody protein is purified in the conventional way. The V_H and V_L genes in the M13 can also be used to build a full-sized humanized IgG molecule using genetic engineering techniques

resulting "bicycles" can be further modified in a variety of ways to form multiple ring systems and drug conjugates (Bicycle Therapeutics 2020).

Nomenclature of Antibody Drugs

Therapeutic antibodies have a nonproprietary or generic name assigned to them by the same organizations that control the nomenclature of small molecule drugs (Chap. 3). They also have a brand name, like Humira® or Remicade®, described earlier. The generic name is built up from a group of word elements that describe the

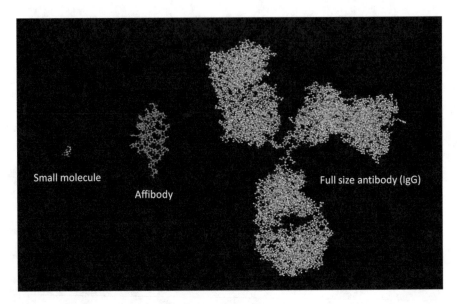

Fig. 8.8 Relative size of engineered affibody compared with IgG molecule. Typical small drug molecule shown for comparison

Table 8.2 Examples of therapeutic antibody nomenclature. The first part of the generic name is specific for the drug, the second for the condition (e.g., "im" or "lim" for immune system), the third for type of antibody, "xi" for chimeric, and "umab" for humanized

Generic name	Brand name	Derivation
Infliximab	Remicade®	Chimeric antibody for immune system
Bevacizumab	Avastin®	Humanized for cancer
Adalimumab	Humira®	Fully humanized for immune system
Efungumab	Mycograb®	Humanized for fungal infection

type of antibody (e.g., chimeric, humanized) and the indication that it is used for (e.g., cancer, respiratory) as shown in Table 8.2.

Finally, a four-letter code (with no meaning) is assigned at the end of the nonproprietary name as a unique identifier to avoid confusion between the range of products available (i.e., biosimilars) so, for example, adalimumab is presented as adalimumab, adaz; adalimumab, adbm; adalimumab, bwwd; and others.

8.1.5   Other Large Molecule Drugs

Peptides, Nucleic Acids, Lipids, and Carbohydrates
Although most biological drugs are proteins (this includes vaccines), other large molecule drugs are being developed. Peptides range from 2 to about 50 amino acids,

after which they are called polypeptides or proteins. Naturally occurring peptides act on dedicated receptors to control a wide range of physiological processes, including blood pressure, appetite, and glucose levels. Mimetics of these peptide ligands are therefore sought after as potential drug molecules through modifications to improve stability in the bloodstream and cell permeability (Otvos and Wade 2014, Stein 2018). It is even possible to formulate peptides to be orally available so the drug can be administered by mouth (see Chap. 10).

Chap. 2 contained some basic background to nucleic acids, so these will not be discussed much further, except to say that they are designed to alter drug targets at the level of the genome. Both use technology described earlier, i.e., bioinformatics to choose the correct DNA sequences, oligonucleotide synthesis and medicinal chemistry to create the siRNA or antisense drugs, and vector engineering for gene therapy.

Apart from proteins and nucleic acids, some carbohydrates and lipids are molecules with large molecular weights. Lipids are used more for drug delivery than as drugs per se, something which will be covered briefly in Chap. 10. Carbohydrates are molecules consisting of individual saccharide units that link together to form complex polymers with a variety of functions. The anticoagulant drug heparin, which has been in use for many years, is an example of a large molecule carbohydrate drug. It is related to the glycosaminoglycan (GAG) family of molecules built up from the small molecules N-acetylglucosamine or N-acetylgalactosamine.

One of the functions of carbohydrates is to create adhesion between viruses or bacteria and the surfaces of cells. This makes them interesting from the point of view of infectious diseases, where carbohydrates may be present either on the cell being infected, or else on the virus or bacterium. The anti-influenza drugs Relenza® and Tamiflu® are both small molecule inhibitors of the sialidase enzyme that breaks down a carbohydrate molecule on the surface of infected cells. Inhibiting this enzyme then prevents viral entry into the cell and reduces the duration of flu symptoms. Carbohydrates are also involved in the adhesion of white blood cells to blood vessels during acute inflammation and have been under consideration as potential drug targets for many years. Unfortunately, for technical reasons, carbohydrates are not easy to turn into inhibitor drugs with high affinity for their protein target; this, however, has not discouraged researchers from trying to make breakthroughs in this area.

8.1.6   Cell Therapy

The use of cells as medicinal products has already been described in Chap. 2. This section gives some more detail about the isolation and manipulation of human cells for tissue repair and enhancing the immune response against tumors.

8.1.6.1 Stem Cells

In an ideal world, tissue regeneration would be achieved by swallowing an orally active drug to stimulate the growth of new tissues in specific places, like the brain. Unfortunately, tissue growth during development depends on many complex interactions between growth and differentiating factors (like cytokines), signaling pathways, cell adhesion molecules, and the programmed death of unwanted cells. Since it is hard to envisage any way of repairing tissues other than through using stem cells, there is an enormous worldwide effort to understand what makes these cells special and how they can be coaxed into forming new tissues to order. The starting point is the population of pluripotent stem cells that arise in the early cell division stages of the fertilized egg (described in Chap. 2).

The problem with natural stem cells is that they must be isolated from human embryos that have been discarded during in vitro fertilization (IVF) procedures. This is both ethically controversial and impractical if these cells are going to be developed and marketed by the biopharmaceutical industry. There is also the major problem of the donor cells being rejected by the immune system of the recipient patient. The only practicable way of getting around this situation is somehow to create stem cells from a patient's own adult cells. There is no easy answer to this as embryonic stem cells (ES cells) have all the correct molecular systems in place to perform the task of creating different tissues; alternative sources will almost certainly differ in subtle ways that could lead to clinical problems emerging in the longer term. Despite this caveat, there is an enormous amount of research directed at finding stem cells that do not require the use of embryos. A breakthrough occurred in 2006 when Shinya Yamanaka and colleagues in Japan succeeded in creating stem cells from adult mouse cells (Takahashi and Yamanaka 2006). These were named induced pluripotent stem cells (iPS) and created a great deal of excitement in the field of regenerative medicine. The process of creating them required the introduction of a series of genes that control other genes critical for the formation of ES cells in the embryo. Since the initial discovery, the process has been refined by optimizing the cocktail of agents (including small molecules) to increase the efficiency of transforming adult cells into stem cells. Progress in using iPS cells to treat disease has been slow, however. Despite intense activity (notably in Japan), progress in developing these agents for the clinic has been disappointing. This is in part due to the oncogenic potential of iPS cells (see Nagoshi et al. 2019 for a discussion of these issues as part of a therapy for spinal cord injury).

Despite the above caveats, iPS cells are moving into the clinic. At the time of writing (2020), the first USA-based clinical trial of these cells is being established for geographic atrophy, the "dry" form of age-related macular degeneration (AMD), a leading cause of vision loss among people age 65 and older (National Institutes of Health 2019).

The origin of pluripotent stem cells during early embryogenesis and the generation of iPS cells are illustrated in Fig. 8.9.

8.1.6.2 Genetically Engineered T Cells

Cytotoxic T lymphocytes are responsible for killing virus infected or tumor cells after physical recognition of target antigens through the T-cell receptor protein complex (TCR). Interest in using T cells for tumor therapy has been around for many years, but the development of genetically engineered cells in the recent years that has reignited the field. Chimeric antigen receptor-modified (CAR) T cells whose activity has led to remissions in previously intractable tumors were described in Chap. 2. There are over 500 clinical trials for CAR therapies (MacKay et al. 2020) for a wide range of tumor antigen targets. Despite this activity there are serious issues with toxicity, for example, as a result of inflammation through cytokine release. The CAR T cells may also show reduced efficacy due to competition between the engineered receptors and endogenous TCRs present in the patient's white blood cells. Other changes can occur to reduce the potency of the T cell response, for example, the induction of the PD-1 receptor that functions to naturally downregulate T cells after activation[5]. Finally, there is the problem of immune rejection through using allogeneic cells (i.e., from donors other than the patient) as a means of creating large numbers of CAR T cells as an off-the-shelf therapy[6]. All the above situations can be addressed by using gene editing to selectively remove the endogenous TCRs, PD-1 receptor, and the MHC molecules on allogeneic cells. Preliminary clinical studies of T cells edited with CRISPR-Cas9 to remove endogenous TCRs and the PD-1 receptor show that the gene editing appears to be safe thereby leading the way for further development.

In addition to CAR T cells, other cell therapies are being developed for cancer. These include T cells modified with cancer antigen specific TCRs and CAR-NK cells which are natural killer cells engineered with CAR receptors (EbioMedicine 2019). NK cells are important in attacking tumors and therefore have potential as another line of attack.

The basic design of CAR T cells is shown in Fig. 8.10.

Summary of Key Points

Many drug targets are too large to be effectively modified by small molecules. Large molecule drugs like proteins are proving to be effective and commercially successful medicines.

Therapeutic antibodies (immunoglobulins) are large proteins that are providing new treatments for cancer, arthritis, and other major conditions.

Antibody engineering is used to create humanized versions of immunoglobulins as well as produce antibody fragments using phage display technology.

Stem cell therapy, particularly with iPS cells, shows great promise for regenerative medicine, but there are many technical obstacles to overcome before it can become routine in clinical practice.

Genetically modified cells such as CAR T cells and CAR-NK-cells are being developed for a wide range of tumors once issues of safety and efficacy can be resolved.

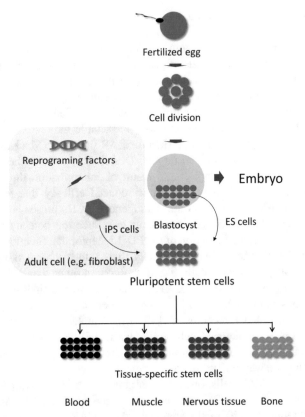

Fig. 8.9 Pluripotent stem cells. After fertilization, the egg (ovum) keeps dividing until the blasto-cyst stage is reached. In normal development, this ball of cells will implant into the uterus and develop into an embryo. All the different tissues in the embryo (and adult) derive from the pluripo-tent stem cells present in the blastocyst. If these embryonic stem cells (ES cells) are removed in the laboratory and cultured under special conditions, they can be turned into tissue-specific stem cells and ultimately new tissues such as blood, bone, muscle, etc. in a way that mimics embryonic devel-opment. Shaded box: normal adult cells can be reprogrammed to revert to an embryonic state and form pluripotent stem cells (iPS cells). Reprograming factors may include genes expressing the transcription factors Oct4, Sox2, Klf4, and c-Myc or many variants including small molecule drugs

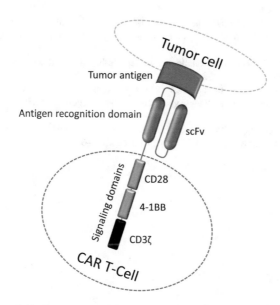

Fig. 8.10 Design of third-generation CAR T cells. Patient-derived cells genetically modified to express a protein containing different domains. The extracellular scFV region binds antigen in the same way as the phage display antibodies described earlier; this substitutes for the T-cell receptor in recognizing a specific antigen present on the surface of the tumor cell. The intracellular domains CD28, 4-1BB, and CD3ζ are involved in cell signaling to stimulate the T cells into dividing and exerting their cytotoxic action on the tumor. Further generations of CAR T cells are under development to enhance activity and reduce unwanted effects such as cytokine release syndrome

References

Ablynx website http://www.ablynx.com/research/ Accessed 17 Mar 2020

Bicycle Therapeutics. https://www.bicycletherapeutics.com Accessed 18 Mar 2020

EbioMedicine (2019) Natural killer cells for cancer immunotherapy: a new CAR is catching up. https://doi.org/10.1016/j.ebiom.2019.01.018

Frejd FY, Kim K-T (2017) Affibody molecules as engineered protein drugs. Exp Mol Med 49:e306. https://doi.org/10.1038/emm.2017.35

Garfall AL, June CH (2019) Trispecific antibodies offer a third way forward for anticancer immunotherapy. Nature 575:450–451

Kim SJ et al (2005) Antibody engineering for the development of therapeutic antibodies. Mol Cells 20:17–29

Leslie M (2018) Mini-antibodies discovered in sharks and camels could lead to drugs for cancer and other diseases. Science. https://doi.org/10.1126/science.aau1288

Lu R-M et al (2020) Development of therapeutic antibodies for the treatment of diseases. J Biomed Sci 27:1. https://doi.org/10.1186/s12929-019-0592-z

MacKay M et al (2020) The therapeutic landscape for cells engineered with chimeric antigen receptors. Nat Biotechnol 38:233–244

McCafferty J (1990) Phage antibodies: filamentous phage displaying antibody variable domains. Nature 348:552–554

Nagoshi N et al (2019) Cell therapy for spinal cord injury using induced pluripotent stem cells. Regen Ther 11:75–80

National Institutes of Health (2019) https://www.nih.gov/news-events/news-releases/nih-launches-
first-us-clinical-trial-patient-derived-stem-cell-therapy-replace-dying-cells-retina Accessed 19
Mar 2020

Otvos L, Wade JD (2014) Current challenges in peptide-based drug discovery. Front. Chem. doi.
org/10.3389/fchem.2014.00062

Stein RAA (2018) Peptide therapeutics near the sweetest spots. Genetic Engineering and
Biotechnology News https://www.genengnews.com/category/magazine/312/ Accessed 16
Mar 2020

Takahashi K, Yamanaka S (2006) Induction of pluripotent stem cells from mouse embryonic and
adult fibroblast cultures by defined factors. Cell 126:663–676

Chapter 9
Screening for Hits

Abstract This chapter describes the process of screening that is undertaken once the target for a drug has been identified and the compounds or biologicals have been produced to test against it. Different laboratory assays are used to measure interactions with the target and, depending on the throughput of the screen, may require robotics to automate the process. These technologies are described in the chapter along with the relevant terminology used in this part of the drug discovery process.

9.1 Introduction

The identification of drug targets, small molecule synthesis, and the production of large molecules has already been covered. This chapter introduces the screening assay (or just screen) that is used to identify either compounds or biologicals that interact with a chosen target. Screening is the primary strategy for small molecule drug discovery by the biopharmaceutical industry despite the element of chance associated with it. Random collections of small molecules (synthetic or natural products) are screened against targets to identify pharmacophores which can then be modified by medicinal chemists to create drug candidates for full clinical development. Biological products, such as antibodies, are screened in a similar way to small molecules, but the lead candidates are progressed to development along a different path (e.g., antibody humanization).

9.1.1 Terminology

Screening has its own terminology, which may vary somewhat from company to company:

- Active.

The first stage of screening produces an "active" that influences the target in question. This active could be a false positive; in other words it has affected the target in a non-specific way. This is a common problem, as many compounds stick

E. D. Zanders, *The Science and Business of Drug Discovery*,
https://doi.org/10.1007/978-3-030-57814-5_9

non-specifically to proteins or act like detergents; there may therefore be a high rate of attrition of such compounds in a screen.

- Hit.

If the active effect is repeated in a second screen (i.e., is reproducible), the compound becomes a "hit." Before a hit can become a lead, it must be selective against the target and must have a clearly defined Structure Activity Relationship (SAR) with related compounds (see Chap. 7). Quite often a screening collection that produces an active will contain several related compounds which can be tested at the same time to provide some initial SAR information.

- Lead.

The desired outcome of a screening campaign is the identification of a lead compound (or lead) that can be further evaluated in vitro and in animal models of disease. Leads are hits that have been chemically modified in the laboratory to optimize their biological properties.

- Primary, secondary, and tertiary screens.

The primary screen is the initial survey of compounds using the drug target alone. Active compounds are then tested in secondary screens to measure the selectivity of the compound by using different protein targets or cell types. This is crucial for ensuring that the active compound is working on the target and not through some non-specific disruptive effect. It is particularly important for secondary screening on whole cells since many compounds may simply be toxic. Further (tertiary) screens may be performed to ensure that the active is a real hit.

Phenotypic screens are designed to find agents that alter the overall properties of a cell or tissue rather than a purified molecular target. This approach has the advantage of exploring one or more targets in their natural environment, something that is lost when the target protein is purified away from the cell or is produced using genetic engineering. It also eliminates compounds that cannot pass through the cell membrane to bind to intracellular targets. However, once compounds have been selected by phenotypic screening, it is not always straightforward to identify their molecular targets. This requires a process of "deconvolution" to make sense of the data. This is made more difficult if there are multiple targets as is often the case with small molecule drugs.

The size of a screen is determined by its throughput, i.e., the number of compounds that are tested in a particular campaign. The throughput is determined by the type of screen being undertaken and is summarized as follows:

- Low throughput.

A maximum of around 200 compounds tested. Often these are not small molecules picked at random but highly biased towards a defined chemical structure such as those used for focused compound libraries or for fragment-based screening (Chap. 7). It is also possible that the screening target is difficult to set up for more than a few tests, a situation that arises when screening whole tissues or organs. Because the number of tests is low, there is normally no requirement for automation.

- Medium throughput.

Compounds (and microbial extracts) are screened in the low thousands. This is used when the target is not easily set up in large numbers, for example, with live model organisms like zebrafish embryos (Chap. 6). Even with several thousand compounds, it is possible to perform the screen without automation, but as I know from bitter experience, the process becomes extremely tedious if performed manually.

- High throughput.

This has been the mainstay of biopharmaceutical screening and is now increasingly being taken up by academic institutions. Large collections of compounds (in the many thousands) are screened with the assistance of robotic workstations to manipulate compounds and target samples (see later).

- Ultrahigh throughput (uHTS).

The problem with screening random compounds is that the chance of finding a useful hit is much lower than if the input were carefully preselected to eliminate poor candidates. Obviously, the more compounds screened, the more chances of finding something useful; however, the adage "rubbish in, rubbish out" definitely applies to screening (actually this is usually described in more "robust" language by drug discovery scientists). Ultrahigh-throughput screening of several million compounds certainly increases the input, but only large companies such as GlaxoSmithKline and Pfizer have the resources to pay for the materials and robotics required for such a huge enterprise.

- Virtual (or in silico) screening.

Screening compounds in the computer is the ultimate high-throughput method since modern processing power can handle millions of interactions within a short period of time. This topic and the application of machine learning in this area have been covered in Chap. 7.

The basic outline of a screening strategy is shown in Fig. 9.1.

9.1.2   Screening in Practice

The different classes of drug targets detailed in Chap. 5 have their own characteristics which determine how a screen (at any throughput) is configured. For example, screens for inhibitors of enzymes will be different to those designed for receptor agonists or antagonists. The target itself may be a purified protein or might remain in the environment of the cell in which it is normally expressed. In both cases there is a need to use as little biological and chemical material as possible to keep the cost

Fig. 9.1 Screening flowchart. Compounds or extracts of natural products (cultured bacteria or fungi or plant/animal extracts) added to drug target protein and the effect measured (see details in text). If the desired effect is observed, the compound is retested against the same target. If the effect is reproducible, the compound is tested against an irrelevant target. If no effect is observed, the hit compound is modified by medicinal chemistry to generate a series of compounds for retesting on the target. If a clear Structure Activity Relationship (SAR) is observed between the different compounds, the most promising examples are selected as lead compounds for further development into drugs

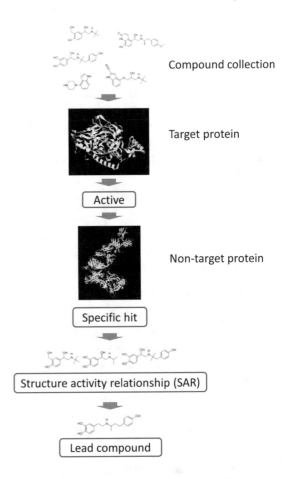

Compound collection

Target protein

Active

Non-target protein

Specific hit

Structure activity relationship (SAR)

Lead compound

of each screen to a minimum. This is achieved by using disposable plastic vessels, known as microplates, which hold small volumes of liquid. Screening plates are molded with 96, 384, and 1536 wells which are suited to different degrees of throughput. The amount of plastic used for a large screen is daunting; screening one million compounds without any duplication or extra samples would require about 650 x1536-well plates.

Compounds to be screened are dissolved in small volumes of liquid and added to the primary assay at a fixed concentration (normally 10 μM). This figure is a compromise between having too many false positives because the concentration is too high and not picking up useful starting points if the concentration is set too low. The storage and manipulation of compounds can be extremely challenging, particularly when as many as 100,000 are being screened each day, so robotic workstations are needed to automate the process. An example of a 96 well microtiter plate and a screening robot is shown in Fig. 9.2.

Large collections of compounds have been built up by large pharmaceutical companies over periods of many years and include those bought from academic

Fig. 9.2 (**a**) 96 well plastic microtiter plate with colored liquid. These are used as vessels to hold the screening reactions in small volumes (in this case up to about 250 µl). (**b**) A screening robot for automating the addition of liquids to screening plates and measuring responses to different compounds. Three robot arms that manipulate the plates (in racks left side of image) and position them are shown

a

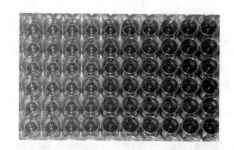

b

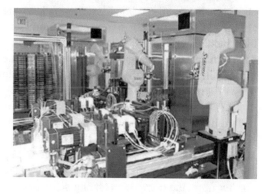

institutions and specialist collections. The majority, however, emerge from internal company research programs; this has tended to bias the screening results towards the type of target in which the company has had a historical interest. I discovered this with my screens for immunosuppressant compounds, where most of the actives turned out to be steroid-like molecules that were produced for other programs in the past. Unfortunately, they were of no interest and had to be weeded out. Modern compound collections have been pre-screened to remove compounds likely to produce non-specific or undesirable effects. A more recent development is the use of highly focused screening collections consisting of small molecule libraries built up from pharmacophores discovered in nature or designed in the computer (see Chap. 7). This chemoinformatics approach ensures that the chance of finding a useful hit is much greater than through searching at random. However, the creation of massive compound collections such as DNA-encoded libraries may prove to an advantage in finding novel chemotypes (Chap. 7).

9.1.2.1 What Is Actually Measured?

The main requirement for a screen is a simple and robust system that generates a signal in a reproducible way. Depending on the target being screened, this signal may be anything from a color change to the growth of cells. Antibiotics were discovered by measuring the inhibition of growth of whole bacteria when screening natural product extracts. Other cell-based screens use mammalian cell cultures to measure growth or some specific aspect of their function. The immunosuppressant drug cyclosporine A, for example, was discovered by inhibiting the function of whole lymphocytes which had been artificially activated in the screening assay to mimic an immune reaction. There are different techniques to measure these cell effects, some of which involve the uptake of small amounts of radioactivity using labeled nucleotides or amino acids that will be incorporated into DNA and proteins, respectively. Radioactive compounds, however, have to be handled with great care under strict regulations for monitoring and disposal; this has prompted the search for nonradioactive alternatives such as a range of small organic compounds that change color when modified by cells undergoing division. The intensity of a color change can be quantified in a spectrophotometer, which measures the absorbance of light by a sample and expresses this absorbance as a number. Spectrophotometers are the mainstay of the screening (and biochemistry) laboratory and can be designed to measure the color absorbance of many samples in parallel such as those dispensed into multi-well screening plates.

The use of live cells to screen drugs should, in principle, provide a more realistic picture of how an active compound is likely to behave in a patient (see phenotypic screens earlier). Unfortunately, the cell-associated target is also in the presence of many other molecules which can interfere with a screening reaction through creating false-positive results or through producing results that are difficult to interpret. Furthermore, the compounds being screened may not penetrate the cell membrane and get to their target (if it resides inside the cell) so that potentially interesting hits may be missed. The alternative to whole cell screens is to use the isolated target protein in varying degrees of purity. Screens for the GPCR class of receptors, for example, are often set up using cell membranes to express the protein that is then mixed with radioactive ligands. The introduction of scintillation proximity assays (SPA) has had a big impact upon this type of screen as it avoids the need to filter a radioligand away from a receptor or other binding protein, unlike the example given in Chap. 4, Fig. 4.9. For this reason, SPA is known as a homogeneous assay; small plastic beads coated with receptor protein are mixed in the same tube or plate with the radioligand to allow binding. The beads enclose a scintillant chemical that only emits light when the radioactivity is close by, i.e., when the compound binds the receptor, hence the term scintillation proximity. The emitted light is detected in a scintillation counter as with other detection methods for isotopes like ^{3}H or ^{14}C. An overview of SPA is given in scintillation proximity assays 2020.

It is now quite routine to express human receptor proteins in animal cells (such as CHO or COS cells, Chap. 8) by using recombinant DNA technology. This means that almost any GPCR, or other receptor type, can in principle be screened, so long as its DNA sequence is available.

Screens for both inhibitors and activators of enzymes are often set up with the purified protein dissolved in a solution of water and various salts, known as a buffer. Both enzymes and other biological molecules are sensitive to changes in pH (acidity and alkalinity; see Chap. 4), temperature, and concentrations of salts such as sodium chloride. Buffers are designed to maintain the solution at a constant pH, hence their use for any biochemical reaction that has been set up in vitro, such as an enzyme screen. The activity of an enzyme is determined by the conversion of a substrate into a product. The substrate for a protease enzyme (e.g., pepsin in the stomach) is itself a protein, and the product a collection of smaller peptides formed by breaking this protein's chains. In a screen for protease inhibitors, an artificial substrate is produced in the form of a small peptide synthesized in the laboratory with a colored dye molecule attached. When the peptide is cut by the enzyme, the dye molecule is released and registers as a color that can be accurately measured in a spectrophotometer. Any screened compound that inhibits the enzyme will stop the color increase and therefore register as an active.

9.1.2.2 Advanced Screening Technologies

Modern screening operations are constrained by cost and safety issues, particularly when applied at very high throughput. The unit cost of a screen can be reduced by using smaller amounts of precious reagents in each well. Safety is a particularly important issue with radioactivity, because experiments must be performed in dedicated facilities, and everything must be monitored for accidental spillage. This has encouraged the search for labels that have equal or greater sensitivity to radioisotopes without the need for special precautions in their use. These labels and detection systems exploit known physical phenomena related to the interaction between radiation (of all types) and matter. Some of these detection systems are listed as follows:

- Fluorescence.

This occurs when light with one wavelength (color) stimulates a fluorescent molecule to emit light of another wavelength. Fluorescent labeling is sensitive and widely used in screening and diagnostics (Chap. 14). A related technology, fluorescence resonance energy transfer (FRET), is widely used to measure the interactions between large molecules; positive contact between molecules labeled with certain elements (e.g., europium or terbium) results in a reduction of the fluorescent signal.

- Phosphorescence.

This is similar to fluorescence in that light of a specific wavelength is emitted from phosphorescent materials, but the duration and nature of the emission are different. Phosphorescence is used in specialized assays and for imaging radioactivity.

- Chemiluminescence.

The wonderful glow of fireflies at night results from the conversion of chemical energy to light, aided by the enzyme luciferase. The biochemical mechanism behind this has been exploited in many screening formats which use recombinant firefly luciferase in conjunction with light emitting compounds. The sensitivity of these labels is extremely high and can exceed that of some radiolabels.

- Surface Plasmon Resonance (SPR).

This rather complex phenomenon is extremely useful for measuring the interaction between different molecules since there is no need to tag them with any sort of label. The instrumentation required for SPR is expensive and unsuited for high-throughput applications but is being increasingly used in the screening laboratory (see GE Healthcare 2020).

High content screening is a screening technology that involves the use of sophisticated microscopes and imaging systems to simultaneously visualize multiple events occurring inside cells that have been treated with different compounds. A screening compound can be evaluated for its effect on many different processes occurring inside cells in addition to those caused by the primary drug target. This has an advantage over simpler screens in that mechanisms of action or possible toxicity can be evaluated at an early stage. In vivo screening offers the ultimate test of a compound's activity but is normally impractical to perform on more than a few animals. This has begun to change, with the introduction of the small model organisms discussed in Chap. 6. Prominent among these is the zebrafish whose embryos are small enough to be screened in 384 well microplates. The embryos can also be manipulated using recombinant DNA technology to express fluorescent marker proteins in specific body structures to observe the effects of compounds on their appearance and function. This approach has been used to highlight blood vessels in zebrafish by using genetic engineering to express a gene called green reef coral fluorescent protein (GRCFP) which is derived from a marine organism. The vessels shine green under fluorescent light and can be imaged with a microscope and CCD device to show blood vessel growth (angiogenesis) over several days. A medium throughput screen of a compound library produced known inhibitors of angiogenesis plus a novel compound that also inhibited human angiogenesis in secondary assays (Tran et al. 2007). Inhibitors of this type have the potential to treat cancer by preventing the growth of new blood vessels, thereby starving tumors of their blood supply.

9.1.3   Chemical Genomics and Proteomics

The interaction between small molecules and the proteins encoded by the genes of organisms is not only important in drug discovery but also in the understanding of fundamental biological processes hence the development of chemical genetics or chemical genomics. Chemical genetics is simply the process of altering biological function with small molecule probes instead of through genetic mutations. Just as genomics is genetics on a genome-wide scale, chemical genomics (and the associated chemical proteomics) is the search for small molecules that interact with all the proteins encoded by the genome (MacBeath 2001). This makes chemical genomics a central part of small molecule drug discovery as it overlaps considerably with target discovery, compound synthesis, and screening. Although most proteins are not likely to be drug targets, chemical probes made for each of them can provide valuable information about basic human cell biology. This could be exploited in pharmaceutical research for target identification as well as to identify proteins that are the targets for toxic drugs, as exemplified later in this chapter. The success of the Human Genome Project in identifying human genes has prompted similar initiatives in chemical genomics/proteomics with the aim of generating small molecules that interact with all the protein products of the human genome. Some of these initiatives have not been maintained, leading to proposals in 2018 for a serious assault on the problem of creating a probe for every human protein in the form of Target 2035 (Carter et al. 2019). Most of the targets lie within the so-called dark proteome and therefore have received relatively little attention from drug discovery and basic scientists. The natural bias against such proteins must therefore be resisted by covering all the genome, albeit at a projected cost of billions of dollars.

Some of the characteristic tools of genomics such as microarrays have been exploited in chemical genomics research. For example, it is possible to immobilize small molecules at high densities on specially treated glass slides in a microarray format. A protein under investigation can then be labeled with a fluorescent tag and passed over the chemical microarray to generate a signal where binding occurs. Since the identity of the compound is known at each position on the slide, it is possible to determine which one(s) is responsible for binding. The miniaturization of this binding assay makes it possible to screen many thousands of compounds simultaneously.

9.1.3.1 Working with Chemical Probes

When a purified protein is used in a screen for small molecules, the results are relatively straightforward to interpret; if the protein is a drug target, then molecules that bind to it will be of interest as potential drugs. The situation is quite different with screens that use the function of live cells as readout. Sometimes the precise molecular target may not be known, and there is no guarantee that compounds are active just because they interact with it. In this case, the screen has to be deconvoluted; in other words, the complex web of proteins in the cell has to be unraveled to reveal the true drug target. Although modern understanding of the chemistry and biology of the cell has accelerated the process of deconvolution, it is still a major undertaking for any research group. There are, however, techniques based on affinity selection for identifying the protein targets of chemical probes (or hits from phenotypic screens). One approach is to bind the drug target protein to the probe that has been immobilized on a solid support. The latter may a bead of cellulose or similar inert material that can be packed into a chromatography column. When human cells are broken up in mild detergents, their proteins are dissolved in a solution that can be poured onto the top of the column and allowed to pass slowly through into a collection vessel. Most of the proteins will not bind to the compound in the column, but those that do can be recovered and identified using the mass spectrometry approach outlined in Chap. 6. One way of recovering the protein from the column is to add a large excess of the same compound that is also immobilized on the beads. This has the effect of competing for the protein binding sites and makes it possible to screen other compounds by assessing their ability to displace the protein. This approach has been used for target discovery by academic groups and commercialized by companies such as Cellzome (now part of GSK). A variant of this is affinity-based protein profiling (ABPP) in which probes are immobilized to capture all members of a specific protein family, such as the serine hydrolases (Moellering and Cravatt 2012).

The second approach involves binding tagged small molecules to the protein targets in live cells prior to extracting the complex and identifying the binding partners by mass spectrometry or visualization on electrophoresis gels. This has been used for determining the binding partner of natural product drugs where the cellular target is unknown (Wright and Sieber 2016).

Examples with Known Drugs

Affinity selection has been used to analyze so called off-target effects where compounds act on proteins that are not their primary target, an example being the drug thalidomide. This was prescribed in the 1960s as a sedative to pregnant women and for preventing morning sickness but gained notoriety because of its teratogenicity, i.e., birth defects characterized by limb malformations. However, the thalidomide analogue lenalidomide is effective in treating multiple myeloma and has been approved by the FDA for this indication. The protein target responsible for the birth defects and clinical effects turns out to be cereblon (CRBN) which was isolated using thalidomide attached to beads in a column (Ito et al. 2010). The biological effects of CRBN can be explained by its involvement with the ubiquitin system that selectively degrades proteins (Chap. 6).

Finally, even the venerable drug aspirin has been shown to bind to multiple protein targets, in this case using affinity-based protein profiling (Wang et al. 2015). A total of 523 protein targets were identified with a variety of biological functions, the majority being enzymes. Aspirin (acetylsalicylic acid) covalently modifies the amino acid serine in the active site of the cyclooxygenase enzymes thereby blocking the synthesis of the prostaglandin PGE_2 which is responsible for fever and inflammation. However, the amino acids targeted by aspirin in the 523 proteins show a much broader picture, revealing a considerably more complex action of aspirin on human cells than may have been expected at first sight (Table 9.1).

Table 9.1 Amino acids modified by aspirin in 523 target proteins identified by ABPP (Wang et al. 2015)

Amino acid	Number modified
Lysine	2282
Serine	167
Threonine	135
Arginine	82
Histidine	59
Tyrosine	47
Tryptophan	12
Cysteine	2

Summary of Key Points

Screening is used to select small and large molecules as leads for further development.

The number of samples tested can range from a few hundred to millions, with virtual compounds numbered in the billions.

Different physical phenomena are used to detect the binding or modifications of molecules in screening assays; these include radioactivity and fluorescence.

Chemical genomics focuses on the interactions between small molecules and proteins on a genomic scale.

Affinity selection is a method of purifying a target protein from a complex mixture using immobilized small molecules or chemically modified probes.

References

Carter AJ et al (2019) Target 2035: probing the human proteome. DrugDiscToday 24:2111–2115

GE Healthcare (2020) https://www.gelifesciences.com/en/us/solutions/protein-research/knowledge-center/surface-plasmon-resonance/surface-plasmon-resonance Accessed 27th March 2020

Ito T et al (2010) Identification of a primary target of thalidomide teratogenicity. Science 327:1345–1349

MacBeath G (2001) Chemical genomics: what will it take and who gets to play? Genome Biol. https://doi.org/10.1186/gb-2001-2-6-comment2005

Moellering RE, Cravatt BF (2012) How chemoproteomics can enable drug discovery and development. Chem Biol 19:11–22

Scintillation Proximity Assays (2020) https://www.perkinelmer.com/uk/lab-products-and-services/application-support-knowledgebase/radiometric/scintillation-proximity.html#Scintillationproximityassays-Scintillationproximitytechnologies Accessed 30 Mar 2020

Tran TC et al (2007) Automated quantitative screening assay for antiangiogenic compounds using transgenic zebrafish. Cancer Res 67:11386–11392

Wang et al (2015) Mapping sites of aspirin-induced acetylations in live cells by quantitative acid-cleavable activity-based protein profiling (QA-ABPP). Sci Rep. https://doi.org/10.1038/srep07896

Wright MH, Sieber SA (2016) Chemical proteomics approaches for identifying the cellular targets of natural products. Nat Prod Rep 33:681–708

Chapter 10
Manufacturing and Formulation

Abstract This chapter covers the scale up and manufacturing of drug molecules, using process chemistry for small molecules, large-scale expression systems for antibodies and other proteins and the specialized production of gene therapy vectors and engineered cells. The second part deals with formulation of active pharmaceutical ingredients into a dosage form that is suitable for administration to patients. The chapter is concluded with a brief review of nanotechnology and its role in drug delivery.

10.1 Introduction

Once a compound or biotherapeutic molecule has been selected as a lead for further development, it is ready to leave the discovery phase and move into preclinical development. The end of the discovery phase signifies a major transition between unregulated, sometimes "blue sky" research and the heavily regulated process of producing and marketing a medicine that may be sold to millions of people worldwide. Decisions and actions taken at these latter stages now have major legal, moral, and financial implications for a biopharmaceutical company.

There has always been a gulf between discovery scientists and those involved in development, because the working ethos of each group is driven by different requirements. The discovery scientist, almost by definition, is more freewheeling and experimental, in keeping with the "discovery" in their job title. There is nothing more frustrating to such a scientist than having to perform a prescribed set of experimental procedures that must be rigorously followed to the letter. The history of drug discovery, and indeed science in general, is full of people who have broken the rules, or exploited a chance observation that had nothing to do with what they were supposed to be working on. Development scientists possess the same experimental skills but use them in very clearly defined processes must be performed in a uniform and consistent manner. In the past, this cultural gap was enormous, with each side understanding their part of the drug development process but having little inclination to learn from each other's experiences. Discovery scientists would metaphorically throw the compounds over a wall to development teams and then forget about them; equally there was no real appetite in development for exploiting the latest

E. D. Zanders, *The Science and Business of Drug Discovery*,
https://doi.org/10.1007/978-3-030-57814-5_10

findings in basic research to improve their processes and procedures. This all began to change in the 1990s, the result being a much closer integration between early discovery and the later stages of drug development. Included in these later stages are the manufacture and formulation of compounds and biologicals. These are critical parts of the development process since whole programs have failed simply because the drug was impossible to formulate or was too expensive or hazardous to manufacture.

This chapter describes some of the basic processes used in the scale up and manufacture of small molecules and biotech products and then covers their formulation into the products that are sold in the pharmacy. These development functions are encompassed by the term "pharmaceutical development."

10.1.1   Scale-Up and Manufacture

Compounds and biologicals (like antibodies) are normally produced in milligram to gram quantities for primary testing and evaluation in disease models. Once a lead compound has been identified, its production must be scaled up from grams to kilograms to provide enough material for preclinical evaluation in animals as well as for formulation.[1] Clinical trials and commercial production require multi-kilogram quantities produced under strict guidelines collectively known as Good Manufacturing Practice or GMP (also CGMP or current GMP). The regulatory authorities responsible for its administration include the American Food and Drug Administration (FDA), the European Medicines Agency (EMA), and the Ministry of Health, Labour and Welfare in Japan. These agencies will be covered in more detail in later chapters, along with a description of the regulatory submissions process. Each country must also conform to regulations applied by their competent authorities, for example, the Medicines and Healthcare products Regulatory Agency (MHRA) in the UK. Manufacturers are now encouraged by the FDA and others to follow the principles of Quality by Design (QbD) using Process Analytical Technology (PAT) to ensure that the quality of the drug is optimized during process development and not subsequently at the production stage (Barshikar 2019). Manufacturing guidelines can be found on the International Council for Harmonisation of Technical Requirements for Pharmaceuticals for Human Use (ICH) website (ICH Quality Guidelines 2020).

---

[1] To put these amounts into perspective, a daily 10 milligram dose of an averaged-sized small molecule drug administered to one million patients chronically over 1 year would require about 4 tons of material (Marti and Siegel 2006).

10.1.1.1 Small Molecule Process Chemistry

Medicinal chemists have a reasonable amount of freedom in the choice of synthetic routes and materials used to produce the small quantities of compounds required for initial screening. This freedom is lost when the compound must be scaled up in a pilot plant during chemical development since the number of synthetic steps has to be kept within reasonable bounds (normally 5 to 10 steps). The choice of raw materials for the synthesis is also important, since these must be readily available, inexpensive, and compatible with environmental regulations. The term "green chemistry" is becoming more widely spread as the biopharmaceutical industry responds to the changing expectations of the outside world (Dunn et al. 2010).

In practice, the initial scale up of compounds is undertaken by the process chemist who later partners with chemical engineers to design manufacturing schemes for the marketed drug. Since the manufactured compounds must be as pure as possible, analytical chemists are employed to liaise closely with the chemists and engineers in the process teams. A great deal of creative effort goes into developing new synthetic routes for compounds using novel catalysts or other techniques. One of the main concerns of the process chemist is the behavior of chemical reactions on a large scale. For example, a small reaction may give off some heat as it proceeds; on a larger scale, however, this heat output could be highly dangerous, so this must be anticipated in advance.

Process chemistry also must be as efficient as possible to keep costs down to a level that the pharmaceuticals market will bear. The sum of all the manufacturing and licensing costs of a compound is known as the cost of goods (CoG). This must be reduced once a drug comes off patent and is taken up by generic manufacturers, since the cost per dose must be at a level to make it commercially viable. Therefore, pressure is put on the generic companies to design synthetic routes that are cheaper than those used initially for the branded medicine. The painkiller ibuprofen is a case in point. This was produced by Boots in the late 1960s using a chemical synthesis involving six individual reactions. Although the process created enough material for the marketplace at the time, it required large amounts of chemical reagents and produced an excess of aluminum salts. Later, the BHC process was developed by the Boots-Hoechst-Celanese company and used in 1992 in a Texas plant to produce around 4000 tons of the drug per annum. The synthesis employed only three steps and used a solvent that was itself a part of the reaction process and which could be recycled with minimal waste as shown in Fig. 10.1.

The cost of goods is a key pressure point in the commercial development of a branded medicine (see Jermini and Polastro 2020 for an overview). For example, Roche developed enfuvirtide (Fuzeon®), a 36 amino acid synthetic peptide discovered by Trimeris for AIDS therapy and approved by the FDA in 2003. Unfortunately, the cost of goods was substantial due to the high daily doses required (180 milligrams). The result was a higher launch cost ($25,000) than healthcare providers were prepared to pay, while cheaper alternatives were available. Fuzeon® was a potential blockbuster drug (>$1billion in sales) but is now used only as a salvage therapy due to the cost of producing the medicine.

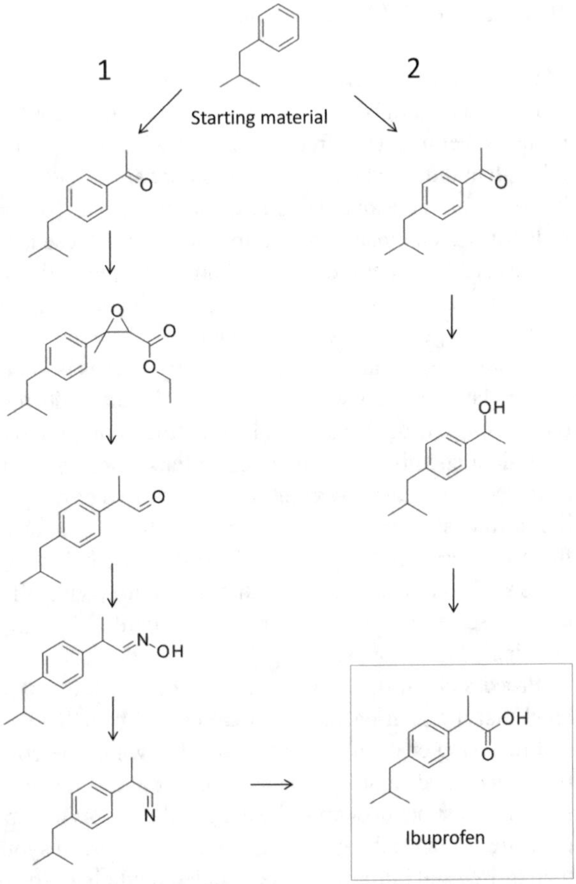

Fig. 10.1 Process chemistry for ibuprofen. Two schemes for manufacturing the painkiller from a common starting material are shown. The first involves six reactions which were reduced to three using Scheme 2. (Figure adapted from Sheldon 2010) (where reaction conditions are also shown) with kind permission of Wiley-VCH Verlag GmbH & Co. KGaA

10.1.1.2   Manufacture of Biologicals: Bioprocessing

Drug compounds and biologicals are created in different ways, because the synthesis of small molecules and large molecules requires separate processes. As described in earlier chapters, small molecules can be synthesized in a laboratory, or factory, using quite harsh conditions and can then be purified to an extremely high degree. In addition, the level of purity can easily be monitored using sensitive analytical methods. On the other hand, most biological products are produced in living organisms and are far less forgiving of mistreatments, such as variations in temperature during manufacture or storage. There are also two major issues with biological sources: the consistency of production and the purity of the final product. Even cloned (genetically uniform) cell lines are liable to variations in their yield of recombinant protein and cannot be kept growing continuously. They must be grown from batches of cells that have been stored frozen in liquid nitrogen. Batch-to-sbatch variation is a major concern, since a sudden change in protein expression during any phase of the development process could have a significant impact on the approval process by the regulatory authorities. The problem with purity stems from the fact that cells contain materials that are difficult to fully remove in the finished product. Bacteria contain molecules called pyrogens which cause fever in humans and are responsible for the fatal sepsis that can occur after infection with gram negative bacteria like *E.coli*. The mammalian cells used to produce recombinant proteins, including antibodies, contain viruses that might conceivably integrate with the patient's DNA with unknown consequences. The downstream processing of biological molecules produced in large batch cultures is therefore of considerable importance.

Antibodies

Since antibodies are the fastest-growing class of protein therapeutics, a great deal of experience has been gained in their scale-up and manufacture by the biopharmaceutical industry. Despite this, producing antibodies for clinical trials and beyond is a complex process involving many different steps that impose great demands on a company's manufacturing capacity. This means that the cost of goods is significantly higher than that of small molecules, although process optimization is bringing the cost down from around $300 per gram to less that $100 (Kelley 2009).

A typical production scheme for a humanized therapeutic antibody is shown in Fig. 10.2. The starting point is a master cell bank (MCB) of mammalian cells[2] expressing the H and L chains of the antibody. These cells are carefully selected for their ability to produce high levels of antibody protein in simple growth media that are free of animal-derived material. The working cell bank (WCB) is taken from this master stock and grown in successively higher volumes of growth medium in flasks, then a rocking platform. The resulting large-scale inoculum is used to seed antibody production in a bioreactor containing up to 25,000 liters of medium. Once the cells have produced enough antibody in the culture medium (after 10–14 days), they are removed using a centrifuge, and the culture supernatant cleared by depth filtration through course filters followed by a very fine membrane filter (0.2 microns, or one twenty thousandth of a millimeter). These are the first of the downstream processing steps that lead to the final purified antibody.

The antibody is purified from the rest of the material in the culture fluid using an immobilized protein called protein A, which has the property of binding tightly to human IgG antibodies. This selective binding is exploited by immobilizing the protein in a column on an inert resin and then passing the culture fluid through it to pull out the antibody. Once most of the host cell proteins (HCPs) have been washed off the column, the antibody is recovered by lowering the pH (thereby increasing the acidity). A low pH step is also used to inactivate any viruses that may have been carried through from the CHO cells. A set of polishing procedures is then used to

---

[2] CHO cells derived from Chinese hamster ovary are most used.

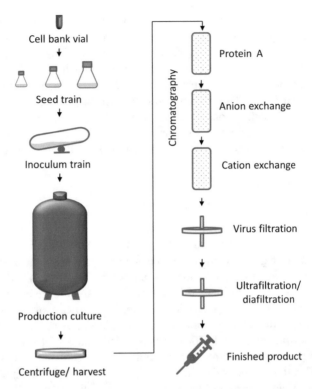

Fig. 10.2 A highly schematic illustration showing the principle of antibody manufacturing adapted from Kelley 2009 (not to scale). CHO cells producing antibody grown in laboratory to produce enough cells to store as a master cell bank in liquid nitrogen. A tube of cells from the working cell bank is thawed and used to start the manufacturing process. Cells grown in successively larger volumes until there are enough to add to a large bioreactor. Recombinant antibody is produced in the bioreactor growth medium which is separated from cells and pumped off into the purification chain. A series of filters and chromatography columns are used to purify the IgG from the growth medium and to remove viruses and other hazardous material. The result is sterile antibody in the correct formulation and ready for clinical use

reduce virus, host cell proteins, and DNA to levels that are acceptable to the FDA and others. This is achieved by using anion exchange then cation exchange chromatography followed by filtration. The antibody is then concentrated and formulated prior to sterilization by further filtration.

Among the technicalities associated with antibody manufacture is the environmental impact analysis (E-factor). For example, typical monoclonal antibody manufacture consumes 3000–7000 kilograms of water per kilogram of product, with cell culture taking up 20–25% and the chromatographic steps over 50% (from Pollock et al. 2017).

10.1.1.3 Production of Proteins in Animals and Plants

Although cultures of genetically modified cells are currently the method of choice for producing therapeutic proteins, alternative sources include transgenic animals or plants, since both can be engineered to produce large amounts of correctly glycosylated human protein. This area has become known as pharming; it is still in its infancy because of regulatory, safety, and political issues. Nevertheless, the human clotting factor antithrombin III has been purified from the milk of genetically engineered goats and marketed by GTC Biotherapeutics (now LFB SA) as ATryn® for use in patients with an inherited clotting disorder. Other animal-derived recombinant expression systems are under development, but the rate of progress is slow (overview by Sullivan et al. 2014).

Plants are, on the face of it, an attractive source of therapeutic proteins since they can be cultivated in large numbers, and the necessary expression systems are well characterized (Burnett and Burnett 2019). In addition, they lack the mammalian viruses and other contaminants that must be eliminated during downstream bioprocessing. The main obstacle to the commercial production of plant-derived proteins is the regulatory environment for genetically modified crops. However, there are situations like the Ebola outbreak in which transient expression in tobacco plants was used to rapidly produce a candidate vaccine (ZMapp) although this failed to show benefit in a clinical trial. At the time of writing, similar efforts are underway to produce a tobacco-derived vaccine for the SARS-CoV-2 coronavirus responsible for the COVID-19 pandemic.

10.1.1.4 Nucleic Acids

Nucleic acid drugs take the form of chemically modified oligonucleotides and gene therapy vectors. The former, such as the antisense oligonucleotide nusinersen (Spinraza®) approved for spinal muscular atrophy, are manufactured by solid phase synthesis using synthetic chemistry. Gene therapy vectors, such as modified viruses, must be produced using genetic engineering techniques and bioprocessing. Figure 10.3 shows the process for manufacturing a vector consisting of viral (AAV) DNA engineered with the gene to be expressed in human cells for example, to correct mutations (adapted from Smith et al. 2018). Since these products have been produced in both bacteria and mammalian cells, they must undergo stringent quality control to remove contaminating viral material and pyrogens as for therapeutic protein production. Note the timelines for each part of the process.

10.1.1.5 Cell Manufacture

Cell therapy using stem cells or genetically modified CAR-T or NK cells, for example, requires highly specialized manufacturing conditions. The most extreme form requires cells to be taken from a patient, manipulated under GMP conditions, then

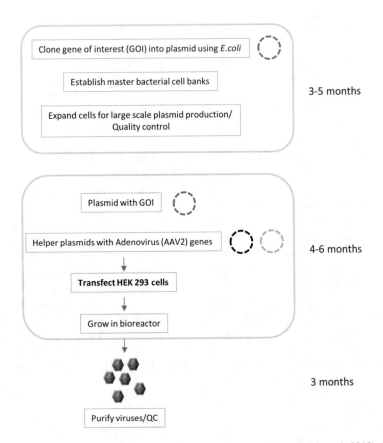

Fig. 10.3 Manufacture of AAV2 gene therapy vectors. (Adapted from Smith et al. 2018). *E.coli* bacteria are used to clone the therapeutic gene into a plasmid that will form part of the recombinant AAV2 virus. Scale-up and quality control under GMP conditions. Human embryonic kidney (HEK293) cells in suspension are transfected with the plasmid and two helper plasmids containing all the components necessary for AAV2 production as virus particles. After growth in a bioreactor, the virus particles are purified by ultracentrifugation and chromatography steps

reintroduced via the circulation. The growth phase of this process where CAR T cells are expanded in culture can take up to 11 days. The general process for creating CAR T cells for clinical use (Levine et al. 2017) is summarized in Fig. 10.4.

10.1.2   Formulation

A small molecule or biological drug is never administered to patients on its own but is mixed with other ingredients in a dosage form that reflects the intended route of administration. Dosage forms for compounds taken by mouth are tablets or capsules and make up most currently prescribed medicines. Dosage form design uses the

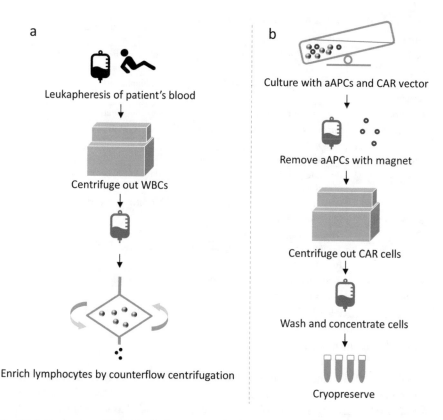

Fig. 10.4 Production of CAR T cells for clinical use (adapted from Levine et al., 2017). (**a**) White blood cells (WBCs)/leukocytes containing lymphocytes taken from patient by leukapheresis. After centrifuging to remove plasma, etc. lymphocytes enriched on basis of size by counterflow elutriation centrifugation to remove monocytes, platelets, and residual red blood cells. (**b**) Lymphocytes cultured with artificial antigen presenting cells (aAPCs) and the virus vector containing the chimeric antigen receptor (Chap. 7). aAPCs are necessary for stimulating T-cell growth, a common type being magnetic beads coated with antibodies to CD3 and CD28 that allow removal with a magnet at the end of the culture period. The resulting CAR T cells are harvested from the culture, washed, and concentrated prior to freezing in liquid nitrogen

formulation process to ensure that the medicine can enter the circulation in a controlled way with maximal bioavailability and in a form that is acceptable to the patient (particularly important in the case of injectable or inhaled products). There are challenges, such as the increased number of compounds with poor solubility in water, so there is plenty of room for innovation.

Table 10.1 Different classes of formulation ingredients

Class	Function	Examples
Fillers	"Bulk up" tablet or capsule	Cellulose, lactose, vegetable oils (diluents for capsules)
Solubilizing agents	Help solubilize lipophilic drugs	Cremophor EL
Binders	Hold tablet together	Sorbitol, gelatin, PVP
Disintegrants	Encourage breakup and dissolution of drug in the stomach	Carboxymethyl cellulose
Lubricants	Stop powder sticking to tablet die	Talc, silica
Glidants	Promote powder flow during manufacture	Magnesium carbonate
Antiadherents	Prevent tablet sticking to punch	Magnesium stearate
Sweeteners	Mask unpleasant taste of API	
Coloring	Visual appeal, drug identification	FD&C red no. 40
Sweeteners	Mask unpleasant taste in liquid formulations	Aspartame
Preservatives	Resist microbial contamination	Benzoic acid, vitamin E
Coating agents	Resist moisture in the air, hide unpleasant flavors	Hydroxypropyl methyl cellulose, gelatin (capsules)

10.1.2.1 Terminology

The drug part of a formulation is called the active pharmaceutical ingredient (API), while the remainder is formed with inactive compounds called excipients. The latter are classed by the regulators as GRAS (generally regarded as safe) and are required to ensure that the API is bioavailable. Excipients are also important for the actual manufacturing process since the tablet must be formed in a machine in a consistent way. Information and industry standards on excipients are available from organizations such as the International Pharmaceutical Excipient Councils Federation (IPEC Federation 2020).

The various types of formulation ingredients and some representative examples are listed in Table 10.1.

10.1.2.2 Formulation Development

Formulation development is undertaken at several different phases of the development process. The first significant involvement of formulation scientists/pharmaceutical chemists really begins after a lead compound or biological has been chosen for further development. At this exploratory stage, there are some key criteria that must be established to produce a workable formulation. These are:

- Solubility and Lipophilicity

Most drug candidates are lipophilic (fat soluble, Chap. 3), since highly charged molecules do not generally penetrate the cell membrane. This can cause problems with formulation since the drug may not readily dissolve in the stomach.

- Solubility at Different pHs

This is an important property, as the degree of acidity and alkalinity can vary significantly in different body compartments. For example, the stomach has a low pH between 1.0 and 3.0, while the duodenum is at pH 6.0–6.5. This means that an orally available drug could experience a major change in solubility and as it transits (migrates) through the stomach and into the small intestine.

- Physical Form of Drug

Chemical substances exist in different forms (polymorphisms) that can have profound effects on their physical properties. Carbon, for example, is found as a crystal (diamond), powder (soot), and layered sheet (graphite), along with exotic forms such as the fullerenes, graphene, and nanotubes. Pharmaceutical compounds are also polymorphic, being either amorphous, like a powder, or crystalline. Two important factors in the formulation of a small molecule drug are particle size and crystal habit (i.e., the shape of the crystal). Formulations used in dry powder inhalers (DPIs), for example, require careful control of particle size. When crystals are used in drug formulations, they are micronized (ground into a fine powder using a mill). Whether a drug is developed as a powder or in crystalline form will have profound consequences on drug development, since the different forms will determine the solubility in the body and the ease of manufacture. Since polymorphic forms can appear and disappear in different preparations of the same compound, there is a danger that this could happen in the middle of a clinical trial; it would then be impossible to know the exact dose of drug administered to patients, and the trial data could be completely invalidated. The experience of Abbott Laboratories with their drug Ritonavir (marketed as Norvir® for HIV infection) is an example of how polymorphisms can adversely affect the manufacturing stage as well. Ritonavir was produced as a semisolid form in capsules because the drug was not bioavailable in a tablet. Since it was in liquid form (dissolved in alcohol/water), there was no regulatory requirement for checking the crystalline form. Two years after launch in 1996, a new crystal form appeared in manufacturing batches that greatly reduced the solubility of the compound; this form rendered the drug un-manufacturable and held up supplies until a new formulation was devised to overcome the problem (Bauer et al. 2001).

- Physical and Chemical Stability

The physical form of a medicine must confer stability as well as resistance to breakdown during storage. Some compounds, particularly in amorphous form, are liable to absorb water from the atmosphere (i.e., they are hygroscopic) which can affect the shelf life of the compound as well as the ease of manufacture. A drug product requires a shelf life at room temperature of at least 18 months, so stability testing is an important factor in the overall development process.

- Salt Selection

Different salt forms (and hydrates) of small molecules have already been described in Chap. 3. If a salt form is required at all, each type will cause the above properties (stability, physical form, etc.) to vary, so evaluation of different salt forms is important at an early stage.

Once the above evaluations have been made and suitable formulations identified, the experimental drug can be submitted to the preclinical development process that involves toxicology and other animal studies (Chap. 11). The final formulation into a tablet or capsule is not completed until well into the clinical development phase; prior to that, the compound is administered to phase I clinical trial volunteers in a capsule that is filled by hand. One challenge for formulation scientists is the preparation of tablets and capsules containing placebos for clinical trials. These formulations must have an identical appearance and taste to those containing the drug. Another, unrelated, concern is the possible chemical interaction between plastic in the bottle used to store the medicine and the medicine itself. This even extends to concerns about the adhesive in the label; this highlights the level of detail required for the regulation of drug development and the emotional stability required by the pharmaceutical scientists concerned.

10.1.2.3   Formulation of Biologicals

Mention has already been made in Chap. 2 and elsewhere, about the lack of oral bioavailability of nearly all protein and nucleic acid-based drugs and vaccines. This means that the formulation must be compatible with parenteral administration, i.e., administration through the skin or into the airways. Nucleic acids, such as the siR-NAs, present additional challenges, as they must penetrate the cell membrane to reach their target. Since antibody proteins are administered at high concentrations, these large molecules can stick together (aggregate), thereby reducing stability and creating unwanted stimulation of the immune system (immunogenicity). In fact, all protein therapeutics must be checked carefully for aggregation and immunogenicity during both the clinical and manufacturing phases of development.

10.1.2.4   Protein Formulation

The formulation of proteins brings its own challenges; the product must be stable for at least 18 months and, in the case of monoclonal antibodies, be concentrated enough to administer in small volumes subcutaneously (Wang 2015). Unlike tablets that are swallowed, proteins are directly introduced into the normal blood circulation so the pH encountered will be around 7.0, and the concentration of salts in the plasma will be at the normal physiological level. Administration of the drug at a different pH will cause the patient to experience pain at the site of injection, so proteins are formulated to maintain physiological levels of pH and salts. Further ingredients, including amino acids, mannitol, glycerol, and trehalose, are used to stabilize the protein during the process of freeze drying (for storage) and reconstituting in buffer prior to injection. This is because proteins are susceptible to denaturation, the process by which their chains of amino acids are unfolded by heat or mechanical agitation into an inactive form. Boiled eggs and melted cheese represent an extreme example of this, but therapeutic (and other) proteins dissolved in liquids can readily denature at the interface between the liquid and air. Special detergents such as the polysorbates are added to reduce this specific problem.

Proteins tend to stick to containers and seemingly disappear completely if formulated in low concentrations. This has proved to be the downfall of many a biochemist who may have spent weeks isolating small amounts of a protein of interest only to wonder where it went to, once it was stored in a glass bottle. The loss of material can be a problem with proteins like erythropoietin or interferon beta that are injected in small amounts; in these cases, gelatin, or human serum albumen (the main protein in blood), is used as a bulking agent.

Another way of enhancing the stability of protein drugs in the circulation is to attach small molecules of polyethylene glycol (PEG). This process of PEGylation has been used in marketed drugs, such as PEGASYS interferon alpha produced by Hoffmann-la-Roche for treating hepatitis C infection.

10.1.3 Drug Delivery

One aspect of drug delivery has already been hinted at, namely, the oral delivery of proteins, a "Holy Grail" of pharmaceutical development. The field of drug delivery requires engineering and microfabrication skills as well as the chemistry and biology outlined earlier, for example, in the delivery of insulin to diabetics via electronic pumps. Research is ongoing into drug implants that release small molecules over many months thus avoiding the need for regular oral dosing. Merck have obtained promising results with MK-8591, a replication inhibitor of the HIV virus that can deliver therapeutic levels of the drug for over a year (Matthews 2019). An overview of some ongoing research into drug delivery is given in Lawrence 2017.

Another important aspect of drug delivery is nanotechnology, which is described below.

"There's Plenty of Room at the Bottom"
This was the title of a talk given by the Nobel Laureate Richard Feynman to the American Physical Society in 1959 where he introduced the basic ideas behind nanotechnology (Feynman 1960). The name is derived from the Greek for dwarf and is concerned with producing materials and devices at nanometer (one billionth of a meter) scales; Feynman speculated that it would be quite feasible to print the entire edition of the Encyclopedia Britannica on the head of a pin. Since then, remarkable strides have been made in micro-fabricating machines and devices at the near atomic scale. This has relevance to pharmaceuticals since there is a great deal

of interest in using nanotechnology in formulations that enhance the delivery of drugs to target tissues. A common formulation is based on liposomes; these consist of minute phospholipid spheres that have encapsulated small molecule drugs and then fuse with cell membranes to allow the compound to enter the cell. Liposomes are also used for delivery of nucleic acids; this is of increasing importance, given the current interest in gene therapy and RNA therapeutics.

One drug delivery problem is the introduction of small molecules into the brain to treat CNS diseases. The blood-brain barrier (BBB) is a physical structure of vessels that keep the general circulation of blood separate from the fluid in the brain and spinal cord (cerebrospinal fluid). It acts as an important barrier to bacteria and molecules that would otherwise be harmful if introduced into the brain. These molecules have a relatively high charge which places restrictions on the structures of drugs that can target the brain because they must be lipophilic, unless a way is found to bypass the BBB using liposomes or other nanoparticles.

10.1.3.1 Oral Delivery of Proteins

The gastrointestinal (GI) tract presents a formidable obstacle to proteins due to the low pH and thick layer of mucus in the stomach, plus the digestive enzymes present in the small intestine. Nevertheless, it is possible to deliver proteins orally, most successfully in the form of oral vaccines like the attenuated polio virus. For protein therapeutics, there are several strategies for circumventing the above barriers, including encapsulation in nanoparticles, and co-administering with protease inhibitors. Novo Nordisk co-developed an oral insulin with MIT, but the requirement for large amounts of insulin in the formulation meant that it was not commercially viable. However, the company has introduced semaglutide, marketed as Rybelsus®, a glucagon-like peptide 1 (GLP-1) receptor agonist as a daily tablet. This was approved by the FDA in 2019 for use in addition to the current injectable form Ozempic®. Lastly, insulin and similar peptides are much smaller than the monoclonal antibodies that comprise most biological therapies. Despite the seeming impossibility of delivering a 150,000 molecular weight via the oral route, it might be possible to achieve this if the protein is protected until released in the large intestine where there is a relative lack of interfering digestive enzymes and transporters. For a review of these different aspects of oral protein delivery, see Brazil (2019).

Delivering Microbial Therapeutics

Finally, the human microbiome presents therapeutic opportunities as the importance of individual microbes in health and disease has become recognized (see Chap. 2.) Live microbial therapeutics are unlike other biotech medicines in that they will (or should) self-replicate in the GI tract and form a stable niche that provides a continuous source of active molecules. Ensuring that this happens and is kept under control is one of the main challenges with this approach (Jimenez et al. 2019).

Summary of Key Points

Scale-up and manufacture of small molecule drugs require the input of process chemists to devise safe, efficient, and environmentally friendly reaction schemes.

The cost of goods is an important factor in deciding whether a drug discovery project is commercially viable.

Manufacture of gene therapy vectors and recombinant proteins employs large bioreactors and downstream processing to remove contaminants from cells and culture medium.

Transgenic animals and plants are being investigated for production of drugs by "pharming."

Personalized cell therapy (e.g., CAR T cells) requires highly specialized manufacturing facilities.

Drug formulation of the active pharmaceutical ingredient (API) requires a range of inert excipients.

Nanotechnology is used to enhance drug delivery.

References

Barshikar R (2019) Quality by Design (QbD) and its implementation in Pharma Industry. Express Pharma https://www.expresspharma.in/business-strategies/quality-by-design-qbd-and-its-implementation-in-pharma-industry/ Accessed 31 Mar 2020

Bauer J et al (2001) Ritonavir: an extraordinary example of conformational polymorphism. Pharm Res 18:859–866

Brazil R (2019) Binning the sharps: the quest for oral insulin. Pharmaceutical Journal doi. https://doi.org/10.1211/PJ.2019.20207045

Burnett JB, Burnett AC (2019) Therapeutic recombinant protein production in plants: Challenges and opportunities. New Phytologist Trust https://doi.org/10.1002/ppp3.10073 Accessed 2 Apr 2020

Dunn PJ, Wells AS, Williams MT (eds) (2010) Green Chemistry in the Pharmaceutical Industry Wiley-VCH Verlag gmbH, KgaA Weinheim

Feynman RP (1960) Plenty of room at the bottom. Caltech Magazine http://calteches.library.caltech.edu/47/ Accessed 3 Apr 2020

ICH Quality Guidelines (2020) https://www.ich.org/page/quality-guidelines Accessed 31 Mar 2020

IPEC Federation (2020) https://ipec-federation.org/ Accessed 2 Apr 2020

Jermini M, Polastro E (2020) Drug Substance Cost: A Non-Issue? Insight and implications of drug substance cost analysis. Contract Pharma https://www.contractpharma.com/issues/2020-03-01/view_features/drug-substance-cost-a-non-issue/ Accessed 1 Apr 2020

Jimenez M et al (2019) Microbial therapeutics: new opportunities for drug delivery. J Exp Med 216:1005–1009

Kelley B (2009) Industrialization of mAb production technology - the bioprocessing industry at a crossroads. MAbs 1:443–452

Lawrence J (2017) Making drugs work better: four new drug delivery methods. Pharmaceutical Journal doi. https://doi.org/10.1211/PJ.2017.20203530

Levine BL et al (2017) Global manufacturing of CAR T cell therapy. Molecular Therapy: Methods & Clinical Development 4:92–101

Marti HR, Siegel JS (2006) Process chemistry in API development. Chimia 60:516

Matthews R (2019) First-in-Human Trial of MK-8591-Eluting Implants Demonstrates Concentrations Suitable for HIV Prophylaxis for at Least One Year. http://www.natap.org/2019/IAS/IAS_19.htm Accessed 6th April 2020

Pollock J et al (2017) Integrated continuous bioprocessing: economic, operational, and environmental feasibility for clinical and commercial antibody manufacture. BiotechnolProg 33:854–866

Sheldon R (2010) Introduction to green chemistry, organic synthesis and pharmaceuticals. In:

Smith J et al (2018) Overcoming bottlenecks in AAV manufacturing for gene therapy. Cell Gene Therapy Insights 4:815–827

Sullivan EJ et al (2014) Commercializing genetically engineered cloned cattle. https://doi.org/10.1016/B978-0-12-386541-0.00027-8

Wang W (2015) Advanced protein formulations. Protein Sci 24:1031–1039

Chapter 11
Preclinical Development

Abstract This chapter describes the preclinical studies that must be performed on drug candidates before they can be administered to human volunteers. Investigations are made into how the drug affects the body (pharmacodynamics) and how the body affects the drug (pharmacokinetics), as well as safety pharmacology and toxicology. All these investigations are laid out in the chapter, which concludes with a description of how the preclinical information is used to estimate the dose of drug that will be administered in first time in human (FTIH) trials.

11.1 Introduction

The point in the drug discovery pipeline has now been reached where animal studies must be used to gather the information required to plan clinical trials in human subjects. Moving towards this stage is sometimes referred to as "entering the valley of death." This is not meant to convey sinister undertones about the outcomes of clinical trials but simply conveys the fact that many drugs fail to make it into full clinical development, for reasons that will become clear in this chapter.

Before discussing these reasons, it is assumed that a drug candidate is potent (ideally at the nanomolar level) and specific for its target. It should also have been subjected to in vitro and in vivo tests that show that the target is affected, preferably in a model of the disease. Manufacturing and formulation issues may also have been resolved at this early stage, but none of the preceding work will guarantee that the drug candidate will become a medicine. For this to happen, the candidate must do the following:

- Get to the target.
- Stay there for the required time.
- Be eliminated sufficiently quickly to allow next dose to be taken.
- Be sufficiently stable, i.e., resistant to metabolism.
- Be safe.

The procedures used to check whether the drug fulfils these criteria, come under the umbrella term preclinical development (which includes the scale-up and formulation reviewed in the previous chapter). Preclinical development is used to gather

E. D. Zanders, *The Science and Business of Drug Discovery*,
https://doi.org/10.1007/978-3-030-57814-5_11

all the information needed to support a regulatory application for administering a drug to humans for the first time. This goal is known as first time in humans or FTIH.[1] The applications themselves are termed IND (Investigational New Drug) and CTA (Clinical Trial Authorization), for the FDA and EMA, respectively.

Different animal species are used in preclinical development to examine the following:

- Pharmacodynamics.

  The effect of the drug on the body.

- Pharmacokinetics.

  The effect of the body on the drug.

- Safety pharmacology.

  Any undesirable effects of drugs on body functions

- Toxicology.

  Any undesirable effects of drugs on body structures.

11.1.1   Basic Requirements

11.1.1.1   Good Laboratory Practice

Once a decision has been made to move a compound or biological into full development, the precise demands of the regulatory agencies must be met. Good manufacturing practice (GMP) has been mentioned already in this context, but laboratory procedures are also regulated. These come under the term good laboratory practice or GLP. This is employed at the preclinical phase to ensure that clear validated operating procedures are followed in the laboratory and that management structures have legal accountability. GLP procedures help to anticipate and respond to the accidental mislabeling of samples and other routine errors that can occur in any laboratory. Site inspections are carried out by the regulators to enforce GLP regulations, and the condition of all the apparatus and materials used in the lab is monitored and recorded to a much higher degree than would be considered normal in a research laboratory. This high level of regulation exists because of the high stakes involved, both legal and ethical.

---

[1] This used to be "first time in man" but presumably changed to reflect modern social attitudes; more trials are, in fact, performed in men than in women (see later chapters).

11.1.1.2 Harmonization of Procedures

The need to coordinate drug development activities in a globally consistent way has prompted governments and regulators to collaborate on a series of industry guidelines. The result of this collaboration is the International Council for Harmonisation of Technical Requirements for Pharmaceuticals for Human Use (ICH) based in Geneva (ICH 2020). Some of the extensive documentation available on the ICH website is referred to in this and subsequent chapters; these guidelines provide detailed information on required procedures and a useful indication of current thinking in the regulatory field. Despite these harmonization initiatives, there are still procedural variations between the different regulators. It is not possible to cover these differences in any detail, but the websites of two ICH founding members, the FDA and EMA, contain a great deal of information about the entire regulatory process, from preclinical development to marketing authorization. Another founding regulatory member is Japan's Pharmaceuticals and Medical Devices Agency (PMDA), part of the Ministry of Health, Labour and Welfare (MHLW). China's National Medical Products Administration (NMPA) represents an important pharmaceuticals market and is also involved in ICH harmonization.

11.1.2 Preclinical Testing in Animals

All experimental drugs must, by law, be tested on animals prior to administration to humans. This is an area of great controversy outside the drug discovery industry since animal experimentation has tarnished the image of scientists for over 100 years. Many drug discovery scientists have become used to checking for bombs under their cars and working in highly secure environments. However, if we want to continue to produce drugs with any kind of safety profile, there is currently no alternative to using animals. Despite all the sophisticated knowledge and technology at our disposal, it is still scientifically impossible to make exact predictions about how drugs will interact with the human body. While most, if not all, pharmaceutical industry scientists would be more than happy to dispense with animals altogether, the much-discussed alternatives of in vitro tests and computer simulations are simply not acceptable substitutes for live animals at the present time. I have had personal experience with compounds that appeared to be safe and effective in vitro, being lethal when given to rodents. Obviously, this animal test had to be in place to ensure that the compound did not reach the clinic. In summary, few people in their right mind would risk taking a drug that had not been through at least some basic screening in animals.

Despite this somewhat forthright statement of current realities, there is undoubtedly a genuine need to minimize or eliminate the use of animals in preclinical development. This is partly to make predicting a drug's effect on humans as accurate as possible but also to avoid ethical dilemmas and to save costs.

One problem with testing drugs in animals is that the results do not always correlate with those obtained later in human subjects. This issue of concordance was formally reviewed by Olson et al. (2000) in a survey of human toxicities relating to 150 compounds. The toxicities were picked up in 71% of tests using rodents and non-rodents (63% non-rodents alone, 43% rodents alone). Of all toxicities detected in animals, 94% occurred within 1 month of testing. Since then, other surveys have analyzed more data, with Clark and Steger-Hartmann (2018) analyzing no fewer than 3290 drugs and 1,637,449 adverse events in humans and registered over 70 years. In this case, the highest concordance between human subjects and animals related to cardiovascular events such as arrhythmias. These figures offer support for the continued use of animals in toxicology, but nobody can doubt that there is enormous room for improvement. Research into predictive toxicology is recognized as a high priority, by both academia and industry. Model organisms like the zebrafish (described in Chap. 6) offer the possibility of a genuine compromise between the desire to avoid experiments on higher mammals and the need to use living organisms to screen for organ toxicity, and a range of new technologies are under evaluation, including the application of machine learning. These will be discussed later in the chapter.

11.1.2.1 Which Species Are Used?

The following table lists the animal species normally used for drug development. In the case of antibodies and other biologicals, which do not bind to non-primate tissues, there is no advantage to be gained by using animals like mice and rats, so these species are not used (Table 11.1).

Table 11.1 List of animal species used in preclinical development

Animal	Comment
Rat	Commonly used small animal species
Mouse	Alternative to rats and used when compound limiting
Dog	Commonly used large animal species
Rabbit	Used for reproductive toxicology
Hamster	Large rodent model used for oral dosing and carcinogenicity
Guinea pig	Used to test for allergic reactions
Pig (minipig)	Similar organ layout and skin to humans
Non-human primates	
Cynomolgus monkey	Most used macaque for compounds/biologicals
Rhesus monkey	Used for biologicals, HIV drugs
Marmoset	Small primate but more distant relative of humans

11.1.3   The Preclinical Development Process

The commitment to take a lead compound or a biological through preclinical evaluation towards the first exposure to humans is not taken lightly. Even with a standard set of test packages, it can take up to 1 year to complete and show little change from two million dollars; this is assuming that no problems occur on the way. The tests are designed to establish the dose of drug will affect the target in patients while at the same time being safe enough to administer to human volunteers. The preclinical development process will now be considered in some detail.

11.1.3.1   Pharmacodynamics

Initial investigations of the effect of the drug on the body will have already been undertaken in animal models of the disease. This should, in principle, give investigators some idea of the dose of drug that will work against the target in patients. The amount of drug required for optimal activity in an animal model depends on different factors, including the number of drug targets in the animal and the strength of binding. To estimate how much drug would be required for human use, the dose must be scaled up from the dose in animals in proportion to the weight of the patient. A rat weighs about 150 g, while the average human is 60Kg (and rising). This allometric scaling is not necessarily straightforward, however, as the pharmacodynamic response of the drug will depend on its fate in the body (pharmacokinetics, see below) and other factors that influence the responses of human patients. These factors include the age of the patient and preexisting conditions which may be unrelated to the disease to be treated. Pharmacodynamics becomes an important issue in clinical trials and subsequent marketing, since different groups of patients can exhibit different responses to the drug (see Chap. 14).

11.1.3.2   Pharmacokinetics

The effect of the body on the drug is one of the most import go/no go areas in drug development. Pharmacokinetic analysis looks at the way in which compounds or biologicals are processed after they have been administered to test animals (and later humans) which will give some idea of the bioavailability of the compound.

Terminology

- Absorption, Distribution, Metabolism, Excretion (commonly referred to as ADME).

  This covers the uptake of a drug into the circulation and its distribution within the body tissues, followed by its metabolism and then removal from the body.

- ADMET or ADME-Tox.

  ADME packaged together with toxicology.

- Drug metabolism pharmacokinetics (commonly referred to as DMPK).

  This covers both pharmacokinetics and metabolism, i.e., the chemical reactions in the body that modify an administered drug (see later).

11.1.3.3   How ADME Studies Are Carried out

Drug candidates are given to experimental animals intravenously and by mouth (if the compound is designed to be orally active). Biologicals are normally injected. Samples of blood are taken from the animal before drug administration and then at several time points thereafter, so that the speed at which the drug enters and leaves the circulation can be measured. The laboratory procedures involve adding an anti-coagulant to blood, spinning out the cells in a centrifuge, and analyzing the straw-colored plasma by chromatography coupled with mass spectrometry (LC-MS). The result is a precise measurement of the concentration of drug in the sample (more details are given later in Drug Metabolism). A simple pharmacokinetic experiment is shown in Fig. 11.1 where a compound is administered, either orally or intrave-nously, and plasma samples taken over different time points to measure the amount of drug in the circulation. Not surprisingly, intravenous administration gives the highest level of drug early on, but with oral administration, there is a lag time before the drug enters the circulation.

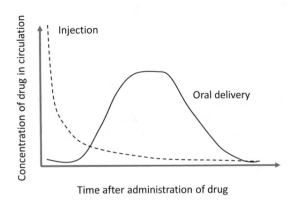

Fig. 11.1 Pharmacokinetics: changes of drug concentration in the circulation over time. Blood samples are taken at regular time intervals after the drug is administered to the animal by oral and parenteral routes. The concentration of drug is measured in the plasma and plotted against time. The curves are characteristic of the different routes of administration and reflect both the uptake and elimination of the drug in vivo

The size and shape of these concentration curves provide a set of numbers that describe the ADME properties of the administered drug. Some of the important numbers are as follows:

- Cmax.

  Maximum plasma concentration of the drug.

- Tmax.

  Time taken to achieve Cmax

- $T_{1/2}$.

  Half-life of drug or the time taken for the drug to be reduced to 50% of its starting concentration

- AUC area under the curve.

AUC is used in non-compartmental analyses[2] to estimate the total amount of drug administered. The area is calculated by using integration (mathematical formulae based on integral calculus). Compartmental analysis is potentially more accurate over different time points and uses terms such as rate constants and other kinetic parameters. More detailed descriptions of the pharmacokinetic equations and other topics described in this chapter can be found in Han et al. (2010). The information gained after measuring the above values in an animal model will determine whether a drug candidate can progress through the remainder of the preclinical evaluation process. A candidate should ideally be well absorbed (by the mouth if relevant), distributed evenly throughout the body, and then eliminated in a timely way. The factors that influence each of these actions are now covered in turn:

Absorption

If a drug is to be taken by the mouth, it must be absorbed through the stomach and small intestine. Drugs that are designed to treat CNS diseases must also cross the blood-brain barrier; alternatively, some medicines (like the non-sedating antihistamines) are tested to ensure that they do not enter the brain.

Absorption through the gut is assessed in laboratory models that mimic the cells which control the movement of compounds across the human intestine. The industry standard model uses a human cell line (Caco-2), originally derived from a cancer of the colon. This line is grown as single sheets (monolayers) in a special apparatus that measures the amount of drug moving from one side of the layer to the other. A compound that is likely to be well absorbed in the living animal will move freely across the cell layer. These layers (and the human gut) are not just passive barriers;

---

[2] These models treat PK data as if the body compartments (brain, lungs, heart, etc.) in the animal species do not have separate characteristics, as opposed to compartmental models that give a more quantitative picture of drug disposition.

some compounds are driven through specialized protein assemblies (transporters) such as the ABC and SLC families, the former using energy provided by the cells in a process called active transport. The transporters can have a strong influence on the absorption of drugs and therefore their pharmacokinetics and clinical efficacy. For example, the multidrug resistance proteins (MRPs), of which P-glycoprotein (P-gp) is a major example, are part of a defense mechanism that evolved to protect the body from noxious chemicals produced by outside sources (xenobiotics). They can reduce intestinal absorption by literally pumping the drug back through the cellular barrier and are problematic with cancer therapeutics. Conversely, drugs that inhibit MDR proteins may create problems with excessive absorption of other medicines taken concurrently.

Protein Binding

Drugs normally exert their effects in the tissues of the body after leaving the circulation, which is rich in plasma proteins such as albumen (sometimes called human serum albumen or HSA) and alpha-1-acid glycoprotein. Many compounds bind to these plasma proteins thereby reducing the amount of material that is available for interaction with the drug target. Protein binding studies are therefore carried out to guide the selection of drug doses for later studies, since it may be necessary to use larger amounts of drug in animals and the clinic if protein binding is high. Protein binding is measured in a laboratory process called ultrafiltration, which employs a plastic filter to separate small molecules bound to plasma proteins from those that are free in solution. High levels of protein binding do not necessarily preclude further development of a compound, however, since marketed medicines, such as diazepam and warfarin, show greater than 98% binding (Brunton et al. 2005).

Drug Metabolism

Cellular metabolism is controlled by enzymes which conduct a wide range of chemical reactions that build up or break down molecules of all sizes. This allows the cell to function as a living entity. Drug metabolism also uses cellular metabolism, but in a series of specialized chemical reactions that are designed to clear xenobiotic compounds from the body. As already outlined in Chap. 2, small molecule drugs pass from the stomach into the liver via the hepatic portal vein prior to reaching the general circulation through the hepatic vein. The liver is the primary site for drug metabolism in which chemical transformations occur to produce metabolites from the administered drug structure. The number and types of metabolites must be considered, as these may show toxicity or themselves be the active drug entity. In the latter case, the administered drug, inactive until subjected to metabolism, is known as a prodrug; one example is acyclovir, which is used to treat *Herpes simplex* infections.

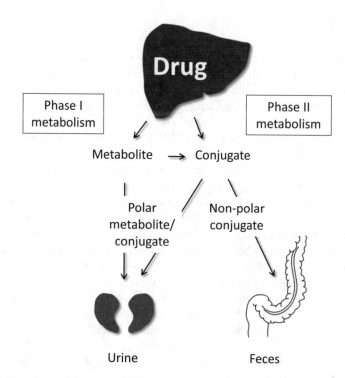

Fig. 11.2 Two phases of drug metabolism in the liver. Compounds entering the liver are subjected to a series of chemical reactions catalyzed by specialized enzymes including the CYP450 system. Drugs and metabolites from phase I may be further modified by phase II enzymes that add chemical groups to produce conjugates. Polar (water soluble) metabolites and conjugates are excreted in the urine, while non-polar compounds (soluble in the detergent-like bile produced by the gall bladder) are excreted in the feces (The choice of animal species for study is important as rodents display a much higher rate of biliary excretion than humans)

The general scheme of drug metabolism is illustrated in Fig. 11.2. Compounds entering the liver are subjected to phase I and phase II metabolism by separate groups of enzymes.

Phase I reactions break the compounds down by oxidation or other chemical reactions. The enzymes that perform this task are from the cytochrome P450 family (shortened to CYP450).[3] Drug conjugates are produced in phase II reactions, where chemical groups are added to drugs as a type of sorting code that allows the compound to be directed towards elimination in either the feces or the urine. The enzymes have complex names and exist in multiple forms (isoforms) within a large family. The main examples are UDP-dependent glucuronosyltransferase (UGT), glutathione-S-transferase (GST), and phenol sulfotransferase (PST).

---

[3] So called because a complex between this heme-containing enzyme and carbon monoxide absorbs light maximally at a wavelength of 450 nm.

CYP450 Enzymes

The importance of these enzymes in drug development cannot be over-emphasized, as they can have a profound impact upon the way different people respond to the same medicine (covered in Chap. 14). CYP450 enzymes are also responsible for the interactions that may occur when drugs are taken with some foods or when more than one medicine is taken at the same time. Figure 11.3 shows the possible outcomes of a single drug interacting with CYP450 enzymes in the presence of another medicine or foodstuff.

In the first case, drug 1 inhibits a specific CYP450 enzyme that metabolizes drug 2, leading to a rise in the level of drug 2. This may be either beneficial or potentially dangerous due to toxicity. Alternatively, drug 1 might increase the level of the CYP450 in the liver (induces it) and therefore will reduce the level of drug 2, again with potential benefits or downsides. The inducibility of CYP450s is largely through the pregnane X receptor which recognizes a wide variety of ligands and itself is a useful target for screening compounds during preclinical evaluation.

Some examples of drug or food interactions are given in Table 11.2.

There are plenty of examples of toxicities caused by drug-drug interactions via the CYP450 system and cases of commercial advantage. In the latter case, the histamine H2 receptor antagonist cimetidine inhibits several CYP450s where ranitidine (Zantac®) does not, thereby showing an improved safety profile. The result was that Zantac® superseded cimetidine (Tagamet®) in the marketplace (see also Chap. 4).

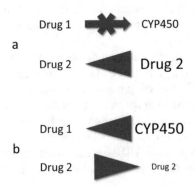

Fig. 11.3 The importance of CYP450 enzymes in drug interactions. (**a**) Drug 1 inhibits a CYP450 making it less active on drug 2, so the latter is poorly metabolized. (**b**) Drug 1 induces more CYP450 activity which then breaks down drug 2. The induction of CYP450 enzymes in the liver occurs over several days in contrast to inhibition which occurs more rapidly

Table 11.2 Major CYP450 isoforms and some agents that affect them. More examples are given in Ogu and Maxa (2000)

CYP450	Comments
1A2	Inhibited by tobacco smoke
2C9	Inhibited by fluvastatin (Lescol®)
2C19	Induced by norethindrone (oral contraceptive)
2D6	Inhibited by fluoxetine (Prozac®)
3A4	Inhibited by grapefruit juice

Laboratory Evaluation of Drug Metabolism

Bioanalysis

The process of ADME evaluation depends upon being able to measure the concentrations of drugs and their metabolites in plasma and other body fluids, such as urine. The range of techniques available for this, which includes chromatography and mass spectrometry, come under the heading of bioanalysis.

Chromatography is used to separate molecules from complex mixtures. The name is derived from the Greek for color writing, so called because the first chromatography experiments, performed early in the twentieth century, separated mixtures of colored substances, such as plant dyes. To separate compounds that are dissolved in a solution, the mixture is poured into a column (i.e., a tube made of glass, plastic, or metal), which contains an insoluble material called a chromatography matrix. The matrix is chemically modified in such a way that different classes of molecules (both large and small) will bind to it. This binding is reversible and dependent on the chemistry of the individual compounds, so each one can be successively eluted from the column and collected in separate tubes. The elution process can be monitored by continuously measuring a property, such as light absorbance, using a spectrophotometer (Chap. 9). The result is the display of a series of peaks on a graph, each peak representing a purified molecule. If the separated compounds must be identified, they are subjected to standard analytical chemistry techniques, such as mass spectrometry. Normal liquid chromatography relies upon gravity to force the samples through simple columns for the separation. While this is acceptable for many applications, this method really lacks the sensitivity needed for the detection of minute quantities of drugs and metabolites in plasma. Instead, high-pressure liquid chromatography (HPLC) is used for drug analysis. The principle is the same as normal chromatography, but high pressures are used to force small volumes of plasma through strong steel columns. Different types of matrix are packed into each column, according to the types of compounds to be analyzed. These matrices come under the general headings of ion exchange, reverse phase, size exclusion, or hydrophobic supports. An example of an HPLC analysis is shown in Fig. 11.4.

Fig. 11.4 An HPLC trace showing individual compounds, separated from a mixture, as sharp peaks on a graph. Each peak elutes at a specific and reproducible time after the mixture has been applied to the HPLC column, so that each peak can be identified by comparing the migration of standards, i.e., compounds whose identity is known

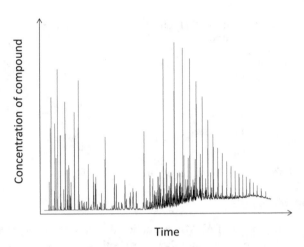

In Vitro Metabolism

While the first metabolism studies on compounds are conducted on live animals in vivo, it is normal practice to assess the metabolic stability of drugs in vitro using human liver cells (hepatocytes). Although it is preferable to use isolated hepatocytes of human, rather than animal origin, these are not always readily available, in which case, liver microsomes are used. These are isolated from whole liver and contain the relevant enzymes for drug metabolism; unlike hepatocytes, they can be stored frozen in batches for use at any time. Microsomes from rat liver are often used for a preliminary metabolic screen, as human liver is less readily available.

Human microsomes are also used as a source of the different CYP450 isoforms which vary in expression according to the genetic background of the donor. These isoforms contribute to the individual responses that patients have to a given medicine (see pharmacogenetics, Chap. 14). Some CYP450 enzymes are available as purified proteins, produced using genetic engineering technology, thus obviating the need to use enzymes derived from human liver donors.

Drug Clearance

The final part of the ADME analysis is elimination, in which drugs are removed from the body via the liver and kidneys. The time taken for this to occur and the amount of compound that is cleared are both measured in at least two animal species. Drug clearance (Cl) is a measure of the efficiency by which a drug is removed from the plasma. Hepatic and renal clearance rates are defined as the volume of plasma that is cleared of drug after flowing for one minute through the liver and kidney, respectively. These rates depend on several factors, including plasma flow and protein binding. The clearance rates of experimental compounds are classified as high, medium, or low; those with high clearance rates are unlikely to be useful medicines, as they will not remain in the body for long enough to provide a useful therapeutic effect. On the other hand, drugs with low clearance rates could be difficult to control during regular dosing, as levels could exceed the required margins of safety.

11.1.4 Drug Safety

At the same time as ADME studies are being performed, the safety assessments will be initiated and carried on through the rest of preclinical development. The main objective is to ensure that the amount of drug required to have a therapeutic effect is significantly lower than the amount that provokes a toxic reaction in the body. This difference, known as the therapeutic window, is illustrated in Fig. 11.5. The amount of drug that produces a toxic response must be determined in order to establish an important value, the no-observed-adverse-effect level or NOAEL. This value is used to calculate the dose of drug that will be given to human volunteers (see later).

The results of drug safety studies will provide information on effects on the following:

- The organ systems of the body.
- The relationship between toxicity and exposure to drug.
- The dose dependence of the toxic effect.

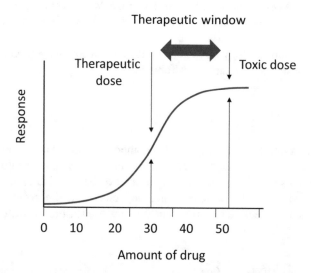

Fig. 11.5 Therapeutic window between drug efficacy and toxicity. Increasing amounts of drug are administered to an animal or human to measure the maximum therapeutic responses (i.e., desired change in disease symptoms, etc.) and then the point at which serious toxicity occurs. The difference in dose between the two gives a measure of the therapeutic window. It is obviously desirable to develop drugs with as wide a window as possible, although all drugs are toxic at some level

- The potential reversibility of the toxic effect.

The remainder of this chapter will cover the procedures used to assess drug toxicity and conclude by showing how all the ADME-Tox information is used to estimate the dose of a drug that will be administered to human volunteers. Further information about each of the safety pharmacology and toxicology procedures is available from the ICH, through the detailed guidance documents on their website (ICH 2020).

11.1.4.1 Safety Pharmacology

Drug candidates are normally tested for their effects on whole organ systems in experimental animals under the umbrella term safety pharmacology (also known as secondary or general pharmacology). This is conducted under GLP guidelines where possible and is run separately or as part of a general toxicology program. Formal safety pharmacology is not required with drugs that are administered through the skin (dermally) and into the eye (ocular) and with cytotoxic cancer drugs and biologicals that do not interact with non-primate tissues. Safety pharmacology studies are normally conducted in rats (or sometimes mice) and a large animal species, like dogs. In some cases, it may be necessary to use a non-human primate such as the cynomolgus monkey. Compounds are administered by the route expected to be used the clinic, i.e., oral or intravenously, depending on the drug's bioavailability. The amount of drug delivered is normally a multiple of the anticipated clinical dose (ACD), or human equivalent dose (HED), which has been estimated from the ADME data and earlier disease models. In practice, the investigations may use 10, 30, and 100 times the anticipated clinical dose to create a wide margin of safety for the final dose estimation.

The safety pharmacology package is divided into a core battery, plus follow-up, or secondary tests, as summarized in Table 11.3.

Core Battery

This provides a first look at the effect of a compound on vital body functions, such as heart rate, respiration, blood pressure, and nervous system function. This last category requires the use of a functional observation battery (FOB) or Irwin profile that is designed to monitor basic coordination, behavior, and nervous reflexes. The core battery studies require electronic instruments to measure the different parameters under investigation.

Table 11.3 Safety pharmacology tests according to ICH Safety Guidelines 2020

Core battery	Follow-up and supplemental studies
Central nervous system	Renal/urinary system
Cardiovascular system	Autonomic nervous system
Respiratory system	Gastrointestinal system

Follow-Up and Supplementary Tests

Sometimes a class of compound or biological will have safety issues associated with it; in this case, the core battery may be extended with follow-up studies. These will provide more details about functions such as blood pH, cardiac output, and learning ability. Supplemental studies are also designed to observe the effects of drugs on organ systems that are not already covered by the core battery. These include a detailed analysis of urine, responses to nerve stimulation, and secretion/transit times in the stomach and intestine.

Cardiac Rhythm and Q-T Interval

One of the most important safety issues relates to the effect of drugs on the rhythm of the heart. This is a complex area but one that must be carefully examined, as particular classes of drugs can induce *torsade de pointes*, a potentially fatal heart arrhythmia. This also occurs in a genetic condition, long Q-T syndrome, so called because the trace on an electrocardiogram (ECG) shows an increase in the distance between the Q and T peaks resulting from a delayed repolarization in the heart ventricle. The heart muscle is controlled by electrical signals that are conducted through ion channels (Chap. 6), one of which, a potassium (K^+) channel (called hERG), is implicated in the Q-T effect.[4] Safety pharmacology should therefore include in vivo evaluation of heart function using ECG, as well as in vitro testing for the effect of drugs on the hERG channel in cell lines. Detailed guidance is given in ICH document S7B (ICH Safety Guidelines 2020).

11.1.4.2 Toxicology Package

While safety pharmacology covers the effect of drugs on major organ function, there are some defined procedures that must be carried out as part of a toxicology package (M3(R2) section of ICH Multidisciplinary Procedures 2020). These include detailed analyses of animal tissues taken *postmortem*, as well as studies on the potential of drugs to cause birth defects or cancer. Firstly, however, there is a need to determine the toxicokinetics of a drug by dosing up to the maximum level tolerated by the animal.

---

[4]The naming of hERG is a good example of the sense of humor exhibited by scientists working with model organisms. The name means human ether-à-go-go-related gene originally found in the fruit fly *Drosophila* (Kaplan and Trout 1969). When mutant flies were given ether (as part of a normal laboratory analysis), their legs began to shake in the same way as dancers at the Whisky à Go Go nightclub in Los Angeles.

Table 11.4 ICH recommendations for duration of repeat dose toxicity trials in animals according to the clinical trial design

Maximum duration of clinical trial	Minimum duration of repeated dose toxicity studies to support clinical trials	
	Rodents	Non-rodents
Up to 2 weeks	2 weeks	2 weeks
2 weeks to 6 months	Same as clinical trial	Same as clinical trial
> 6 months	6 months	9 months

Acute and Repeat Dose Toxicity Studies

Acute dosing of different amounts of compound will already have been performed in rats as part of the safety pharmacology process. This will help to determine the maximum tolerated dose (MTD) of drug, which could be administered at up to 1000 mg/Kg per animal. While acute toxicology is not seen to be particularly valuable if chronic studies are going to be undertaken anyway, they can used on their own for testing drugs that are going to be administered to volunteers at very low doses in exploratory clinical trials (Chap. 12).

A chronic repeat dose study is undertaken in two species (rat, dog, or rat, primate), run under GLP conditions, to provide enough regulatory information for human trials. In addition to taking measurements of body weight and food consumption, investigators must analyze the body tissues and fluids of each dosed animal in detail. Over 40 tissues are examined using histology (Chap. 6), adding to the enormous amount of experimental data that will already have been accumulated during preclinical development.

The exact duration of the repeat dose study depends on the envisaged duration of the clinical trial. These ICH guidelines are shown in Table 11.4.

Genetic Toxicology (Genotox)

If a drug causes physical damage to DNA, this may result harmful mutations and possibly cancer. Since this is a real concern in drug development, a key part of the toxicology package involves screening drug candidates to identify those compounds which may be genotoxic or mutagenic. A "standard battery" of tests is employed, involving the use of bacteria, animal cells, and whole animals, as described in S2(R1) from ICH Safety Guidelines 2020.

Bacterial Ames Test

This assay was devised by the American scientist Bruce Ames to monitor damage to DNA in *Salmonella typhimurium*. A strain of this bacterium is selected to grow on agar plates (Chap. 4) only when the amino acid histidine is present in the growth medium. This is a result of a genetic mutation in the bacterial DNA that prevents it from producing its own histidine. To test for genotoxicity, the bacteria are spread onto plates that lack histidine and then incubated with test compounds. If a compound is mutagenic, it will alter the bacterial DNA and reverse the histidine mutation, so the bacteria can then grow independently of this amino acid. The number of bacterial colonies that appear on the plate is related to the degree of genotoxicity. This assay is a useful primary screen for compounds and is also used to check for drug metabolites that may themselves be mutagenic. In this case, rat liver microsomes (see earlier) are added to the bacteria, along with the test compound, to metabolize the drug in situ.

Mouse Lymphoma Assay

While the Ames test is an important first screen for genotoxic compounds, it is important to confirm any positive results using animal cells, the most widely used being the mouse lymphoma L5178Y$^{tk+/-}$ cell line. This line has a mutation in the thymidine kinase (TK) gene that causes the cells to die if they are grown in the presence of the synthetic compound trifluorothymidine. Using the same principle as the Ames test, any drugs that damage DNA will reverse the TK mutation and allow the cells to grow in the presence of the toxic compound. Instead of counting bacterial colonies on a plate, the mouse cells are counted in a liquid culture medium.

11.1.4.3   In Vivo Micronucleus Assay

In addition to causing DNA damage, genotoxic compounds also damage chromosomes (i.e., they are clastogenic). This can be seen directly by examining chromosomes under a microscope but also indirectly by using the micronucleus assay. A test drug is administered to rats or mice and their bone marrow cells removed for examination under a microscope. Under normal conditions, blood cells are produced in the bone marrow by cell division, to then become mature red blood cells after expelling a nucleus (these cells are the only ones in the body that do not have this structure). If any damage to chromosomes occurs during this division process, the cells release small fragments, called micronuclei, which can then be counted under a microscope.

Carcinogenicity

While the tests outlined in the previous sections are useful for screening for potentially genotoxic compounds, they are conducted over a short time frame and cannot fully mimic the chronic administration of a drug to humans. Long-term carcinogenicity studies (18 to 24 months) are therefore undertaken in mice and rats of both sexes, using compounds at the maximum tolerated dose. Since this time frame is longer than the preclinical development phase, these "lifetime studies" provide further safety information, while the drug candidate is proceeding through clinical trials. There is some debate about the usefulness of using both mice and rats over this time period; as a result, a number of more sophisticated models of carcinogenesis are being evaluated by the regulators (see ICH S1A, S1B, S1C(R2) accessed through ICH Safety Guidelines 2020).

Reproductive Toxicology (Reprotox)

The thalidomide tragedy focused minds on the teratogenic potential of drugs, i.e., their potential to cause birth defects. This area of reproductive toxicology is clearly important for women of childbearing age but also for men, whose fertility may be affected by experimental drugs. As with carcinogenicity studies, Reprotox is conducted in parallel with clinical development and requires only some basic information on embryonic development in animal studies prior to first human trials.

Studies of the effects of drugs on embryo development (embryotoxicology) are often undertaken in rats and rabbits, where all aspects of embryonic development are studied from conception to birth. For male fertility, sperm samples are obtained from rodents during chronic toxicology studies.

Safety Tests for Biologicals

Most of the drug candidates passing through safety pharmacology and toxicology are small molecules, rather than biotherapeutics such as antibodies, nucleic acids, and vaccines. Safety assessment is of paramount importance for these latter drug types as well, but many, like antibodies, are agents which interact specifically with human targets. This means that many biotherapeutics will not bind to body components in animals other than those derived from the primate family (which consists of monkeys, apes, and humans). The degree of cross-reactivity between species is assessed at an early stage in order to select the appropriate animal species for toxicology. The scale of these tests is often limited, however, by the need to restrict the use of non-human primates. The ICH guidelines S6 and S6 (R1) (ICH Safety Guidelines 2020) cover the safety evaluation of biologicals and include the requirement to test for immunogenicity and local tolerance. Immunogenicity is the degree to which the drug provokes an antibody response in the animal (which has already been mentioned in Chap. 7) and is covered in ICH S8 (ICH Safety Guidelines 2020). Local tolerance is the response to administration of the biotherapeutic at the site of injection. Some inflammation may occur, depending on the formulation used or through the possibility that drugs designed to modify the immune system may provoke inflammation through some unanticipated mechanism. Reproductive toxicology may have to be undertaken in a non-human primate, but, for practical reasons, these will be more limited than the equivalent rodent or rabbit studies. Finally, genotoxicity and carcinogenicity are evaluated according to practical circumstances.

Some biologicals may be inherently unable to damage DNA or cause cancer, but others, like nucleic acids, may do so. These must be assessed on a case-by-case basis.

11.1.4.4 New Technology

Preclinical safety evaluation is time-consuming, expensive, and not always predictive of later problems. The case of the monoclonal antibody TGN1412 is a stark example of how misleading preclinical data supported a disastrous phase I evaluation in which six volunteers suffered multiorgan failure and nearly died (Attarwala 2010). In addition, several drugs have been withdrawn from sale over recent years because of fatal adverse reactions in patients; examples include the antidiabetic drug Rezulin® and the statin Baycol®. These considerations, including the need to reduce and replace animals where possible, have driven research into finding better ways of predicting toxicity. There is no shortage of ideas in this field, but the demands of formal regulation make it difficult to introduce new technologies into the standard battery of procedures. Any variation on existing methodology must be rigorously tested before being acceptable to the regulatory authorities. Nevertheless, as an example, the FDA has published a roadmap of predictive toxicology designed to identify new technologies that may be incorporated into their regulatory procedures (FDA's Predictive Toxicology Roadmap 2020). Brief mention will be now made of some promising new approaches to toxicology.

Zebrafish

While the small model organism *Danio rerio* has already been described as a model organism for drug target discovery, it is attracting a great deal of attention as a possible adjunct to large animal toxicology. The fish has a small transparent embryo that rapidly develops the organs found in conventional test species. The zebrafish heart, for example, has a similar structure to that found in mammals and can even be monitored by ECG. Compounds that damage the mammalian heart also damage this organ in zebrafish embryos; this damage can be visualized directly under the microscope even while the fish is alive. Despite many unanswered questions about the suitability of zebrafish for toxicology screening in a regulated drug development environment, this organism appears to be a useful bridge between in vitro testing and larger animal toxicology (Cassar et al. 2020).

Toxicogenomics

Genomics and microarrays have already been introduced in Chap. 6. The technology allows researchers to measure the (mRNA) expression of many thousands of genes in parallel and provides information about cellular function in health and disease. Toxicogenomics (TGx) uses the same technologies (high-throughput sequencing and microarrays) to measure gene expression in the liver, kidneys, and other target organs of animal models after drug administration. The objective is to generate a characteristic pattern of gene expression (gene expression signature) that relates to the type of compound administered. Signatures from known toxins can then be identified and used to flag up a warning if an experimental drug shows a similar pattern. One example is the induction of messenger RNA for KIM-1, a marker of kidney damage, by specific nephrotoxicants. This proved to be a highly sensitive assay in rats where a change in KIM-1 expression could be detected in the absence of observable tissue damage (Chiusolo 2010).

This toxicogenomics approach encompasses protein and metabolite expression, as well as mRNA, but it is vital that the laboratory model accurately reflects the situation in human beings. For example, rats are used for many of these experiments, including the example above, but they still belong to a different species. A discussion of these issues and an assessment of the advantages and pitfalls of toxicogenomics can be found in Qin et al. (2016).

Organ-on-a-Chip

Even if human cells are used for in vitro tests, they are still cultured under highly artificial conditions. In real life, a given cell type is associated with other cell types within a three-dimensional matrix, so attempts are being made to recreate this complexity in the laboratory as a step towards producing a fully predictive and reproducible test system. For example, HEPATOPAC cultures, consisting of primary hepatocytes from the species of interest in contact with a supporting matrix of stromal cells, can mimic important liver functions for over 4 weeks (BioIVT 2020). Going a step further, it is possible to culture tissues and cells on microfluidic devices to create an "organ-on-a-chip" or "tissue chip" as a single entity or a series of interconnected organs to mimic the human body. The National Center for Advancing Translational Sciences (NCATS), as part of the US NIH, in collaboration with the FDA and others, supports several programs aimed at using tissue chips for pharmaceutical and clinical research (NIH 2020). This close association with the US regulators suggests that validated tissue chip systems will one day be approved for use in formal drug development.

Finally, stem cells from a variety of sources can be used generate mini organs or "organoids" that retain some of the features of the adult organ. This has enormous potential for basic science as well as pharmaceutical research and of course drug screening. In the latter case, Takasato et al. have produced a human kidney organoid from iPS cells (Chap. 8) that successfully mimic the toxic response to cisplatin seen in vivo (Takasato 2015).

11.1.5   Estimating the Dose for First Time in Humans

The data gained from the studies described in this chapter are used to estimate the maximum recommended starting dose (MRSD) for human volunteers. This is not a straightforward area, since drugs will differ from each other in their potential toxicity and other properties, so a single method of calculating this dose is not ideal. There is a danger that the starting dose could be underestimated, so there is little chance of a therapeutic effect once the drug enters patients; equally of course, there is a danger that unforeseen toxicities will emerge in humans if the dose is set too high. Therefore, biopharmaceutical companies are investigating different ways of making the MRSD prediction as accurate as possible for the full range of compounds and biologicals that are being submitted for clinical trials. One approach is to use mathematical techniques to provide more refined estimates of pharmacokinetic and pharmacodynamic data (this is known as pharmacometric analysis). The FDA document "Estimating the Maximum Safe Starting Dose in Initial Clinical Trials for Therapeutics in Adult Healthy Volunteers" 2005 provides guidance on how to select the dosage for humans (FDA Guidance for Industry 2005). The process is summarized as a flowchart in Fig. 11.6.

The first step is the calculation of a human equivalent dose (HED) based on the dose used in animal models for safety pharmacology and toxicology. The animal dose is determined as the NOAEL (no-observed-adverse-effect level) or the maximum amount of drug (in mg/kg) that produces no adverse effect (a NOEL is a no effect level, which is different).[5] It has been suggested that a comparison between human and animal doses is best made by comparing body surface area rather than

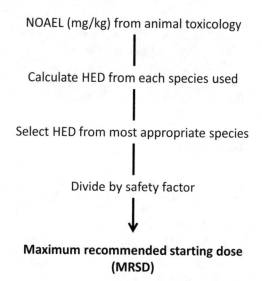

Fig. 11.6 Calculation of the maximum recommended starting dose (MRSD) for human clinical trials based on preclinical animal data. The no-observed-adverse-effect level (NOAEL) is used as the starting point to generate a human equivalent dose (HED) that is then divided by a safety factor to derive the MRSD

NOAEL (mg/kg) from animal toxicology

Calculate HED from each species used

Select HED from most appropriate species

Divide by safety factor

Maximum recommended starting dose (MRSD)

---

[5] The acronyms can be quite amusing, for example, SNARL "suggested no adverse response level."

Table 11.5 Figure used to divide NOAEL value to obtain an equivalent dose in humans

Species	Conversion factor
Mouse	12.3
Hamster	7.4
Rat	6.2
Rabbit	3.1
Dog	1.8
Cynomolgus monkey	3.1

weight, so NOAEL can also be converted to mg/m^2. Each animal species requires a different conversion factor as shown in Table 11.5.

Once an HED value has been calculated for each animal species used (e.g., rat and dog), the lowest value is used in combination with a safety factor to calculate the final dose. The safety factor is normally 1/10 of the HED, although this can vary according to the class of drug under investigation. Once the maximum recommended starting dose has been estimated, the time has come for preclinical development to end and to make way for the lengthy clinical development process.

Summary of Key Points

Preclinical development is used to assess the potential suitability of a compound or biological as a medicine prior to testing in humans.

Pharmacodynamics and pharmacokinetics relate to the effect of the drug on the body and the body on the drug, respectively.

Pharmacokinetics are reflected in the ADME properties of the drug.

Drug metabolism studies include the assessment of possible drug-drug interactions.

Safety pharmacology assesses the effect of the drug on body functions.

Toxicology assesses the effect of the drug on body structures.

Pharmacokinetic and safety data are used to determine a maximum dose of drug that can be given to human volunteers.

References

Attarwala H (2010) TGN1412: from discovery to disaster. J Young Pharm 2:332–336

BioIVT (2020) https://bioivt.com/about/technologies/hepatopactechnology Accessed 15 Apr 2020

Brunton L, Lazo J, Parker K (2005) Goodman & Gilman's the pharmacological basis of therapeutics. McGraw-Hill, Columbus

Cassar S et al (2020) Use of zebrafish in drug discovery toxicology. Chem Res Toxicol 33:95–118

Chiusolo et al (2010) Kidney injury Molecule-1 expression in rat proximal tubule after treatment with segment-specific Nephrotoxicants: a tool for early screening of potential kidney toxicity. Toxicol Pathol 38:338–345

Clark M, Steger-Hartmann T (2018) A big data approach to the concordance of the toxicity of pharmaceuticals in animals and humans. Regul Toxicol Pharmacol 96:94–105

FDA Guidance for Industry (2005) Estimating the Maximum Safe Starting Dose in Initial Clinical Trials for Therapeutics in Adult Healthy Volunteers. https://www.fda.gov/regulatory-information/search-fda-guidance-documents/estimating-maximum-safe-starting-dose-initial-clinical-trials-therapeutics-adult-healthy-volunteers Accessed 14 Apr 2020

FDA's Predictive Toxicology Roadmap (2020) https://www.fda.gov/science-research/about-science-research-fda/fdas-predictive-toxicology-roadmap Accessed 15 Apr 2020

Han C, Davis CB, Wang B (eds) (2010) Evaluation of drug candidates for preclinical development. Wiley, Hoboken

ICH (2020) www.ich.org Accessed 14 Apr 2020

ICH Multidisciplinary Guidelines (2020) https://www.ich.org/page/multidisciplinary-guidelines Accessed 14 Apr 2020

ICH Safety Guidelines (2020) https://www.ich.org/page/safety-guidelines Accessed 14 Apr 2020

Kaplan WD, Trout WE (1969) The behavior of four neurological mutants of Drosophila. Genetics 61:399–409

NIH (2020) Tissue Chip for drug screening. https://ncatsnihgov/tissuechip Accessed 15 Apr 2020

Ogu GC, Maxa JL (2000) Drug interactions due to cytochrome P450. Proc Bayl Univ Med Cent 13:421–423

Olson H et al (2000) Concordance of the toxicity of Pharmaceuticals in Humans and in animals. Regul Toxicol Pharmacol 32:56–67

Qin C et al (2016) Toxicogenomics in drug development: a match made in heaven? Expert Opin Drug Metab Toxicol 12:847–849

Takasato M (2015) Kidney organoids from human iPS cells contain multiple lineages and model human nephrogenesis. Nature 526:564–568

Part III
The Drug Development Pipeline – Clinical Trials to Marketing Authorization

Chapter 12
Clinical Trials

Abstract This chapter outlines the different phases of clinical trials, starting with phase I human volunteer studies through to phase III studies in large numbers of patients. The overview includes case histories of a phase I and a phase III trial to illustrate their design and implementation. Also covered are the regulatory applications for first time in human studies.

12.1 Introduction

The drug discovery process moves towards its culmination in the clinic. This is the ultimate proof of concept for a new drug; does it work in patients? This chapter describes the different clinical trial phases that lead from initial studies of a drug in volunteers to its evaluation in patients with a specific disease. Also covered are the key regulatory applications that must be approved before a drug can be administered to humans for the first time (later applications to market the drug are covered in the next chapter). Like many other aspects of drug development, there are variations in procedures that apply to individual types of medicines, diseases, or patient populations. This means that some trials do not readily fit into a specific phase; late-stage cancer patients, for example, may be treated in a phase I study that looks for efficacy, as well as for drug safety and tolerability.

A great deal of information about the conduct and regulation of clinical trials is available on the websites of the FDA, EMA, and ICH (FDA clinical trials links 2020; EMA clinical trials in human medicines 2020; ICH efficacy guidelines 2020). Further information is available from trials managed by the NIH's ClinicalTrials. gov 2020 and the World Health Organization's International Clinical Trials Registry Platform (ICTRP) 2020.

Clinical trials are conducted in different phases as listed in Table 12.1.

E. D. Zanders, *The Science and Business of Drug Discovery*,
https://doi.org/10.1007/978-3-030-57814-5_12

Table 12.1 Phases of clinical testing human subjects. Applications are documents required by the regulatory authorities at different stages of clinical development

Clinical phase	Comment	Timescale
Phase 0	Preclinical pharmacokinetics/dynamics using humans instead of animals	Weeks
Phase I	Dose ranging study in human volunteers	Weeks
Phase II	Testing drug in up to approximately 100 patients for proof of concept	Months
Phase III	Testing drug in 100s to 1000s of patients over longer period	Years
Phase IV	Post-marketing studies	Years
Phase V	Post-marketing surveillance	Years
Application		
IND	Investigational New Drug – FDA	Pre-phase I
CTA	Clinical Trial Application – EMA	Pre-phase I
NDA	New Drug Application – FDA	During phase III
MAA	Marketing Authorisation Application – MAA	During phase III
REMS	Risk Evaluation and Mitigation Strategy – FDA	During phase III

12.2 Preparation for Clinical Testing

Information about the pharmacokinetics, pharmacodynamics, and safety of experimental drugs will already have been gained from animal models; this information is used to guide the selection of a dose that will be used in first time in human studies (FTIH studies). ADME studies are now repeated by clinical pharmacologists, using human volunteers instead of animals, to provide information that will support further clinical development of the drug. The clinical department is responsible for establishing the clinical endpoint of a trial. Some diseases, like migraine, have a clear manifestation that can be measured, in this case headache and visual disturbances. Other conditions, however, may have to be measured indirectly, using surrogate endpoints such as lowering of blood glucose for diabetes, or reduction of pathogen levels in infectious diseases. This area of surrogates and biomarkers will be discussed later in the book.

12.2.1 Good Clinical Practice (GCP)

Unsurprisingly, clinical trials are highly regulated, so they are subject to harmonization and quality control rules in the same way as manufacturing and laboratory analysis (see ICH E6 (R2) 2020). Good clinical practice is required for the design, implementation, and reporting of drug trials on human subjects, who must be treated with the ethical considerations required by the Declaration of Helsinki. This declaration was drawn up by the World Medical Association (WMA) in 1964 to act as a set of guidelines for physicians wishing to conduct research on human subjects. Some of the main requirements for GCP as applied to clinical trials are listed below:

- Adherence to ethical standards
- Careful analysis of risk/benefit
- Informed consent of patients
- Fully qualified personnel
- Review by Institutional Review Board or Independent Ethics Committee
- Clear reporting of data
- Materials produced under GMP conditions

12.2.2 Terminology

Some basic terms are now listed, before moving on to the design and implementation of clinical trials:

- Active comparator

  Instead of using a placebo, which should give no benefit to the patient, a medicine that is commonly used for the condition under investigation may be used instead. This active comparator, or active control, is being increasingly used in clinical trials which would previously have been designed with placebo arms.

- Adaptive platform trial

  Adaptive trials are a comparatively recent initiative, in which a study may be modified in midstream because of early clinical results. These are covered in more detail later in this chapter.

- Adverse drug reaction (ADR)

  All unintended and noxious responses to a drug administered at any dose. A serious adverse drug reaction may result in death or major disability.

- Adverse event (AE)

  An untoward symptom, or laboratory finding, that occurs after drug administration and which may not necessarily be caused by the treatment. A serious adverse event (SAE) may result in death or major disability. The Medical Dictionary for Regulatory Activities (MeDRA) is used as a standard reference for describing adverse events.

- Blinded trial

  - Single: a trial in which the subject does not know whether they are taking drug or placebo, but the clinical investigator does.

– Double: a trial in which both the subject and the investigator do not know which is placebo and which is drug (obviously a third party must know!).

• Bridging study

A study performed in a new marketing region or population (e.g., pediatric or geriatric) to provide data on safety and efficacy that can be related to preexisting clinical trial information. This may be necessary if there are significant differences in pharmacokinetics between these different groups.

• Crossover trial

A trial in which one group receives placebo and the other the drug for a certain period, after which the treatment is reversed. In this way, placebo-controlled groups should improve after transfer to drug and the clinical improvement should cease when drug group is given the placebo. These trials can be double blinded to add further objectivity to the clinical assessment.

• Investigator

The individuals performing the trial, whose leader will be called the principal investigator.

• MAD – multiple ascending dose
• Open-label study

An unblinded trial in which patients know which medicine they are taking. This can be used for comparing different doses and similar drugs, or for situations where placebos would be inappropriate (e.g., in advanced disease).

• Placebo

Literally "I shall please." A dummy pill, capsule, or injectable formulation designed to look and taste identical to the drug compound. Placebos are used to counteract the powerful placebo effect that can profoundly influence some trials.

• Randomization

The random assignment of patients into different treatment groups in a random-ized controlled trial (RCT).

• SAD – single ascending dose
• Sponsor

Any company, or organization such as clinician networks, that finances and orga-nizes a clinical trial. If a trial is sponsored by a contract research organization (CRO) working under contract (e.g., to a biopharmaceutical company), then ultimate legal responsibility lies with the latter organization. The sponsor is also required to take out insurance as an indemnity against claims due to the trial itself, but this does not cover medical malpractice.

• Stratification

The separation of patient types into categories (e.g., >60 years, <40 years).

- Washout period

Before new drugs are administered to clinical trial subjects, any preexisting medicines should be cleared from the body (unless part of the trial). The time set by the clinical trial organizers is called the washout period.

12.2.3   Regulatory Requirements Before First Human Exposure

Before a drug can be administered to humans for the first time, a set of documents must be drawn up for submission to the regulatory authorities. In the USA these are specified in Section 21, Part 312 of the Code of Federal Regulations (CFR) and form the Investigational New Drug (IND) application 2020. In the European Union, the regulations come under Directive 2001/20/EC of the European Commission and form the Clinical Trial Application (CTA). In case there is any doubt about the seriousness with which these regulations are enforced, a few dishonest clinicians have found themselves in prison for misleading the authorities.

Although there are some differences between the IND and CTA applications, essentially the same type of information is required for both, so the IND will be used here as a representative example of a formal application for FTIH trials.

12.2.3.1   The Investigational New Drug (IND) Application

This application (submitted to the FDA) is a key set of documents which support the case for trialing an experimental drug in human subjects. It may also support the use of an existing drug for another indication (i.e., disease). Since the application will consist of over two thousand pages, it will take several months to compile prior to submission. The main categories (apart from administrative details) are listed as follows:

> **Introductory Statement and General Investigational Plan**
> This includes the name, formula, and nature of the drug, as well as its formulation, routes of administration, and dosage. Also submitted is a general plan for how the trial is going to be conducted and whether there are any known risks associated with the drug.

Investigator's Brochure
This contains a summary of the drug and its formulation as well as the phar-macokinetic, pharmacodynamic, and toxicology data gathered to date. It also describes possible side effects and any special monitoring required during the trial.

Protocols
Descriptions of how the studies are going to be carried out, with more detail required as the trials progress through the different phases. All aspects of patient recruitment, monitoring procedures, and flexibility in study design are covered, including the formal qualifications of the proposed investigators.

Chemistry, Manufacturing, and Control (CMC) Information
This contains detailed information about the chemical nature of the drug and its purity, as well as the associated excipients and the placebos used in the trial. The drug must also be tested for stability over at least the duration of the trials. Modifications to the method of preparation of the new drug substance and dosage form and changes in the dosage form itself are likely as the inves-tigation progresses, so final specifications for the drug substance and drug product are not expected until the end of the investigational process.

Pharmacology and Toxicology Information
This not only requires the relevant ADMET and other laboratory data but also the names, qualifications, and locations of the individuals who evaluated it.

Previous Human Experience with the Investigational Drug
The drug may already have been marketed, in the USA or elsewhere, and have background information on safety which should be submitted.

Additional Information
Information on abuse potential, use in children, or use of radioactive drugs all fall into this category.

The application is normally reviewed within 1 or 2 months, and, if approved, the first clinical trial can then commence.

12.3 The Clinical Trial Phases

12.3.1 Phase 0 (Microdosing)

As mentioned in Chap. 11, the animal models used for ADMET in preclinical development are not always predictive of human reactions to the drug. One possibility, therefore, is to administer minute doses of compound or biological to human volunteers and obtain valuable pharmacokinetic data before committing large resources to a phase I trial. This is called microdosing, or phase 0. A microdose is a maximum of 100 micrograms of drug, or less than one hundredth of the dose required to generate a pharmacological effect. Preclinical toxicology requirements are therefore limited to single-dose administration to animals without the need for genetic toxicology; this means that preclinical development times can be cut from 1 year to about 4 months and involve considerably less effort. Since the drug is present at an extremely low concentration in the human plasma, it is tagged with trace amounts of carbon 14 (^{14}C) that can be detected and measured using accelerator mass spectrometry (AMS), a sensitive technique also used for the carbon dating of archaeological samples. The potential value of microdosing in early clinical development has been recognized by the FDA and EMA, and the regulatory requirements are in place through guidance on using radioactive tracers for ADME studies (ICH M3 (R2) guidelines 2020). Although microdosing is not yet a routine part of drug development, there is no doubt that it will become more commonplace as biopharmaceutical companies gain confidence in the ability of the technique to provide an accurate picture of drug disposition in the human body.

Lastly, microdosing can be used to explore drug pharmacodynamics as illustrated by Presage Biosciences' CIVO (Comparative In Vivo Oncology) injection technology for dosing arrays of anticancer drugs into human tumors (Presagebio. com 2020). The tumors are then resected (removed) and the effect of the drugs noted through histology, etc. The promise of this approach as a means of avoiding unpredictive animal models of cancer has led to interest from companies such as Celgene (now part of BMS).

12.3.2 Phase I: Human Pharmacology

A typical phase I trial is conducted with up to 100 human volunteers assembled at a hospital or specialized contract research unit. The objective of such trials is to assess the maximum tolerated dose (MTD) of experimental drug and the recommended phase II dose (RP2D) for studies in patients. These volunteer trials are therefore not usually designed to look for clinical effects (they generally use healthy people after all), but there may be a pharmacodynamic effect that can provide the encouragement to commit resources to phase II trials. This is particularly true of oncology trials where phase I studies are performed using heavily pretreated patients with few therapeutic options. To gain maximum value from such trials, their design may include so-called expansion cohorts where further subjects are recruited after the RP2D has been established (Manji et al. 2013). Trial designs may involve an initial cohort of three individuals which moves on to the next three according to toxicities observed (3 + 3 design) or the "rolling six" design using groups of six.

Phase I trials can be subdivided into phase 1a and 1b, the former with single ascending doses in small groups and the latter with multiple ascending doses. Studies will also be undertaken to look at food interactions. Some trials of vaccines, anticancer agents, and other situations where efficacy must be rapidly determined can be reported as phase I/II trials.

12.3.2.1 Phase I: Case History

To give an idea of how phase I trials operate in practice, this example is taken from the trial of a compound designed to increase the number of platelets in the blood of patients with disease or undergoing chemotherapy (Jenkins et al. 2007). Platelets are the small cell-like bodies in blood that aggregate together to form clots, so reduced platelet counts can increase bleeding times. The GlaxoSmithKline drug eltrombopag or SB-497115[1] is a small molecule agonist for the thrombopoietin receptor on platelets. Preclinical work established that the compound worked with human cells in vitro and increased the number of platelets when given orally to chimpanzees.

Phase I studies were performed with male volunteers in a London hospital after full ethical approval had been obtained in advance. Volunteers were pre-screened for existing conditions that might affect blood clotting or related phenomena; the mean age, height, and weight of the volunteers were 27.5 years, 1.8 meters, and 79.8 kg, respectively. The placebo-controlled trial design is illustrated in Fig. 12.1.

Six groups of 12 subjects were assigned daily oral doses of eltrombopag, from 5 to 75 milligrams in capsules; 9 received the drug and three a placebo. The study was single blinded, so that only the subjects did not know whether they were receiving drug or placebo. Safety assessments (ECGs, blood and urine chemistry, and reporting of adverse events) were made throughout the trial. Blood and urine were taken at regular intervals during treatment and analyzed for drug levels, while platelet numbers were monitored in the blood. In this way, the phase I study was used to study pharmacokinetics (by looking at drug levels) as well as pharmacodynamics

---

[1] The code can indicate where the compound was developed originally. In this case, SB means SmithKline Beecham.

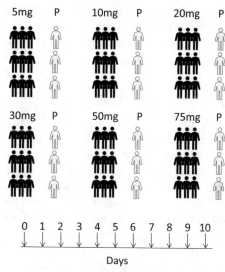

Fig. 12.1 Study design of phase I clinical trial of eltrombopag. Each group of nine volunteers was given the drug at the indicated doses (in mg/kg). Three individuals receiving placebos were added to each group. Drugs or placebos were given daily over 10 days with blood and urine sampled at regular intervals. Pharmacokinetics: C_{max}, maximum concentration; AUC, area under the curve; $T_{1/2}$, half-life. Safety and tolerability: ECG, electrocardiogram; BP, blood pressure; HR, heart rate; AEs, adverse events

(increases in platelet numbers). The results of the trial indicted that the pharmacokinetics were consistent with the animal data, since the drug was orally bioavailable and increased in concentration in proportion to the dose given. Furthermore, the dose-dependent rise in the number of platelets showed that the drug was working in the expected way. No safety issues were reported that could be related to the medication, and the small numbers of adverse events (AEs) such as headache or tiredness were independent of drug dose or placebo.

12.3.2.2   Dealing with the Unexpected

Although 20–40% of phase I trials fail to progress into phase II because of safety issues, the vast majority that do proceed do so without any serious adverse reactions. The rare exceptions are the disastrous phase I trial of the antibody TGN1412 in which six volunteers experienced multiple organ failure and nearly died (see Chap. 11). The drug was immediately withdrawn from further development by its maker TeGenero (which folded), and the regulators laid down stringent rules and guidelines for the safe conduct of phase I trials (Duff 2006). Unfortunately, a second disaster occurred when BIA 10–2474, a fatty acid amide hydrolase (FAAH) inhibitor developed by the Portuguese company Bial Portela, resulted in the death of one subject and neurological damage to four others (Kaur et al. 2016). There was no biological reason why this class of drugs should be problematical, as other companies had not experienced issues with their FAAH inhibitors, but BIA 10–2474 may have exhibited significant off-target effects. Furthermore, the dosing schedule was poorly designed so that the possible toxicity observed in one subject did not stop the drug being administered to the next group of four.

12.3.3 Phase II: Exploratory

Once a compound, or biological, has been shown to be safe and well tolerated at doses likely to have a clinical effect, planning for phase II trials can begin. Unless the earlier phase I trial had been specifically designed to look for efficacy, this is the first point at which the proof of concept will be tested on patients to see if there is any hope of moving on to phase III and beyond. Phase II testing is normally undertaken on a closely defined set of 50–100 patients who suffer from the disease that the drug is designed to treat. This relatively small number reduces the risk of unexpected adverse drug reactions or lack of efficacy, both of which could be expensive for the sponsoring organization. Phase IIb trials can extend earlier studies by including placebos and establishing a dose-response relationship prior to selecting the appropriate dose for the much larger phase III studies. Each study is designed to assess the safety and effectiveness of a treatment regime, which could be of the following type:

- A single medicine for a specified disease
- An altered dose of medicine
- A marketed medicine for a new indication
- A new drug compared with a gold standard medicine
- Two or more different medicines

The gold standard medicine, referred to above, is a drug that is generally considered to be the most effective treatment for a particular disease at the time of the trial; this means that an experimental drug has a significant hurdle to overcome before it can show a convincing improvement over an existing medicine. Antibiotics are good examples of gold standard drugs, since existing compounds like penicillin are highly effective in killing bacteria; a new product would have to show significant clinical advantage, perhaps through overcoming the problem of antibiotic resistance. The option of testing two or more different medicines as combination therapies is highly relevant to diseases such as AIDS, where three different drugs are formulated together to reduce the problem of drug resistance to the HIV virus. Current thinking about the cancer treatment strategies is also moving towards multiple drug therapies. While the thought of using drugs in combination may have scientific logic, the regulatory situation is not straightforward. The problem with drug combinations (in the form of one tablet or several) is that they must be evaluated in separate studies for both the combination and the separate drug components (FDA codevelopment guidance 2020). This obviously increases the time and expense of drug development, but the resulting clinical benefits clearly make up for this.

12.3.3.1   How Is Drug Effectiveness Measured?

This question could be rephrased: "How do you know when a drug has worked in patients?" The answer may seem obvious, "the patient gets better," but of course that leads to the question: "How do you define better?" The other problem is the separation between the proposed mechanisms of the drug and the actual clinical effect. In many cases, a drug that has a measurable effect on a protein target in patients may make no difference to the symptoms of the disease in which the target has been implicated. Alternatively, the disease symptoms may be reduced, but the target is unchanged, a consequence of the limits to our understanding of the molecular mechanisms of disease.

The formal design of clinical trials requires the specification of variables that define the desired clinical endpoint of a study as objectively as possible. The variables must be presented in a form that can be analyzed using statistics (see later). More detail about the different types of variables described below is available in ICH E9 guidelines 2020.

• Primary variable

Normally a single measurement that defines the effectiveness of the drug. This can be very wide ranging according to the type of disease or question asked by the investigator. Examples include lowering of blood glucose levels, survival after a certain period, quality of life improvements, number of heart attacks over a defined period, and the slowing down of tissue destruction.

• Composite variable

In many cases (e.g., arthritis or psychiatric diseases) several separate measurements are used to assess the clinical state of the patient, and it may not be possible to identify a single primary variable. In this case, a composite variable consisting of all the different factors in combination is used for the trial analysis.

• Global assessment variable

This is like the composite assessment but includes a more subjective analysis of the disease by both physician and patient as to the effectiveness of the treatment.

• Categorized variable

A requirement that a disease measurement falls into a specified category. One example would be a trial that requires diastolic blood pressure to fall below the fixed value of 90 mm Hg.

• Surrogate variable

There will be cases in which it is not possible to directly observe clinical effects during the clinical trial. In this case, a surrogate marker of the disease will have to be employed (see also Chap. 14). One of the best known of these is the surrogate marker of AIDS used to speed up the development of drugs against the causative HIV virus. The symptoms of AIDS are expressed slowly in different ways, so rather than just showing that drugs reduce levels of HIV virus, investigators use a surrogate of the immune deficiency that is the hallmark of the disease. Immune deficiency is caused initially by depletion of the CD4 T lymphocyte population in white blood cells, since these are the direct target of the HIV virus. The surrogate variable for clinical trials is therefore the level of CD4 lymphocytes in AIDS patients compared to uninfected individuals. Despite cases like this, the choice of surrogate markers must be made extremely carefully as these are not always predictive of clinical outcome. Nevertheless, some diseases such as neurodegeneration are so complex and difficult to assess in an objective way that the search for predictive surrogate markers may offer the best way forward in drug development.

- Secondary variable

As the name implies, secondary variables are considered after the primary variable but may provide new information about a drug that can be exploited in further trials. One example is reduction in the incidence of strokes (secondary) as a consequence of bringing blood pressure below a certain value (primary variable).

12.3.3.2 Basic Clinical Trial Actions

If clinical trials are to be performed on patients, the following actions must be taken:

- Identify hospitals and clinical investigation teams.
- Recruit a suitable cohort of patients.
- Conduct the trial.
- Use biostatistics to determine the effectiveness of the trial.
- Monitor and report adverse drug reactions.
- Present a full report of the trials to the regulatory authorities.

The above items will be covered in the remainder of this chapter and in the next.

12.3.3.3 Geographical Locations

The nature and size of the study will determine whether single-center or multicenter trials are performed. In the latter case, there may be contributions from hospitals in several different countries. The geographical location of the hospital is particularly important, since the ethnic background of the local population of patients may vary widely, along with the potential responsiveness to a drug treatment (see Chap.14). Table 12.2 shows the geographical locations of the clinical studies listed in Clinicaltrials.gov as of April 2020.

Table   12.2 Number   of clinical   studies   listed   in Clinicaltrials.gov   by geographical location

Region	Number of registered studies
North America	145,195
Europe	96,219
East Asia	37,902
Middle East	14,401
South America	10,686
Africa	9839
Pacific region	7850
Southeast Asia	6796
Japan	5875
North Asia	5870
South Asia	5148
Central America	3046

Data from website Accessed April 2020

12.3.3.4   Patient Recruitment and Selection

Once a clinical team and plan are in place, patients must be identified and recruited for trials that may require many months of repeated visits to hospital. Although many patients are willing to submit to repeated interventions in the hope that their contribution will be of use to themselves and others, there are several reasons why they may not be selected for trial or be compliant once enrolled. The inclusion or exclusion criteria used to select patients are defined by the nature of the study. Patients will obviously be included if their condition fits with the trial objective, but they may be excluded for a variety of reasons, including having other diseases that may confound the study. Since patients are also human, they have free will to conduct their lives as best they can. This means that they may choose not to "play ball" with the trial investigators and be noncompliant with their medication or hospital visits. This can be understandable, particularly if unpleasant side effects arise as part of the treatment. There may, however, be logistical problems; for example, some patients with rheumatoid arthritis have such severe disability in their joints that they require most of the morning just to get to a position where they can leave their homes to be taken to the clinic. This limits the number of hours available to the investigator and increases the time taken to complete the trial. Sometimes the reverse occurs, when patients actively look forward to attending the clinic, as one patient said in a trial which I was familiar with: "to get away from the wife for a few days!" Although noncompliant trial participants inevitably reduce the quality of the data, an intention-to-treat principle is applied to subjects assigned to a treatment group. This means that they are assessed as part of that group, regardless of whether they completed the treatment.

The issue of patient recruitment is of great concern to sponsors and investigators and is something that will be discussed further in Sect. 12.4.

12.3.3.5   Biostatistics

People differ from one another in many ways, including their responses to drug treatment. This means that the variability that occurs in clinical trials must be measured and analyzed using statistics. This branch of mathematics has sometimes suffered from an image problem; witness, for example, British statesman Winston Churchill's remark about "lies, damn lies and statistics." Even biologists have come up with the (rather weak) joke that "if the experiment needs statistics, it means it hasn't worked." This, of course, is an exaggeration and perhaps reflects the frustrations that living organisms cause biologists through refusing to behave with the same precision and predictability as the inanimate objects studied by physicists.[2]

The biostatistics department in a company is responsible for the statistical analysis needed prior to the recruitment of patients (or volunteers) and for analyzing the data as the trial proceeds. Statistics deals with likelihoods or probabilities that a certain event or events will occur, with a certainty being 1 and a "no chance" being 0, with all points in between. Probabilities in clinical trials are expressed as decimals, like my chance of winning the UK National Lottery being about 0.0000000714 (or 14 million to one). The main objective of biostatistics in clinical trials is to obtain a measure of the statistical significance of an effect that is measured after drug administration. This measure may compare the responses of the drug treatment to that of the placebo and produce a probability that the observed response to treatment is greater than what would have been observed by chance; this is known as the null hypothesis. Of course, there may be no significant difference, or indeed a worsening of the patients' condition, but whatever the result, the statistical analysis enables the investigators, sponsors, and regulators to assess whether the treatment has worked or not. The accuracy of the significance measurement is related to the sample size; in other words, the more people tested in the trial, the greater the accuracy. This is referred to as the power of a planned study: too low and the conclusions will be ambiguous; too high and resources will have been expended unnecessarily. There is a real danger with underpowered studies that the results will be scientifically meaningless, but sometimes the problem of patient recruitment is so great (e.g., with rare diseases) that investigators may have to proceed with the trial anyhow. As always, a compromise between the desired outcome and the real-world situation must be made.

12.3.3.6   Statistics Terminology

A number of tests of significance are referred to by name in biostatistics reports; while it is not the intention here to offer more than a very brief description, some are listed below and can be studied in detail by accessing the many statistics resources available elsewhere, including ICH E9 guidelines 2020.

---

[2] This is only to make a point. Statistics is fundamental to physics.

- Chi squared, or χ^2 test

This is commonly used to compare variable data between groups by assigning a probability that the observed variability is likely to be true (see p-value below). The variables in a clinical trial designed to assess the ability of a drug to lower blood pressure will be the blood pressure measurements in a group of treated subjects compared with the those in a placebo (or active) control group.

- ANOVA – analysis of variance

This is a statistical method related to the χ^2 test that allows comparisons of multiple sets of data. This is useful for trials with more than just a treatment and placebo arm.

- P-values

A p-value is the probability that there will be a difference between different groups (e.g., drug-treated and placebo) in a situation when no difference exists. In other words, a high p-value (high probability) will imply that the trial has not worked, and a low p-value gives the investigator confidence that any differences between groups are likely to be real. P-values are quoted in clinical trial reports and publications to give the reader a number which allows them to judge whether the drug treatment worked. Typical p-values quoted might be something like $p < 0.05$, or $p < 0.001$. In the last case, this indicates that the chance of the difference between groups not being real is less than 1 in 1000, and therefore the result is considered meaningful.[3]

- Type I and type II errors

Tied up with the comparisons of different sets of variable data is the false positive or type I error, written as the Greek letter α. A false positive might present itself as a lessening of symptoms in a trial for reasons unrelated to the drug. This happens to a surprisingly large degree in patients treated with placebos, possibly related to the extra care and attention they receive in hospital. Type II errors are false negatives written as β. These of course will mislead investigators in thinking that the treatment has not worked when in fact it has. All the above statistical tools (and more) are used to determine the optimal power of the study (number of patients and groups) as well as interpreting whether the drug has really worked in patients and therefore worth pursuing further.

---

[3]The misuse of p-values in science has come under fire, and caution is needed in their use (Nuzzo 2014).

12.3.3.7   The Structure of a Clinical Trial Report

In case there has been some doubt about the amount of paperwork required for drug development, the following ICH (E3) recommendations for the structure of a clinical trial report should correct any misconceptions (ICH efficacy guidelines 2020).

Title Page, Synopsis, Table of Contents, List of Abbreviations Ethics
IRB, Declaration of Helsinki followed, patient information and consent given.
   Investigators and study administrative structure
   Investigators and observers such as nurse, physician's assistant, clinical psychologist, clinical pharmacist, or house staff physician and biostatistician(s)

Introduction, Study Objectives, Investigational Plan, Study Design and Plan Description
Drugs, doses and procedures, patients; blinding, single/double blind, open label; controls, placebo, active drug, etc.; method of patient assignment, randomization, stratification, sequence, and duration of all study periods
   Discussion of study design, including the choice of control groups
   Selection of study population
   Inclusion criteria, exclusion criteria, removal of patients from therapy, or assessment

Treatments
Treatments administered, identity of investigational product(s), method of assigning patients to treatment groups, selection of doses in the study, selection and timing of dose for each patient, blinding, prior and concomitant therapy, treatment compliance

Efficacy and Safety Variables
Efficacy and safety measurements assessed and flow chart, appropriateness of measurements, primary efficacy variable(s), drug concentration measurements

Data Quality Assurance
Statistical methods planned in the protocol and determination of sample size.
  Statistical and analytical plans, determination of sample size
  Changes in the conduct of the study or planned analyses
  Study patients, disposition of patients, protocol deviations
  Efficacy evaluation, datasets analyzed
  Demographic and other baseline characteristics

Measurements of Treatment Compliance
Efficacy results and tabulations of individual patient data
  Analysis of efficacy, statistical/analytical issues, adjustments for covariates, handling of dropouts or missing data, interim analyses and data monitoring, multicenter studies, multiple comparison/multiplicity, use of an "efficacy subset" of patients, active-control studies intended to show equivalence, examination of subgroups, tabulation of individual response data, drug dose, drug concentration, relationships to response, drug-drug and drug-disease interactions, by-patient displays, efficacy conclusions

Safety Evaluation, Extent of Exposure, Adverse Events (AEs)
Brief summary of adverse events, display of adverse events, analysis of adverse events, listing of adverse events by patient
  Deaths, other serious adverse events, and other significant adverse events
  Listing and discussion of deaths, other serious adverse events, and other significant adverse events

Clinical Laboratory Evaluation
Listing of individual laboratory measurements by patient and each abnormal laboratory value, evaluation of each laboratory parameter, laboratory values over time, individual patient changes, individual clinically significant abnormalities

Vital Signs, Physical Findings, and Other Observations Related to Safety
Safety conclusions

Discussion and Overall Conclusions
Data figures, tables, graphs, forms, references, appendices

12.3.4   Phase III: Confirmatory

The following section covers the design and conduct of phase III clinical trials; these are designed to replicate the phase II efficacy data for the drug candidate in a larger number of patients over a longer period.[4] Unfortunately, a trial cannot possibly replicate the numbers of patients who might use the drug after marketing, but it is the only option before the drug reaches the market. The phase III design (generally) ensures that any safety issues that may have been absent in smaller trials can be identified. However, many phase III trials fail to confirm efficacy, or they reveal safety problems to such a degree that the drug development program must be terminated. When phase III trials are designed to support a marketing application to the regulators (regulatory submission, covered in next chapter), they are described as pivotal studies. These trials are large randomized controlled trials in multiple centers that are either coordinated into one study or treated as multiple independent single-center trials with similar objectives. Sometimes a phase III trial is designed to support a marketing approach, or to extend the use of the drug to different types of patients (label expansion), in which case the term phase IIIb is sometimes used.

12.3.4.1   Specialized Patient Groups

The general clinical trials process described up to now is applicable to many patients, but there are specific groups where variations in disease manifestation and responses to drugs can be highly significant. These groups are the elderly, children, and the male and female genders.

Pediatrics

Children have not reached full biological maturity and are still in the development stages that were initiated during embryo formation (embryogenesis). This means that they may respond to medicines in a different way to adults and not just because they are smaller. One much quoted example is the different pharmacodynamics of the antibiotic tetracycline between adults and children where the drug causes permanent staining in children's teeth as they develop. There is, however, specific

---

[4]The phase II and III trials are designed and conducted in a similar way, so there is significant overlap between them.

guidance on pediatric drug development provided by the regulators and the ICH E11 guidelines 2020. In general, studies of many drugs on adults will form the basis of later evaluation on children. If, however, the intention is to treat life-threatening diseases where there is no cure, then children can be used at the outset. There may also be some treatments (such as gene therapy) that are more specific for children, so testing in adults would be inappropriate. Other considerations include formulation, where some excipients may be harmful to developing children, or where it might be necessary to make the product more palatable.

Finally, pediatric cancers often differ from the adult disease through having fewer genomic alterations (see Chap. 14). Trials with drugs aimed at the adult disease may therefore be inappropriate for those under 18 years, so special regulations will be necessary (DuBois et al. 2019).

Geriatric Populations

At the other end of the age spectrum, anyone over the age of 65 is classed as elderly or geriatric, according to the ICH E7 guidelines 2020. This figure does seem rather on the low side, considering the upward trend in life expectancy (see introduction), and think of Mick Jagger. Nevertheless, as a group, the elderly will be differentiated from younger adults because they may have chronic age-related conditions such as reduced kidney function (renal insufficiency) and be prescribed several different medicines at once. The aging population is important to the biopharmaceutical industry, as its scientific and commercial focus is on the chronic diseases that normally occur later in life, such as Alzheimer's and cancer. The ICH E7 guidelines 2020 provide information on specific aspect of clinical trials with geriatric patients that need to be considered by sponsors and investigators. Emphasis is placed on metabolism and drug-drug interactions (see Chap. 11).

Vive la Différence

The biological differences between men and women are generally something to be celebrated, but from the perspective of drug development, these can be a complication. The large changes in hormone levels during monthly cycles and pregnancy, for example, make it important to use female animals or human subjects, yet most preclinical models and human trials use males (unless the intention is to treat females predominantly). There appears to be an increasing realization on the part of clinicians and regulators that gender differences must be taken more seriously when conducting drug development programs (Ballantyne 2019). These differences are manifested as changes in pharmacokinetics and pharmacodynamics, as well as conditions such as cardiovascular disease which may present quite differently between men and women. There has also been a political drive towards increasing the representation of women in clinical trials and paying more attention to specific diseases, such as breast cancer. This shift in emphasis has meant that investigators will have to think carefully about how to conduct trials with pregnant women, or those who may get pregnant, while taking the drug under evaluation. One answer may be to conduct small phase I trials after phase III studies are underway with male patients; in any event, extreme rigor in safety monitoring is required to ensure that no harm is done to mother or fetus. Basic research into the detailed differences between male and female biology and pathology will help to guide future preclinical and clinical trial designs, perhaps through identifying new biomarkers using proteomics or other technologies (Chap. 14).

12.4 The Need to Improve Efficiency

Every time a new drug-related scientific breakthrough is featured in the media, it is nearly always qualified by the phrase "but it will be many years before a treatment is available for patients." This is obviously true because of the preclinical and clinical hurdles involved, but also dispiriting; there is a general feeling among those involved that processes could be improved and regulations made more efficient. The biopharmaceutical industry, regulators, and clinicians are all aware of the problem and are making attempts to streamline processes and even question the value of some types of trial. There is also a strong financial incentive to speed up clinical trials since every month of delay is lost revenue to the sponsoring company as the patent life of the drug ticks away. It is not unreasonable to suggest that the topic of clinical trial efficiency is close to the top of a list of concerns felt by the biopharmaceutical industry. Some of the key issues are discussed below.

12.4.1 Clinical Trial Costs

The planning and executing of clinical trials are a time- and resource-consuming business for a biopharmaceutical company and associated agencies (regulators, clinics, contract research organizations, etc.) It has been thought that the cost of trials is a significant component of the overall cost of drug development. However, according to Moore et al. (2018) in a study of 59 drugs in 138 pivotal trials between 2015 and 2016, the median cost was $19.0 million (interquartile range, $12.2 million–$33.1 million). While not cheap, these figures are much less than the average drug development cost of over $2 billion (see Chap. 16). The costs of individual trials varied 100-fold, and unsurprisingly those with small patient numbers were cheaper than large placebo-controlled studies in, for example, cardiovascular diseases. So, while direct costs may not be so important, the delays in bringing new drugs to market because of issues with clinical testing are certainly a problem, with consequences for both patients and sponsors.

12.4.2   Trial Initiation and Management

The study initiation phase of a clinical trial involves organizing treatment site(s) and patients and dealing with the regulatory groups. This is a highly inefficient process which, according to the Tufts Center for the Study of Drug Development (Tufts CSDD), can last nearly 8 months on average (Getz 2018). One problem is the increase in studies of rare diseases and stratified patient populations, with the numbers of unique FDA-approved clinical principal investigators reaching around 34,000 in 2015, of whom 40 percent dropout each year; this is attributed to stringent regulatory requirements, high workload and staff turnover, and lack of financial incentives.

According to Tufts CSDD, 11% of sites for phase II and III trials fail to enroll any patients, and 37% will under-enroll, which is a problem because these trials would have been activated and must then be conducted regardless of patient numbers. The 37% figure has been adopted by the clinical research company Science37 (science37.com 2020) which is one the emerging organizations aiming to improve efficiency through rebuilding the structure of clinical trials using digital technology (see also Chap. 17).

The complexities of trial management are illustrated by a study of phase III cancer trials organized by the Eastern Cooperative Oncology Group (ECOG) in the USA (Dilts et al. 2008). By studying 16 trials in detail, the group concluded that it took as much or more time to activate a trial (ranging from 435 to 1604 days) as it did to conduct the study itself. Since more than 481 distinct processes were required, including 61 major decision points, this is not too surprising. To make the point even more forcefully, a process diagram capturing all the interactions, if printed out in 8-point type, would measure a staggering 5 by 50 feet. Anyone who contributes to a multidisciplinary project in any field will feel a shudder of recognition when they examine why the timelines were so extended. This was because of the need for review boards and agencies from multiple locations who were all expected to operate with identical procedures and approaches, an almost impossible objective.

12.4.3   Reporting Trial Results

The results of clinical trials, whether positive or negative, provide the knowledge that can be used to support the use of a medicine or to provide a warning about toxicities, etc. In 2018, a law was finally enacted in the USA that required trial sponsors to post results in Clinicaltrials.gov or face financial penalties from the FDA. While most large pharma companies have complied with this, many smaller commercial organizations and academics have been less than compliant, although without incurring any penalty so far (as of January 2020). A survey by *Science Magazine* from more than 4700 clinical trials submitted to Clinicaltrials.gov looked at the number of organizations who reported results as of 2019 and found 31.6% not reported, 23.7% reported late, and 44.7% reported on time or early (Piller 2020). Furthermore, the results of many trials are never published in the biomedical literature, thus depriving the scientific community of basic information. Any change in this situation will no doubt arise from enforcement of the US law and reputational considerations by trial sponsors.

12.4.4   Clinical Trial Design: Placebos

"Drug discovery is not about whether a drug works, it's about whether it works better than a placebo control," Ted Kaptchuk, Director Placebo Lab, Harvard Medical School. The problems arise when the placebo effect is so high as to make it impossible to interpret the results of a drug trial.[5] Furthermore, the adverse consequences of the "nocebo effect" may cause a significant number of patients to dropout of trials because they have perceived adverse effects while on the placebo. So clearly there is a need to understand the biology behind the placebo effect and use the information gained to pre-screen patients enrolled into randomized trials. Some candidate pathways in a so-called placebome have been identified, including the opioid, dopamine, serotonin, and endocannabinoid systems, but this field is in its infancy; detailed validation of these findings would require negative treatment controls (i.e., no placebo) which has ethical implications (Cai and He 2019).

12.4.5   Clinical Trial Design: Synthetic Control Arms

Data from clinical research studies are obviously important but often miss real-world data (RWD) based on the actual experiences of physicians and patients. Real-world data can provide real-world evidence (RWE) whose incorporation into the drug development process is being considered by regulators, such as the FDA (FDA Strategic framework 2020). One way in which this can be done is to create a synthetic control arm for clinical trials to replace the placebo or active comparator using data taken from patient records, laboratory data, etc. The inclusion of a virtual control arm means that patient recruitment is more efficient, and the nocebo effect is eliminated. Although this area is in its early stage, synthetic control arms have been used in oncology trials to successfully fulfill regulatory requirements, for example, with Roche's Alecensa and Amgen's Blincyto (Goldsack 2019).

---

[5] Sometimes the placebo response matches or exceeds that in the treatment group as occurred with Circassia's cat allergy immunotherapy, with the company now repositioned into diagnostics.

12.4.6   Adaptive Platform Designs

The terminology for randomized placebo-controlled trials (RCTs) was highlighted in Sect. 12.2.2.2 along with that of adaptive platform trials (APTs) . The latter are designed to focus on a specific disease while studying multiple (drug) interventions "on the fly" rather than waiting until the end of the planned trial period. Mathematical algorithms are used to decide which interventions to terminate and which to progress to further trials (The Adaptive Platform Trials Coalition 2019). Bayesian probability is commonly used, as it generates probabilities based on prior experience rather than total randomness. The "prior experience" may be, for example, changes in the clinical response as the trial progresses.[6]

A pioneering 4000 patient, phase II adaptive trial, I-SPY 2 (Investigation of Serial Studies to Predict Your Therapeutic Responses with Imaging and Molecular Analysis 2)[7] was launched in 2010 for breast cancer patients. It is sponsored by a charitable consortium, Quantum Leap Healthcare Collaborative, and involves several major pharmaceutical companies, including AbbVie and Merck. I-SPY 2 is designed to compare the efficacy of novel drugs in combination with standard chemotherapy with the efficacy of standard therapy alone (Zipkin 2019). The biomarker

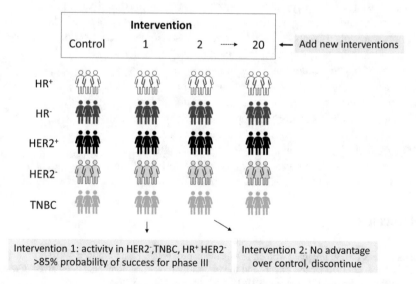

Fig. 12.2 Design of I-SPY2 adaptive trial for breast cancer. Patients subdivided according to tumor subtypes: HR+ or – (estrogen or progesterone receptor), HER2 + or – (epidermal growth factor receptor), TNBC (triple-negative breast cancer). Up to 20 different interventions assessed versus control arm. Intervention 1 is monoclonal antibody checkpoint inhibitor Keytruda (pembrolizumab) in combination with paclitaxel, doxorubicin, cyclophosphamide, and surgery

---

[6] Some oncology designs may be basket trials (patients with a range of cancers enrolled), or umbrella trials where they have the same tumor type.

[7] Clinicians, like space engineers, use catchy acronyms for their major projects.

signatures (see Chap. 14) of patients with breast cancer are used to assign to further trials those with higher Bayesian probability of better responses to the drug than standard therapy. Treatments not offering improved benefits are rejected earlier in the process than would be the case in a randomized controlled trial. The trial itself allows new interventions to be introduced to fill the gap where others have moved on. This is an ambitious trial due to complete in 2025 (Clinicaltrials.gov study NCT01042379, 2020). A summary of the trial design is shown in Fig. 12.2.

In the example with Merck's pembrolizumab, the Bayesian probability of success for this drug in combination with standard therapy was greater than 85% (the threshold for moving into a phase III trial). Evaluation of this drug regimen in I-SPY2 took only 12 months and the results validated in a phase III trial giving pathologic complete responses in patients. These results emphasize the speed in which this type of phase II trial can be conducted and the use of a single control arm for multiple interventions (20 in this case).

Summary of Key Points
Clinical trials are conducted in several distinct phases, firstly to determine the maximum tolerated dose and other clinical pharmacology parameters (phase 0 and phase I).

Phase II trials use patients to check for efficacy and safety and phase III to confirm both in larger numbers of patients over a longer time period.

Most trials use randomized patient groups assigned to a placebo control (or active comparator) and drug treatment arms. These studies are double blinded to avoid bias by patient and investigator.

Biostatistics are used to analyze trial data to determine the effectiveness of a treatment compared to the control group. Bayesian statistics are used in adaptive trial platforms to support or deny progression of a therapy from phase II to phase III.

Clinical trial documentation is complex and extensive.

Patient enrollment is a major factor in the efficient running of clinical trials

References

Ballantyne A (2019) Pregnant women can finally expect better. Hast Cent Rep 49:10–11

Cai L, He L (2019) Placebo effects and the molecular biological components involved. General Psychiatry doi:101136/gpsych-2019-100089. Accessed 23 Apr 2020

Clinicaltrials.gov study NCT01042379 (2020)

Dilts DM et al (2008) Development of clinical trials in a cooperative group setting: the eastern cooperative oncology group. Clin Cancer Res 14:3427–3433

DuBois SG et al (2019) Ushering in the next generation of precision trials for pediatric cancer. Science 363:1175–1181

Duff GW (2006) https://webarchive.nationalarchives.gov.uk/20130105090249/http://www.dh.gov.uk/en/Publicationsandstatistics/Publications/PublicationsPolicyAndGuidance/DH_063117. Accessed 20 Apr 2020

EMA clinical trials in human medicines (2020) https://www.ema.europa.eu/en/human-regulatory/research-development/clinical-trials-human-medicines#new-clinical-trials-regulation-section. Accessed 17 Apr 2020

FDA clinical trials links (2020) https://www.fda.gov/regulatory-information/search-fda-guidance-documents/clinical-trials-guidance-documents. Accessed 17 Apr 2020

FDA Codevelopment guidance (2020) https://www.fda.gov/regulatory-information/search-fda-guidance-documents/codevelopment-two-or-more-new-investigational-drugs-use-combination. Accessed 20 Apr 2020

FDA Strategic framework (2020). https://www.fda.gov/news-events/press-announcements/statement-fda-commissioner-scott-gottlieb-md-fdas-new-strategic-framework-advance-use-real-world. Accessed 23 Apr 2020

Getz K (2018) Trends driving clinical trials into large clinical care settings. Nat. Rev. Drug Disc. 17;703–704

Goldsack J (2019) Synthetic control arms can save time and money in clinical trials. https://www-statnewscom/2019/02/05/synthetic-control-arms-clinical-trials/. Accessed 23 Apr 2020

ICH Efficacy guidelines (2020). https://www.ema.europa.eu/en/human-regulatory/research-development/clinical-trials-human-medicines#new-clinical-trials-regulation-section. Accessed 17 Apr 2020

ICH E6 (R2) guidelines (2020). https://www.ich.org/page/efficacy-guidelines. Accessed 17 Apr 2020

ICH E7 guidelines (2020). https://www.ich.org/page/efficacy-guidelines. Accessed 21 Apr 2020

ICH E11 guidelines (2020). https://www.ich.org/page/efficacy-guidelines. Accessed 21 Apr 2020

ICH E9 guidelines (2020) https://www.ich.org/page/efficacy-guidelines. Accessed 20 Apr 2020

ICH M3 (R2) guidelines (2020) https://www.ich.org/page/multidisciplinary-guidelines. Accessed 20 Apr 2020

Investigational New Drug (IND) application (2020) https://www.ecfr.gov/cgi-bin/text-idx?SID=c89f98db7135bc56111bf0742c1c172f&mc=true&node=se21.5.312_123&rgn=div8. Accessed 17 Apr 2020

Jenkins JM et al (2007) Phase 1 clinical study of eltrombopag, an oral nonpeptide thrombopoietin receptor agonist. Blood 109:4739–4741

Kaur R et al (2016) What failed BIA 10–2474 phase I clinical trial? Global speculations and recommendations for future Phase I trials J Pharmacol Pharmacother 7:120–126

Manji A et al (2013) Evolution of clinical trial Design in Early Drug Development: systematic review of expansion cohort use in single-agent phase I Cancer trials. J Clin Oncol 31:4260–4267

Moore TJ et al (2018) Estimated costs of pivotal trials for novel therapeutic agents approved by the US Food and Drug Administration, 2015–2016. JAMA Intern Med 178:1451–1457

NIH's. Clinicaltrials.gov (2020) Accessed 17 Apr 2020

Nuzzo R (2014) Scientific method: statistical errors. Nature 506:150–152

Piller C (2020) Transparancy on trial. Science 367:240–243

Presagebio.com (2020) https://presagebio.com/civo-platform/. Accessed 20 Apr 2020

Science37.com (2020) https://www.science37.com/clinicians/. Accessed 22 Apr 2020

The Adaptive Platform Trials Coalition (2019) Adaptive Platform Trials: Definition, Design, Conduct and Reporting Considerations. Nat Rev Drug Discov.18:797–807

WHO International Clinical Trials Registry Platform (2020) http://www.who.int/ictrp/en/. Accessed 17 Apr 2020

Zipkin M (2019) Speed SPYing: adaptive clinical trials hit the gas. Nat Biotechnol 37:969–977

Chapter 13
Regulatory Affairs and Marketing Approval

Abstract This chapter covers the main regulatory agencies and provides some detail about the applications required to gain permission to market a drug. The post-marketing surveillance phases are also covered, and the chapter is concluded with a brief overview of marketing to physicians.

13.1 Introduction

The clinical trials process now continues through the late stages of phase III trials and on to the marketing and post-marketing surveillance phases. This chapter will provide some background to the key regulatory agencies and will describe the formal application process needed to gain marketing authorization for a drug. Once a medicine has been approved for sale, the regulatory process does not end there; some high-profile product withdrawals on safety grounds have prompted the demand for increased monitoring of the drug's effects on the population at large. This surveillance by physicians and the producer companies is undertaken in phase IV and V trials, which are conducted once the drug is on the market. Pre- and post-marketing regulation comes under the term "regulatory affairs," which, given their central importance in getting a drug on the market, is a significant business function within a biopharmaceutical company. This chapter marks the final stage of the drug discovery pipeline and concludes with a brief description of the marketing process.

13.1.1 Regulatory Agencies

It should now be clear to the reader that the regulations covering drug discovery and development create huge demands on biopharmaceutical companies, in terms of both time and money. The agencies that issue these regulations for the large markets in the USA, Europe, Japan, and now China exert the most influence on global drug development. The responsibility for authorizing drug sales in individual countries lies with "competent authorities," i.e., national agencies with relevant legal powers.

E. D. Zanders, *The Science and Business of Drug Discovery*,
https://doi.org/10.1007/978-3-030-57814-5_13

13.1.1.1   The Food and Drug Administration

Since the USA is the world's largest market for prescription medicines, the views of the country's regulator, the FDA, are followed closely by the global biopharmaceutical industry. The modern form of this US government agency was born out of tragedy in 1937, when 107 people died of poisoning by *Elixir Sulfanilamide*. The following year, President Roosevelt signed the Food, Drug, and Cosmetic Act to create the FDA. The organization is headed by the Commissioner of Food and Drugs, who is responsible for offices and centers covering both aspects of consumer regulation. The two most important centers for pharmaceuticals are the Center for Drug Evaluation and Research (CDER) and the Center for Biologics Evaluation and Research (CBER), both of which deal with the IND and NDA applications. The agency has had to keep up with advances in science and medicine, as well as respond to political and economic pressures through the introduction of new laws and directives. A series of links to the historical background of US regulatory agencies and laws are available on FDA History 2020.

The FDA must deal with the competing demands of companies wishing to develop and sell products as quickly as possible and patients who expect high levels of drug safety. This requires greater public transparency on the part of the biopharmaceutical companies and acceptance of new technology on the part of the regulators. In response to the latter, the FDA launched the Critical Path Initiative in 2004 with the aim of using modern technology to improve various aspects of the drug development process. These include the conduct of clinical trials, drug manufacturing procedures, and electronic data management. As described in Chap. 12, open access to clinical trial information is provided by a joint FDA-National Institutes of Health (NIH) initiative (ClinicalTrials.gov).

The FDA's list of *Approved Drug Products with Therapeutic Equivalence Evaluations* is commonly known as the "Orange Book."

13.1.1.2   The European Medicines Agency

The creation of the European Union and its expansion has created a pharmaceuticals market to challenge that of the USA. Each member country has its own regulatory body, and, of course, each has its own language. The European Medicines Agency, based in Amsterdam, acts as a hub to coordinate a network of national agencies from all European Union (EU) countries and their political bodies. The EMA is headed by an Executive Director who oversees a group of six committees, made up of experts and representatives from all EU members. CTA and MAA applications are made to the Committee for Medicinal Products for Human Use (CHMP). Despite this centralization, the EMA does not authorize the marketing of every medicine in the EU but leaves this to individual member states. A full list of national competent authorities for each country is available on National Competent Authorities in Europe 2020. As with other agencies, the EMA is concerned with improving the efficiency of the drug development process, so it has published the EMA Regulatory Science Strategy to 2025 (2020) that mirrors the FDA's critical path initiative.

13.1.1.3   Japan's Ministry of Health, Labour and Welfare

Japan is a significant market for pharmaceuticals, partly because of its aging population and a strong biomedical science base. The Ministry of Health, Labour and Welfare (MHLW) is run by a minister in the Japanese government and consists of different bureaux, including the PMDA (Pharmaceuticals and Medical Devices Agency), established in 2004 for drug development authorization (PDMA 2020). The MHLW is also a contributor to the ICH Harmonisation Initiative (ICH.org, 2020), along with the FDA and EMA.

13.1.1.4   China's National Medical Products Administration

The NMPA (NMPA 2020) emerged in 2017 from a reorganization of Chinese regulatory agencies. Given the significant increase in drug development by Chinese companies with ambitions to penetrate global markets, the NMPA is an important national regulatory agency.

13.1.2   The New Drug Application

This major regulatory submission process begins during pivotal phase III clinical trials, where it is hoped that the experimental drug will prove to be safe and effective in patients. The FDA requires a New Drug Application (NDA) and the EMA, a Marketing Authorisation Application (MAA). The purpose of an NDA, as summarized here by the FDA, is to determine:

- Whether the drug is safe and effective in its proposed use(s) and whether the benefits of the drug outweigh the risks
- Whether the drug's proposed labeling (package insert) is appropriate and what it should contain
- Whether the methods used in manufacturing the drug and the controls used to maintain the drug's quality are adequate to preserve the drug's identity, strength, quality, and purity

The tangible product of this process is a package insert (in the USA) that accompanies the medicine to provide prescribing information. This is sometimes referred to as "the label," as in the phrase "off-label" use of medicines. The EMA equivalent is the Patient Information Leaflet (PIL), which is an abbreviated form of the Summary of Product Characteristics (SPC) document written for prescribers.

The main headings required for an FDA-approved package insert are as follows:

- Product Names, Other Required Information
- Boxed Warning[1]

---

[1] These are warnings about serious risks, the most serious being the black box warning, so-called because of the border around the printed text.

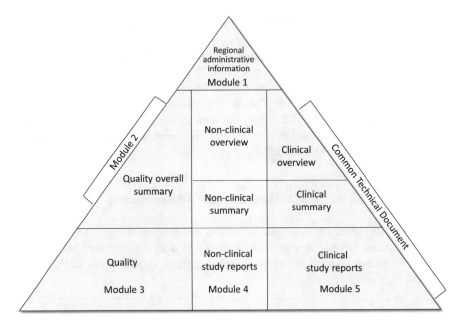

Fig. 13.1 Components of the Common Technical Document (CTD)

- Recent Major Changes
- Indications and Usage
- Dosage and Administration
- Dosage Forms and Strengths
- Contraindications
- Warnings and Precautions
- Adverse Reactions
- Drug Interactions
- Use in Specific Populations

13.1.2.1   The Common Technical Document

The NDA, and its European equivalent, is based on the Common Technical Document (CTD). This consists of separate modules containing the information about drug manufacture, efficacy, and safety that is required for marketing authorization (ICH M4 2020). Each module is so extensive that the whole application may run to over 100,000 pages. To increase efficiency, as well as to save a few forests and acres of storage space, the regulators are introducing a submission process based on electronic CTDs (eCTDs). Although the exact contents of the CTD differ according to the requirements of a given regulator, the ICH M4 documents specify a common set of guidelines, which are summarized in Fig. 13.1.

Each module is broken down as follows:

Module 1: Administrative Information and Prescribing Information
Table of contents of whole submission
    Specific information for each country where drug is to be registered (which means that module 1 is not strictly part of the Common Technical Document)

Module 2: CTD summaries
Table of Contents
    Introduction
    Quality overall summary
    Nonclinical overview
    Clinical overview
    Nonclinical written and tabulated summaries
    Pharmacology, pharmacokinetics, toxicology
    Clinical summary
    Biopharmaceutic studies and associated analytical methods, clinical pharmacology studies, clinical efficacy, clinical safety
    Literature references
    Synopses of individual studies

Module 3: Quality
Table of contents
    Body of data
    Literature references

Module 4: Nonclinical Study Reports
Table of contents
    Study reports
    Literature references

Module 5: Clinical Study Reports
Table of contents
    Tabular listing of all clinical studies
    Clinical study reports
    Literature references

Pharmacovigilance

An important part of the marketing application process is the submission of a pharmacovigilance plan for monitoring the safety of a drug after its introduction to the general population. The ICH E2E guidelines cover the types of material that must be submitted to the regulators, either as part of the CTD or as separate documents (ICH E2E Pharmacovigilance planning 2020). These contain a safety specification and a pharmacovigilance plan for monitoring the drug after launch.

The Safety Specification

This contains some, or all, of the following, depending on the drug in question:

Nonclinical (Animal Models)
Safety pharmacology, toxicology, drug interactions
Clinical
Limitations of the human safety database. Numbers treated so far versus populations likely to receive drug, any new or different safety issues identified. Populations not studied in the pre-approval phase, e.g., children, elderly, different ethnic groups. Adverse events (AEs) /adverse drug reactions (ADRs). Identified and potential risks that require further evaluation. Identified and potential food-drug and drug-drug interactions
Epidemiology
Incidence and prevalence of the disease to be treated in different regions
Pharmacological Class Effects
Particular drug classes may have similar risks (e.g., causing heart arrhythmias)
Summary

Pharmacovigilance Plan

This is structured as follows:

Summary of Ongoing Safety Issues
Routine pharmacovigilance practices. Expedited adverse drug reaction and periodic safety update reports (PSURs). Action plan for safety issues. Action(s) proposed and rationale, monitoring by sponsor and proposed action, milestones for evaluation and reporting
Summary of Actions and Milestones
Pharmacovigilance methods. Design and conduct of observational studies

Pharmacovigilance is undertaken by passive or active surveillance of the patient population taking the drug. Spontaneous reporting of an adverse event to the regulators by patients and doctors is a common example of passive reporting; although clearly useful, it is random and unstructured. Active reporting overcomes this problem by collecting adverse event data from patients enrolled in risk management programs, or who contribute to surveys about the drug. This brief description belies the considerable amount of detailed epidemiology that is undertaken by pharmacovigilance experts in support of the marketing of a new medicine.

The regulatory authorities have specific mechanisms for reporting adverse events, for example, the FDA's Adverse Event Reporting System (FAERS 2020) and the EudraVigilance network for the EMA (EudraVigilance 2020).

The Drug Master File

It is quite normal for the biopharmaceutical company submitting a marketing application to be concerned about revealing proprietary information to the external committees appointed by the regulators. This is prevented by submission of a Drug Master File (DMF) to the regulators; this document is based on the Common Technical Document and contains restricted and nonrestricted information in separate sections. The EMA equivalent is the Active Substance Master File (ASMF).

Variations on the NDA

The marketing application for patented medicines is covered by the NDA, or MAA, but there are variations which cover generic medicines and biologicals, such as vaccines. The Abbreviated New Drug Application (ANDA) is used for generic copies of medicines that are already on the market and therefore have been through a full review process (generic medicines will be covered in Chap. 16). The only requirement on the manufacturer is to be able to demonstrate the bioequivalence of the generic product. In practice, this means showing that the generic product has the same absorption properties as the patented medicine (ANDA 2020). Bioequivalence is relatively straightforward to demonstrate for small molecules, but not at all for biologicals, as mentioned in Chap. 8. The approvals of blood products, vaccines, cell therapy, and gene therapy are made by the FDA's Center for Biologics Evaluation and Research (CBER), with the marketing application being called a Biologics License Application (BLA). This requires essentially the same types of information and clinical trial structure as the NDA. The design and interpretation of vaccine trials is, of course, quite specialized, so the expert review panel will include immunologists and microbiologists.

The Approval Process

The documentation for an NDA is sent to the Centre for Drug Evaluation and Research (CDER) division of the FDA. The MAA is submitted to Committee for Medicinal Products for Human Use (CHMP) of the EMA. The 1992 Prescription Drug User Fee Act (PDUFA) in the USA established either a standard or a priority review by the FDA, the former taking 10 months and the latter 6 months. The submission is reviewed by internal and external experts who require meetings with the sponsors to discuss and clarify any technical points that may arise. When the FDA requires clarification from the sponsor, it sends out an "approvable letter" requesting further information. Once all issues have been clarified to the satisfaction of the regulator, an "approval letter" is sent out to the sponsor to signal that the NDA has been accepted. In the case of EMA approval, the CHMP issues an "assessment report," possibly along with a "request for supplementary information," which will lead to approval, or otherwise. The huge amount of information that must be reviewed inevitably means that the approval process is time-consuming and may take over 1 year. However, the regulators may prioritize the application if the drug is likely to make a significant difference to serious diseases or have a better safety profile than existing drugs. As mentioned above, a priority review shortens the review process to 6 months which is obviously attractive to drug developers. The FDA can issue three types of priority review vouchers (PRVs) to encourage drug development for tropical and pediatric diseases and medical countermeasures (FDA's Priority Review Voucher Programs 2020). The vouchers have intrinsic value for companies and can be sold on to those who need one for their own programs.[2] The PRVs are informally known as "golden tickets" from the story "Willy Wonka and the Chocolate Factory."

Variants of the review process (from FDA Development & Approval Process 2020) are as follows:

Accelerated Approval

Even though phase III clinical trials can take several years, this may still not be enough time to determine whether a drug confers real benefit to patients, through increasing life expectancy, or quality of life. However, in the case of drugs for some cancers, where the main clinical outcome may be patient survival times, the FDA can follow a process of accelerated approval; this system accepts the use of drug-induced changes in biomarkers or surrogates of the disease to provide a more rapid measure of clinical outcome. The markers for cancer, for example, could be tumor shrinkage, or biochemical measures of drug activity, such as target enzyme inhibition. Having passed through accelerated approval, a drug can be marketed, but on the condition that it is reviewed in phase IV studies (below). These studies are

---

[2] For example, AstraZeneca paid Sobi in Sweden $95M for a PRV.

performed with the general patient population and will determine whether the drug makes a long-term difference to the disease; if not, it could lead to the product being withdrawn from the market.

The following examples of FDA-approved cancer drugs illustrate both the advantages and pitfalls of fast-track review. Imatinib (Gleevec®; see Chap. 3) was developed by Novartis as a small molecule drug for a leukemia and fast track reviewed in just 4 months. This meant that the lifesaving medicine was rapidly introduced to patients and provided real clinical benefit, despite the problems with drug resistance that emerged later. The other example is gefitinib, marketed as Iressa® by AstraZeneca, which was granted accelerated approval for treating a type of lung cancer after it completed phase II trials. The small molecule drug was successful in shrinking tumors and given marketing approval in 2003 on condition of running a confirmatory trial. Unfortunately, this trial failed to show survival benefit, so Iressa® was withdrawn from the general patient population, thus casting a shadow over the accelerated approval process. In the meantime, the situation with this drug has improved after the discovery that it only works on patients with a mutation in the gene encoding the Iressa® target protein. Iressa® was approved in 2010 as a first-line treatment for lung cancer patients with the target mutation, so all was not lost with this drug. This example demonstrates the importance of pharmacogenetics and how it can be used to improve the efficiency of clinical trials by selecting only those patients who are likely to respond to the drug (covered further in Chap. 14).

Fast-Track Review

If a drug candidate is likely to fulfill an unmet medical need and treat a serious illness like cancer, the FDA can fast track the review by committing more resources and increasing the number of meetings with the sponsors. It may also be possible to use a rolling review, in which time is saved by reviewing the application section by section as they are submitted, rather than waiting for the whole document. Fast-track review is requested by the sponsor at any time during the drug development process, and a decision as to whether the drug does fulfill an unmet medical need is made by the FDA after 60 days.

Breakthrough Therapy

A breakthrough therapy designation expedites drug development for serious conditions where preliminary clinical evidence indicates that the drug may demonstrate substantial improvement over available therapy. The benefits to the biopharmaceutical company include fast-track designation and assistance with various development issues.

13.1.3   Phase IV: Post-marketing Studies

13.1.3.1   Post-marketing Surveillance

Once a medicine has received marketing authorization and has been approved for reimbursement by healthcare providers, it will start to generate revenues. Although a great deal of money and time will have been spent over the period from the first clinical trials up to marketing authorization, the responsibility of the manufacturer for monitoring safety does not end there. This is because of the risk of serious adverse reactions, which may be rare or undetectable in small numbers of patients, but show up when significant numbers of people are treated with the drug over an extended period. The biopharmaceutical company may be obliged to invest in clinical studies that extend the number of patients to confirm safety and efficacy. These so-called phase IV studies are conducted under GCP conditions in hospitals, or other healthcare centers, and are regulated by the FDA and similar agencies. For example, the FDA issues post-marketing requirements (PMRs) and post-marketing commitments (PMCs), the former being obligatory and the latter optional (FDA Postmarketing Requirements and Commitments 2020). For the EMA, in the case of safety, this is called a post-authorization safety study, or PASS (Post-authorisation safety studies 2020). Sometimes the term phase V is used to describe post-marketing studies undertaken by the originator company, for example, to test the drug in specific populations such as pregnant women, or ethnic groups. Serious adverse drug reactions (SADRs) are reported to the FDA's MedWatch program, or the EU's EudraVigilance (highlighted in Chap. 12), and have led to drugs being withdrawn from the market or given "black box" warnings.

Meta-analysis

Sometimes, adverse drug reactions only come to light after data from several trials are pooled together and carefully examined using statistics. Studies of this type use a technique called "meta-analysis," which can be useful for increasing the power of the analysis by studying data from more patients. For example, the antidiabetic drug rosiglitazone (Avandia®) was pulled from the market by its manufacturers GSK after a meta-analysis of clinical trials by clinicians showed a significant risk of heart attacks due to this medicine (Singh et al, 2007). This was an example of a successful meta-analysis, but they can, however, be subject to bias and inaccuracy, particularly if studies are selectively excluded from the analysis.

Risk Evaluation and Mitigation Strategy

Considering increased concerns about drug safety, the regulators are encouraging drug companies to devise a strategy for monitoring and reporting adverse events that may arise once the medicine is on the market. The FDA's Risk Evaluation and

Mitigation Strategy (REMS), or EMA's EU Risk Management Plan (RMP), is put in place during phase III trials. An example of REMS documentation for the antipsychotic drug clozapine is given in REMS 2020.

13.1.4 Marketing the Drug

Although the commercial aspects of drug development will be covered in more detail later, this chapter is concluded with some comments about marketing a drug to clinicians.

Patients are obviously the end users of prescription medicines, but these products are only available from the medical profession that helped to develop them in the first place. Clinicians make decisions about which drugs are appropriate for certain groups of patients using their personal experience in the clinic and the opinions of their peers. The biopharmaceutical industry therefore devotes considerable resources to informing doctors about its products; this may involve direct visits from sales personnel, published material in journals and magazines, or sponsorship of conferences. Panels of experts, consisting of specialists with senior appointments in hospitals or universities, advise companies about the need for a new medicine and give their professional judgment as to how it should be used in clinical practice. The relationship between biopharmaceutical industry and clinicians has not been without controversy, particularly in terms of financial inducements to promote a product. Whatever the details of cases that emerge in the media from time to time, it must be sensible to form close ties between the medical profession and those who supply the tools to treat their patients. As an industry scientist, who has interacted with many research clinicians over the years, I (nearly) always felt that the relationship was based on a shared interest in the science behind our activities, with the understanding of who held the purse strings kept well in the background.

Scientists and doctors communicate their results by means of publications and oral presentations at conferences and seminars. The publications include peer-reviewed papers in journals, magazine articles, or online blogs, wikis, and webinars. The articles used for the clinical trial examples in the last chapter were all submitted to journals where the editor used external experts to "peer review" the article before agreeing to publication. This is accepted as being the most rigorous method of quality control even though authors, myself included, complain if the referees reject the paper, or make unreasonable demands for more data. The biopharmaceutical industry often employs medical writers to produce scientific communications of all types, including the thousands of pages of regulatory material required during drug development. These medical writers may be directly employed by the company or be part of a specialist agency where the writing is outsourced. Apart from regulatory documentation, this writing covers everything from marketing advertisements to peer-reviewed articles. There is a danger, however; some journal articles have been authored by "ghost writers," and this lack of transparency can lead to a loss of confidence by the scientific and medical profession. Journals are tackling these issues head on by requiring full author transparency and financial disclosures of conflict of interest.

Summary of Key Points
Major regulators are the FDA, EMA, Japan's MHLW, and China's NMPA.
The New Drug Application (NDA) or European Marketing Authorisation Application (MAA) are drawn up and submitted to the regulators during phase III trials.
Depending on the severity of the disease and the novelty of the drug, the application may receive fast-track review and accelerated approval.
Pharmacovigilance is an important part of pharmaceutical development both during and after clinical trials.
A Risk Evaluation and Mitigation Strategy (REMS) plan may be required by the FDA before marketing approval is granted.
Phase IV post-marketing surveillance studies may be conducted once the medicine is on the market to identify adverse reactions in larger patient groups over a longer period than phase III. The term phase V trial is sometimes used when a drug is trialed in different patient groups.
Medical writers are employed to publicize the new medicine using scientific and educational media.

References

ANDA (2020). https://www.fda.gov/drugs/types-applications/abbreviated-new-drug-application-anda. Accessed 27 Apr 2020

EMA Regulatory Science Strategy to 2025 (2020) https://www.ema.europa.eu/en/news/advancing-regulatory-science-eu-new-strategy-adopted. Accessed 27 Apr 2020

FAERS (2020) https://www.fda.gov/drugs/surveillance/questions-and-answers-fdas-adverse-event-reporting-system-faers. Accessed 27 Apr 2020

EudraVigilance (2020) https://www.ema.europa.eu/en/human-regulatory/research-development/pharmacovigilance/eudravigilance. Accessed 29 Apr 2020

FDA Development & Approval Process (2020). https://www.fda.gov/drugs/development-approval-process-drugs. Accessed 27 Apr 2020

FDA History (2020) https://www.fda.gov/about-fda/history-fdas-fight-consumer-protection-and-public-health. Accessed 27 Apr 2020

FDA Postmarketing Requirements and Commitments (2020) https://www.fda.gov/drugs/guidance-compliance-regulatory-information/postmarket-requirements-and-commitments. Accessed 27 Apr 2020

FDA's Priority Review Voucher Programs (2020). https://www.gao.gov/products/GAO-20-251. Accessed 27 Apr 2020

ICH.org (2020) http://www.ich.org. Accessed 27 Apr 2020

ICH E2E Pharmacovigilance planning (2020) https://www.ich.org/page/efficacy-guidelines. Accessed 27 Apr 2020

ICH M4 (2020) https://www.ich.org/page/ctd. Accessed 27 Apr 2020

National Competent Authorities in Europe (2020) http://www.ema.europa.eu/ema/index.jsp?curl=pages/medicines/general/general_content_000155.jsp&murl=menus/partners_and_networks/partners_and_networks.jsp&mid=WC0b01ac0580036d63. Accessed 27 Apr 2020

NMPA (2020) http://english.nmpa.gov.cn/index.html. Accessed 27 Apr 2020

PDMA (2020) https://www.pmda.go.jp/english/about-pmda/outline/0005.html. Accessed 27 Apr 2020

Post-authorisation safety studies (2020). https://www.ema.europa.eu/en/human-regulatory/post-authorisation/pharmacovigilance/post-authorisation-safety-studies-pass-0. Accessed 27 Apr 2020

REMS (2020) https://www.accessdata.fda.gov/scripts/cder/rems/index.cfm?event=RemsDetails.page&REMS=351. Accessed 28 Apr 2020

Singh S et al (2007) Long-term risk of cardiovascular events with rosiglitazone. A meta-analysis. JAMA 298:1189–1195

Chapter 14
Diagnostics and Personalized Medicine

Abstract The drug discovery landscape is changing with the advent of personalized medicine and the companion diagnostics that are used to select patients who will respond to a marketed drug. This chapter provides an overview of the techniques used in the diagnostic laboratory and moves on to discuss personalized medicine in the form of pharmacogenetics and SNP analysis. Mention is also made of the surrogate markers and biomarkers used in clinical trials.

14.1 Introduction

In a word, I consider hospitals only as the entrance to scientific medicine; they are the first field of observation which a physician enters; but the true sanctuary of medical science is in a laboratory.

Claude Bernard (Introduction to the study of experimental medicine 2017 edition).

The previous chapters in this section have covered the drug discovery pipeline from target to marketed medicine, with occasional reference to diagnostics and the differing reactions of patients to the same drug. Diagnostics and personalized medicine are becoming more important to the biopharmaceutical industry for several reasons. Firstly, serious adverse drug reactions, so-called iatrogenic diseases, account for thousands of hospital admissions and deaths annually. A study of data for 2008 from the UK's National Health Service showed that 75,076 out of 6,830,067 emergency admissions in England were drug related. Systemic agents were most implicated followed by analgesics and cardiovascular drugs. Drug-related mortality over the period 1999–2008 was 4.7% of admissions (Wu et al. 2010). Secondly, many drugs only work against a proportion of patients, so it is therefore wasteful and unethical to prescribe such drugs to those who will not respond to them. Thirdly, selecting patients for clinical trials would be made more ethical, scientifically rigorous, and cost-effective if patients were pre-screened to provide a genetically uniform pool of subjects, rather than a random selection who suffer from the same disease. Finally, it is often difficult to make an objective measurement of whether a treatment has really worked in clinical trials; this is particularly true with neurodegenerative conditions, such as Alzheimer's disease, where

E. D. Zanders, *The Science and Business of Drug Discovery*,
https://doi.org/10.1007/978-3-030-57814-5_14

restoration of brain structure is difficult to measure objectively. These various topics, which come under the broad headings of diagnostics and personalized medicine, will be described in some detail in this chapter.   ·

14.1.1   Background to Diagnostics (D$_x$)

Diagnosis of disease is as old as medicine itself; it may be performed during a consultation with a physician, or in specialized hospital laboratories. Strictly speaking, diagnostics development and drug development are quite separate activities, the former being concerned with the identification, rather than the treatment of disease. The diagnostics industry is very much the poorer cousin of the biopharmaceutical industry since the unit cost of a diagnostic kit is considerably less (maybe 100-fold) than that of a patented medicine. The diagnostics market is mostly driven by demand for reagents and instrumentation by hospital laboratories and retailers of home diagnostic kits. Because of this, the products must be robust, cheap, and straightforward to use, which contrasts with the expensive products of the biopharmaceutical industry. This is not, however, to underestimate the importance of diagnostics in supporting public health, or the sophistication of the technology that is used to create the required tests and instruments. These products are also used for the analysis of preclinical and clinical samples during drug development, where tests are conducted in centralized service laboratories.

Routine screening is used in the general population for blood pressure, levels of cholesterol, glucose, and other substances, as well as signs of breast cancer, colon cancer, and aortic aneurisms. While important and potentially lifesaving, these are quite crude techniques; cancer diagnosis, for example, would ideally detect tumor growth before it becomes clinically unmanageable, so sophisticated diagnostics are required that can be applied to large populations. Another important area of diagnostics concerns infectious diseases, where the specific infectious agent must be identified before the appropriate treatment is prescribed. Finally, some diagnostic procedures do not require clinical input, for example, the pregnancy testing kits used at home.

14.1.1.1   Point-of-Care Diagnostics

It is in the obvious interest of the patient that a diagnosis is made rapidly, particularly if it is related to a life-threatening condition. However, even the most efficient hospital laboratory cannot provide the speed or convenience of a diagnosis in the primary care practice, or even the home. Therefore, point-of-care (POC) diagnostics, which can be performed away from the hospital, is such a fast-growing area and one in which there is great scope for innovation.[1] From the drug discovery perspective, POC diagnostics are going to be important for personalized medicine; in this scenario, the primary care physician may have diagnosed a particular disease in a patient and then will perform a further test to see which medicine would be most appropriate for that patient. This is sometimes referred to as "prescribing the right drug to the right patient"; this topic will be discussed later under "personalized medicine."

14.1.1.2 Biomarkers

The term biomarker is being heard more frequently in the biopharmaceutical industry as new molecular and imaging tools are being used to identify changes in the human body that associate with specific diseases. These changes may take the form of increased (or reduced) levels in the blood, or other body compartments, of cells, proteins, nucleic acids, lipids, or small molecules. The ideal biomarker is easily measurable, with minimal inconvenience to the patient and be highly accurate in defining disease status. Biomarkers, such as glucose or cholesterol levels, are routinely measured during drug development programs for cardiovascular diseases and diabetes. The search for biomarkers of complex diseases, such as cancer, arthritis, and Alzheimer's disease, is almost as challenging as the search for treatments. Cancer biomarkers are particularly difficult, as there are large differences in the genetic makeup of tumors taken from different tissues and patients. Because cancer cells rearrange and delete many genes as part of the process of oncogenesis (tumor formation), it is difficult to identify a set of expressed genes or proteins that reliably associate with a tumor. This has not deterred investigators from trying, however; as a result, molecular technologies, such as gene and protein expression microarrays and protein mass spectrometry, have been used to identify interesting biomarkers that may turn out to be clinically useful. These techniques can also identify prognostic markers which are particularly relevant to life-threatening diseases such as cancer; this is because patterns of gene or protein expression may be associated with high or low chances of survival. A list of tumor biomarkers from the US National Cancer Institute is available at cancer.gov 2020. The markers include the HER2/neu gene and estrogen receptor (ER)/progesterone receptor (PR), both used in assessing breast cancer patients for targeted therapies. These are used as biomarkers for the I-SPY2 adaptive clinical trial discussed in Chap. 12. Other examples include programmed death ligand 1 (PD-L1) for patients likely to respond to checkpoint inhibitor therapy (Chap. 2) and UGT1A1*28 variant homozygosity for toxicity to irinotecan in colorectal cancer therapy (see pharmacogenetics, later).

---

[1] Point-of-care diagnostics are also used in hospital emergency rooms, where rapid diagnosis of conditions is clearly important.

14.1.1.3 Companion Diagnostics and Theranostics

According to the FDA, "A companion diagnostic device can be *in vitro* diagnostic device or an imaging tool that provides information that is essential for the safe and effective use of a corresponding therapeutic product" (FDA Companion diagnostics 2020). As an example, the diagnostic kit for the HER2 breast cancer biomarker (previous section) is marketed by Roche alongside Herceptin®, the therapeutic antibody used if the tumor cells express this protein. Some members of the tyrosine kinase family of enzymes are mutated in different tumors and inhibited by small molecule drugs. Companion diagnostics based on nucleic acid detection are used to select those patients who have mutations that match the available drug. A list of companion diagnostics to a range of different drug types is available on FDA companion diagnostics 2020. The technology used in these and other diagnostic procedures are described in the sections below.

Theranostics

This (rather inelegant) term was probably first used in the 1990s and can be defined as the ability to affect therapy or treatment of a disease state. The companion diagnostics examples given above fall into this category, but the term is increasingly being used for imaging and therapy with the same, or similar, radiopharmaceuticals, for example, with ^{131}I and thyroid diseases (Langbein et al. 2019).

14.1.2 Technology

Although it is impossible to describe even a fraction of the diagnostic tests that are available, or under development, it is possible to make some generalizations, since many share common design principles. In broad terms, the diagnostic test (assay) may be based on a physical measurement, such as blood pressure, volume of air breathed out, an image in a CAT scan, or the amount of glucose in the blood. Figure 14.1 shows a schematic overview of some of the most important areas.

14.1.2.1 Imaging

Where diseases are manifested by changes in the living body, such as the appearance of tumors, the obvious way of detecting this is through direct observation using X-rays or other techniques. Although not always easy to quantify, imaging has the potential to rapidly determine the effect of a drug on, for example, tumor size in cancer, or tissue destruction in arthritis. While X-rays are used routinely for many diagnoses, other imaging techniques exist that can visualize soft tissues and even

Fig. 14.1 Main categories of diagnostic tests. Various forms of imaging are used, particularly for examining the brains of living subjects. The other techniques are used on blood, urine, cerebrospinal fluid, or tissue samples

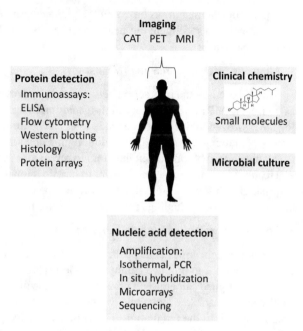

Imaging
CAT   PET   MRI

Protein detection
Immunoassays:
ELISA
Flow cytometry
Western blotting
Histology
Protein arrays

Clinical chemistry

Small molecules

Microbial culture

Nucleic acid detection
Amplification:
Isothermal, PCR
In situ hybridization
Microarrays
Sequencing

the exact area where a drug binds to its target. These techniques are briefly summarized as follows:

- Computerized axial tomography (CAT or CT scans)

Multiple X-ray images are taken of the affected part of the body and the images reconstructed to form a highly detailed view of the bones, blood vessels, and organs. Image contrast agents are often used to provide a clearer distinction between different structures, with much research being undertaken to develop novel agents which can highlight specific tissues and even cells.

- Optical imaging

Fluorescent molecules attached to proteins are visualized by shining light (normally infrared) on the surface of the body. The technique is limited by the poor penetration of light below the body surface, but it is versatile and can be used with genetically engineered animals that express fluorescent proteins (see animal models, Chap. 6).

- Ultrasound and photoacoustic imaging

Ultrasound imaging (sonography) exploits the echo created when sound is reflected off different structures. Specific contrast agents containing microscopic gas bubbles are injected into the circulation to create a high contrast between the blood vessels and surrounding soft tissue. Microbubbles can also be attached to probes which target specific areas of the body.

Photoacoustic imaging is based on the photoacoustic effect, in which light from a laser shines on a specimen; this heats up and emits sound waves which are detected with special sensors.

- Magnetic resonance imaging (MRI scans)

This is based on the interaction between radio waves and certain elements, such as hydrogen. MRI produces very high-resolution images of body tissues by detecting the hydrogen in water molecules. Images can be further enhanced using contrast agents containing elements like gadolinium. Functional MRI (fMRI) is used to detect changes in blood flow in the brain to reveal areas of neuronal activity; it may, as a result, identify the anatomical sites of certain human actions and behaviors. Not surprisingly, this is an area of intense research interest, but is not without some controversy, as some findings relating to behavior may have been overinterpreted.

- Positron emission tomography (PET scans)

Certain radioisotopes emit positively charged electrons (positrons) which interact with electrons to produce gamma radiation. This radiation can be detected using specially designed cameras and processed to create three-dimensional body images. PET scans use tracker molecules which are firstly tagged with radioisotopes and then injected into the circulation. The choice of molecules is almost limitless, but the one most used for medical imaging is ^{18}fluorodeoxyglucose, which is taken up by actively metabolizing tissues. PET scans are also used to image the binding of labeled drugs and probes to precise areas in the body. Although PET scanning has the advantage of high sensitivity and resolution, it comes at the price of producing the radioisotopes used to incorporate into different molecules. The half-life of the commonly used ^{18}F isotope is 2 h, while that of the ^{11}C, ^{15}O, and ^{13}N isotopes is less than 30 min; these very short half-lives mean that the isotopes must be generated in a cyclotron situated very near to the hospital or research facility used for the study.

14.1.2.2   Diagnostic Tests for Small Molecules

The blood levels of small molecules, like glucose, cholesterol, or creatinine, give an indication of the health of the patient in routine clinical practice, or during clinical trials. These molecules are usually detected in a system that generates a colored product, whose intensity is measured by light adsorption (see spectrophotometry, Chap. 9). The system that generates the product is usually an enzyme, which converts the test molecule (analyte) into a product that reacts chemically with another compound to form a colored derivative. The various components of the assay may be bound onto a paper and plastic test strip; this is then dipped into blood or urine to produce a color change if the analyte is present. The strips can then be "read" in small handheld devices that provide a numerical display of the analyte concentration.

14.1.2.3 Antibody-Based Assays (Immunoassays)

Antibodies are exquisitely specific for their targets (antigens) and are ideally suited for diagnostic applications. An antibody created against a hormone in blood, for example, can be used to discriminate between that molecule and the hundreds of other proteins that may be present in the blood sample. Tests that employ antibody-based detection of analytes (which are usually proteins) are called immunoassays. The most important requirements for the assay are sensitivity to low levels of analyte and specificity for it; the assay must also operate over a wide range of analyte concentrations; in other words, it must have a large dynamic range. One example of how this is applied in practice comes from the clinical trial of the therapeutic antibody TGN1412 in phase I volunteers; this trial resulted in a near fatal cytokine storm (described in Chap. 11). The levels of different cytokines were measured in blood samples, before and during the treatment (Suntharalingam et al. 2006), and illustrate the low levels that can be detected in healthy subjects (2.8 pg/ml); this rose to 1760 pg/ml after 1 h and 4675.9 pg/ml after 4 h[2]

Immunoassays are produced in different formats, depending on the specific requirements of the analysis; the main assays are described below.

Enzyme-Linked Immunosorbent Assay

This immunoassay format is more straightforward than the name would imply and is used to detect and measure different hormones, cytokines, and other proteins in solution. It has become the method of choice for screening for diagnostic proteins in blood, urine, or other fluids, replacing the older radioimmunoassay that employed antibodies labeled with radioactive iodine (^{125}I). A typical sandwich ELISA system is illustrated in Fig. 14.2 (the sandwich refers to the two antibodies that form a sandwich with the antigen (analyte) as the filling).

This basic system is often configured to produce greater sensitivity, or improved discrimination between the analyte and interfering substances in blood. Extremely sensitive detection systems can be produced that use antibodies against antibodies, or else by exploiting the high-affinity interaction that exists between the small molecule biotin and the protein streptavidin. These will not be elaborated upon any further, but details can be found in the many external resources that cover immunoassays, e.g., Cell Signaling Technology 2020.

---

[2]This 4-hour level of around 4.6 nanograms TNF-α per milliliter of blood illustrates just how potent and destructive these cytokines can be.

Fig. 14.2 A sandwich ELISA format. (**a**) An antibody protein (Y shape) binds a plastic well to capture (pull out) the analyte protein (diamond shape) out of a complex mixture of antigens as would be the case in blood or other tissues. (**b**) The analyte protein binds to the immobilized antibody, and all other proteins are washed away. (**c**) A second antibody is bound to the analyte; this one is conjugated (tagged) with a label that generates a signal which can be measured in a spectrophotometer. In an ELISA, the label is an enzyme (hence enzyme-linked) that converts a colorless substrate to one with a color, whose intensity is proportional to the amount of enzyme present

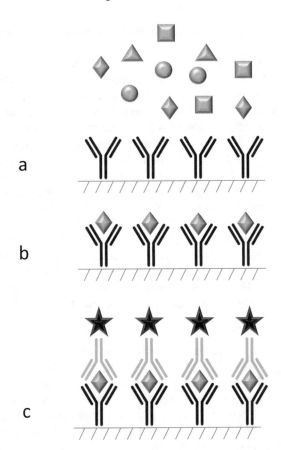

Flow Cytometry

The ELISA system described above is configured to detect hormones and other signaling proteins that are present in solution (e.g., in the blood). However, many important proteins are expressed on the surface of living cells and can act as markers for diseases, like cancer. The CD4 molecule is an example of such a marker protein, in this case for the immunodeficiency that results from infection with the HIV virus. CD4 is expressed on the surface of about 40% of the T lymphocytes circulating in the blood. If an individual is infected with the HIV virus, the levels of CD4 cells are severely reduced, thus compromising the ability of the immune system to fight infection, which eventually results in AIDS. The diagnosis of patients with AIDS, as well as monitoring the effects of anti-HIV drugs, requires a suitable immunoassay that can rapidly count the number of CD4 cells in a sample of blood. This is accomplished using a technique called flow cytometry, in which cells are labeled with an antibody specific for the molecule of interest (CD4 in this example).[3] The cells are then passed through a machine called a flow cytometer. The process of flow cytometry conducted in this machine is illustrated in Fig. 14.3. In this example, white blood cells (leukocytes) are isolated from blood and stained with an antibody to CD4, which is tagged with a fluorescent molecule that glows green when illuminated by a laser. A second antibody, labeled with a fluorescent molecule that glows red under the laser, is used to stain another population of blood cells, called CD8; the process of staining cells using fluorescently labeled antibodies is called immunofluorescence. The cells are forced through a tube into a single stream before passing through the laser beam. The different wavelengths (colors) of the emitted fluorescence are detected by passing through dichroic filters and amplified in photomultipliers. Each color signal is fed into a computer for analysis. Individual fluorescent cells are displayed as a dot on a histogram whose position reflects the color of the label attached via the antibody. It is then straightforward to identify the area of the display where CD4-positive cells have been identified and the areas for CD8 and those blood cells that bear neither molecule. The machine counts the number of cells in each area, so the proportion of blood cells that bear CD4 can be readily determined. Two different fluorescent dyes were used for the above example (two-color immunofluorescence). Now, the introduction of a large range of fluorophores that can be discriminated with lasers and filters in the flow cytometer enables the simultaneous detection of over 20 different antibodies (for more details on flow cytometry and cell sorting, see ThermoFisher 2020).

Mass Cytometry

Despite the power of flow cytometry, there are limits to the number of fluorescent probes that can be used, so an alternative based on mass spectrometry has been introduced. A combination of flow cytometry and mass spectrometry now allows the detection of over 40 different antibodies (Spitzer and Nolan 2016).

Antibodies are coupled to unique stable, heavy-metal isotopes that can be quantified in a time-of-flight mass spectrometer after ionization in an argon plasma. Ion counts are transformed into a data matrix with each column representing a distinct isotope measured and each row a single mass scan of the detector. The matrix can then be used to reconstruct the labeling pattern on antibodies bound to the different cells in the sample.

---

[3] The physical size and shape of cell populations can also be discriminated by measuring the scattering of the laser light.

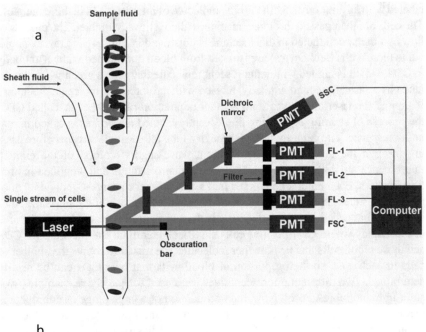

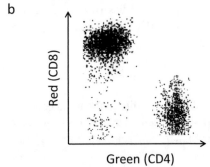

Fig. 14.3 Diagram of flow cytometry used to count the number of CD4 + ve and CD8 + ve T cells in blood. (**a**) White blood cells are isolated from blood and mixed with antibodies to CD4 and CD8 proteins on the cell surface. Each antibody has a molecule attached that emits a signal (red for CD8 and green for CD4). The cells flow down a tube where they are illuminated by a laser to generate a fluorescent signal which is amplified in a photomultiplier before passing through a filter specific for the color and registration in a computer. PMT, photomultiplier; FL-1,2,3, filters; SSC, side scatter; FSC, forward scatter. (**b**) The computer plots each cell as a dot on a two-dimensional display where each axis is the intensity of fluorescence. In this way, cells labeled with green (CD4 + ve) are clearly separated from those labeled with red (CD8 + ve) with a small number being unlabeled by either. Figure (a) courtesy of Bilal Hussain, Creative Commons CC0 1.0

Immunohistochemistry (IHC)

Immunohistochemistry (or immunohistology) is a key part of the drug development process; it is used to examine tissues from animals and human subjects for adverse effects during drug development, as well as for diagnostics. The technique is based on histochemistry (introduced earlier in Chap. 6), where thin sections of tissue are attached to glass microscope slides and stained with dyes to reveal different cell types. Immunohistochemistry is based on similar principles, except that antibodies, tagged with a fluorescent or enzyme label, are used instead of a colored dye. Because antibodies bind to different cells with great selectivity, individual cell types can be readily identified in tissue sections. As with flow cytometry, antibodies used for immunohistochemistry can be tagged with fluorescent molecules (fluorochromes), but they can also be tagged with enzymes that create an insoluble colored product at the site of antibody binding. This is illustrated in Fig. 14.4, where an enzyme-labeled antibody has been applied to a section of breast tissue. The sites of antibody binding are revealed as dark areas in a lighter background.

Modern developments, like the introduction of laser scanning confocal micros-copy, have enormously improved the quality and resolution of cell and tissue images, and, as well as being scientifically useful, they can be real works of art. Diagnostic specimens are normally checked by pathologists who assign a score that depends on the intensity of staining of particular cell types. This is a subjective

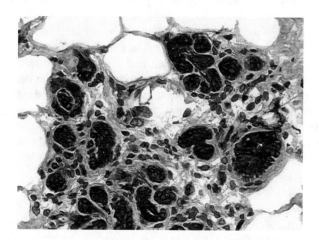

Fig. 14.4 Immunohistochemistry used to visualize cells in breast tissue. A thin section of tissue on a microscope slide is mixed with antibody that can discriminate between fat cells clear struc-tures and cells surrounding ducts. The dark color is produced at the site of antibody binding by a colored product produced by the enzyme attached to the antibody

process, so attempts have been and are being made to automate the process by extracting information from images by using machine learning algorithms (e.g., Komura and Ishikawa 2018). However, while there has been undoubted progress in this area, the fully automated IHC laboratory is not yet in routine use.

Western Blotting

There are some situations where the natural antibodies produced by human subjects can be used for diagnosis. This is particularly relevant for infectious diseases, where the patient will have produced antibodies as a response to the viruses, bacteria, or parasites that may have infected them. The protein targets of these antibodies can be identified using Western blotting,[4] which is based on the deposition of proteins onto a plastic membrane followed by detection using specific antibodies. Samples of bacterial or viral proteins are separated on a gel matrix using an electric charge (SDS-polyacrylamide gel electrophoresis (SDS-PAGE) described in Chap. 6). The Western blotting procedure (see ThermoFisher Overview of Western blotting 2020) is designed to produce an exact copy of the separated proteins on a plastic membrane; this is achieved by layering the membrane over the gel and applying an electric current across its entire surface. The proteins then migrate out of the gel and onto the plastic, where they bind tightly, but still maintain the same pattern as in the original gel. The reason for making this copy is that the plastic membranes are stronger and thinner than the polyacrylamide gel; this means that vigorous washing steps can be performed without risk of breaking the gel and that antibody reactions will occur quite rapidly. Western blotting has an advantage over ELISA assays in that it provides information about the size of each protein, as well as the number that may react with the same antibody; however, it requires specialized apparatus and is less straightforward to perform. The basic principle is illustrated schematically in Fig. 14.5.

Western blotting forms part of the standard test for HIV infection, where the patient has produced antibodies to the virus in the blood; these are used to probe HIV proteins separated on the membranes, and if virus protein bands appear, this means that the donor is HIV positive. The method is also used to detect misfolded proteins, for example, the prion proteins that are responsible for degenerative brain diseases such as Creutzfeldt-Jakob disease. Protein folding diseases are of great interest to the biopharmaceutical industry as they may be responsible for some of the features seen in the brains of patients with Parkinson's and Alzheimer's disease.

Nanoparticles: Beads and Dots

A great deal of ingenuity has been applied to improving the sensitivity of immunoassays using nanotechnology. Magnetic and semiconductor materials created for the electronics industry have been formed into particles with sizes ranging from nanometers to micrometers; these are used for capturing antibodies that are bound to

---

[4] Western blots are named after the compass point – the concept of blotting molecules onto membranes was exploited by Dr. Ed Southern in the form of the southern blot used for analyzing DNA fragments. Someone, with a peculiar brand of humor, subsequently named the transfer of RNA a Northern blot, so Western blots followed for proteins (the nomenclature has got out of hand, with the introduction of Eastern and Southwestern blots).

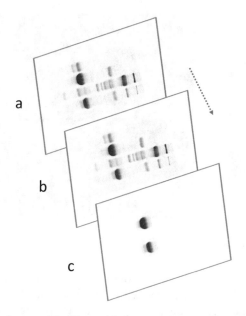

Fig. 14.5 Schematic diagram of the Western blotting procedure. (**a**) Protein samples are separated through an electric field in an SDS-PAGE gel (protein bands would normally be invisible at this stage). (**b**) The gel is overlaid with a plastic (nitrocellulose or PVDF) membrane and an electric current applied to drive the proteins from the gel. The exact distribution and quantity of protein bands are retained on the membrane (these can be visualized using reversible stains). (**c**) The membrane is probed with an antibody specific for the protein of interest. The antibody may be directly labeled with an enzyme to allow visualization, or (more usually) with a labeled secondary antibody that binds to the first (primary) one. The bound antibody and target protein is revealed using a colored reaction or more sensitive chemiluminescence. Standard marker proteins run in parallel to the original can be used to estimate the molecular weight of the target protein

their target, or as supports for fluorescent probes. Antibody capture can be accomplished with magnetic particles (beads) coated with an antibody, or any molecule that binds specifically to the target protein. The beads can then be isolated, to remove everything except the target molecules of interest, by simply applying a magnet. Magnetic particles can also be used to create ultrasensitive assays that reproduce the interactions between a magnetic recording head and a computer hard disc drive. These magnetonanosensors can detect protein levels in the attogram range (10^{-18} grams or one million trillionth of a gram) (Gaster et al. 2009).

Quantum dots are another type of nanoparticle which can be attached to antibodies to produce a signal, in this case emission of fluorescent light. The dots are made of semiconductor materials that emit light of different wavelengths (colors) depending on the size of the particle. This makes quantum dots particularly useful as multicolor probes for different immunoassay formats since a range of sizes can be linked to specific antibodies and used in combination (Qdot probes technology overview 2020).

Fig. 14.6 A microfluidics chip used to perform small-scale reactions produced by Micronit Microfluidics 2020. Note the size of the unit compared with the tweezers. Two channels for pumping in separate reagents are visible on the left-hand side of the chip. Image courtesy of Micronit Microfluidics

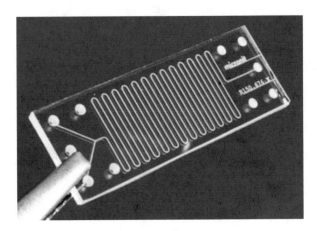

Lab-on-a-Chip (LOC)

The lab-on-a-chip is an example of microfluidics technology that enables biological and chemical reactions to be performed in extremely small volumes (Fig. 14.6). This is ideally suited to immunoassays where different reagents (antibodies, proteins, enzymes, etc.) are reacted together and the waste products removed by washing. Conventional immunoassays can take several hours, both because of the time needed for the antibody to bind sufficiently to its target and because of the need to perform multiple washing steps. LOC devices are under development for many diagnostics and research applications in the biopharmaceutical industry and academia (for review see Microfluidic Reviews 2020 (Fig. 14.6).

Molecular Diagnostics

This term is generally applied to the detection of nucleic acid molecules derived from microbes or human tissues. It is particularly useful for detecting genetic sequences from viruses and bacteria in a much shorter time than would be taken to culture these organisms in a laboratory. Nucleic acid-based diagnostic techniques, for specific cells or organisms, employ an amplification stage from the DNA or RNA present in clinical samples. This is achieved using the polymerase chain reaction (PCR) to amplify DNA, followed by oligonucleotide hybridization to specific sequences in the target DNA (Chap. 6). An example of the need for rapid infectious disease diagnostics is highlighted by the COVID-19 pandemic caused by the RNA coronavirus SARS-CoV-2. The RT-PCR assay design uses reverse transcriptase to convert the viral RNA to DNA which is then amplified by real-time PCR using primers based on the virus sequence. The PCR reaction uses fluorescent dyes (usually SYBR Green 1) to quantify the original target DNA in a specially designed thermocycler (LightCycler Roche and the ABI 7700, Applied Biosystems). See Lyon and Wittwer (2009) for a review of the technology. The instrumentation required for these assays is too expensive and cumbersome to be used "in the field," for example, in countries with poor medical infrastructure. To overcome this problem, the amplified viral DNA can be revealed by a colored reaction on a dipstick. The SAMBA system (Lee et al. 2010) captures the amplified viral DNA on an absorbent matrix and then hybridizes a probe chemically labeled with multiple copies of a small molecule recognized by antibodies (i.e., a hapten). The probe is then recognized by an anti-hapten antibody labeled with colored beads. If the target is present above a certain level, enough antibody will bind to the dipstick to reveal a colored band. The system has been successfully used to detect HIV and has been modified in 2020 for use in the COVID-19 outbreak.

Cancer Diagnostics

DNA amplification can be used for detecting the characteristic DNA rearrangements and mutations that occur in cancer cells. For example, germline mutations in the BRCA1 and 2 genes that confer a high risk of breast and ovarian cancer are routinely tested in DNA from peripheral blood, using NGS sequencing, allowing decisions to be made about preventative treatment. For tumors that have occurred, but perhaps are in difficult anatomical locations (e.g., brain) or are undeveloped, it is possible to consider so-called liquid biopsies in which circulating tumor cells (CTCs) can be detected in the blood. Their concentration can be as little as 1 cell per milliliter of blood, with variation according to the cancer type; being mostly epithelial, these cells can be differentiated from the bulk of the blood cells (Eisenstein 2020). DNA signatures can be obtained using PCR and combined with serum protein biomarkers of disease. The CancerSEEK system uses this combination of DNA and protein markers in blood to detect early malignancy in 5 different cancers (Thrive Earlier Detection Corp.) and is being evaluated in an observational clinical study with 3000 subjects (ASCEND trial 2020).

Lastly, findings such as RNA containing exosomes in blood (Hornung et al. 2020) and microbial signatures associated with cancer (Poore et al. 2020) provide new opportunities for diagnostics development.

14.1.3   Regulation of Diagnostics

Just as the regulation of drug development is designed to ensure that marketed products are safe and effective, so it is with diagnostics. A false diagnosis based on an erroneous positive (or negative) result can have catastrophic consequences for the patient. Some possible scenarios can be envisaged, such as inaccurate HIV tests or prognostic screening tests for cancer that lead to the unnecessary removal of tissues. This potential hazard with unregulated products is the reason why the FDA has a dedicated office of the Center for Devices and Radiological Health (CDRH) to cover diagnostics, namely, the Office of In Vitro Diagnostics and Radiological Health which is part of the Office of Product Evaluation and Quality (OPEQ) (FDA

CDRH 2020). In Europe, a conformity assessment is conducted under regulations for in vitro diagnostics (IVDR). Diagnostic devices, to be sold in the EU, can be self-certified by the manufacturer after reference to notable bodies (i.e., authorized expert institutions) prior to receiving a CE (Conformité Européenne) mark. However, certain tests that fall into higher-risk categories, such as those for infectious agents, must register for a CE mark via a competent authority rather than through self-certification (EMA Medical devices 2020).

14.1.3.1   Classes of Diagnostics Reviewed

Diagnostic devices are assigned to one of three classes for review by the FDA according to potential risk: class I devices include reagents for routine testing, class II for moderate risks (e.g., mutations in blood clotting proteins), and class III for high-risk procedures such as automated analysis of cervical smears for cancer diagnosis (Mansfield et al. 2005).

Class III devices require premarket approval (PMA) by the FDA and will often require data from clinical trials to show effectiveness. Under certain circumstances, however, an investigational device exemption (IDE) may be granted if the test is to be used, for example, to select a specific group of patients with serious diseases for a clinical trial. This scenario is likely to be more common as pharmacogenomics, and biomarker tests become a routine part of clinical trial design.

14.1.4   Personalized Medicine

14.1.4.1   Introduction

If it were not for the great variability among individuals medicine might as well be a science and not an art.

Sir William Osler, *The Principles and Practice of Medicine* (1892). These words by the famous Canadian physician highlight the fact that diseases and the effectiveness of treatments varies between patients. It follows, therefore, that the more this individual variability can be understood, the more scientific the medical intervention, being effectively, the idea of personalized, or precision medicine.

14.1.4.2   Measuring Genetic Variation

The genetic "body plan" of human beings is essentially the same between all members of the species, but with phenotypic[5] differences in anatomy, physiology, and pathology, as well as responses to drugs. Comparisons of DNA sequences from

---

[5] Leaving aside sex chromosome differences.

many thousands of individuals have pinpointed areas of variability that can in many cases be related to these phenotypic differences.[6] Despite advances in next-generation sequencing (Chap. 6), it is currently uneconomical and impractical to sequence the entire genome of every patient who is prescribed a drug by their doctor. In fact, it is not necessary to look at the entire genome sequence, but instead to only look at that part of the genome that is relevant to the disease, or drug, under consideration. Differences between individual genomes come in the form of insertions or deletions (indels), duplications, or substitutions of one nucleotide base by another (SNPs). The term "haplotype" is used to describe groups of these variants carried as blocks on single chromosomes that are passed on from parent to offspring.

Large-scale comparison of human DNA sequences provides a reference set of variations from the so-called human reference genome. Table 14.1 shows the breakdown of variation between individuals from different geographical areas (The 1000 Genomes Project Consortium 2015).

The above table shows that SNPs (single nucleotide polymorphisms, pronounced "snips") and their detection are one of the main ways in which genetic variability is measured. The possible effect of SNPs on genetic function is illustrated in Fig. 14.7, where a sequence of DNA from two individuals is identical except for a change in a single base (nucleotide).

In this example, the T in individual one has been substituted with an A in individual two. If the SNP were present in the vast amount of human genome sequence that does not encode proteins, then it may have no effect on either individual (SNP1). Alternatively, the SNP may affect the DNA involved in regulating the amount of a protein that is made by a cell, and therefore the levels will differ between individuals (SNP2). Finally, the SNP may directly alter the genetic code that specifies the amino acids introduced into the protein (SNP3). This last change may have profound consequences if the function or stability of the protein is altered significantly.

Table 14.1 Median variation in autosomal DNA between African (AFR), American (AMR), East Asian (EAS), European (EUR), and South Asian (SAS) individuals. SNPs, single nucleotide polymorphisms; indels, insertions/deletions; CNVs, copy number variants. M, million; K, thousand

Geographical origin	SNPs	Indels	Large deletions	CNVs
AFR	4.31 M	625 K	1.1 K	170
AMR	3.64 M	557 K	949	153
EAS	3.55 M	546 K	940	158
EUR	3.53 M	546 K	939	157
SAS	3.6 M	556 K	947	165

Adapted from The 1000 Genomes Project Consortium (2015)

---

[6]Variability due to epigenetic changes is also of profound importance (Chap. 6).

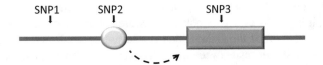

Fig. 14.7 Different types of single nucleotide polymorphisms (SNPs). The oval represents a protein (transcription factor) that binds to DNA and affects the activity of genes further away. In this example, the affected gene (represented by the gray rectangle) produces a protein. SNP1 is in a piece of DNA where mutations have no effect (i.e., they are silent). SNP2 modifies the binding of the transcription factor and therefore modifies the level of expression of the protein. SNP3 directly modifies the amino acid sequence of the protein

Large-Scale Surveys of Genetic Variation

SNP analysis is generally performed at high throughput using microarrays (genotyping chips) and is sufficiently robust to allow accurate genotyping from mouth swabs or sputum samples. If sequence information is required, either from the whole genome or just the protein-coding regions (exome), the cost of the technology has come down to a point where ambitious sequencing projects can be undertaken. For example, the 100,000 Genomes Project in the UK was able to access DNA and data from the large cohort of patients enrolled in the National Health Service. The objective was to sequence 100,000 genomes from 85,000 patients with rare diseases or cancer and to match the data with medical records. According to Genomics England (2020), "To date, actionable findings have been found for 1 in 4 to 1 in 5 rare disease patients, and around 50% of cancer cases contain the potential for a therapy or a clinical trial."

The UK Biobank project has taken a similar approach to the above project, but with a prospective study of normal volunteers (UK Biobank 2020). 500,000 people aged between 40 and 69 years were signed up between 2006 and 2010 to provide medical data and biological samples, including DNA. The information gained is wide ranging, from genetic associations with lifestyle choices, such as food, as well as with a range of diseases. These data are freely available to researchers worldwide and are generating many publications of relevance to personalized medicine.

Surveys of this type are designed to identify genetic factors in disease and will reveal many potential drug targets; unsurprisingly this is of interest to drug developers; as an example, the AstraZeneca and MedImmune Genomics Initiative was established to analyze two million genomes by 2026 (Ledford 2016).

Finally, the SNP genotyping technology which is used by consumer genetics companies like 23andMe for genealogy has potential for healthcare research given the millions of people interested in their family history (23andMe Therapeutics 2020).

14.1.4.3   Pharmacogenetics and Pharmacogenomics

Until recently, for a given disease, drug therapy was undertaken on a "one-size-fits-all" basis, such that everyone gets the same medicine. However, the idea of individualized treatments goes back to antiquity, in Ayurvedic and Chinese medicine, for example (Chap. 4). The fields of pharmacogenetics (PGt) and pharmacogenomics (PGx) have put these ideas on a modern scientific footing and are now a centerpiece of personalized medicine.

For regulatory purposes, the ICH E15 document (ICH E15 guideline 2020) uses the following definitions:

- PGt

The study of variations of DNA and RNA characteristics as related to drug response.

- PGx

A subset of pharmacogenomics defined as the study of variations in DNA sequence as related to drug response.

There is essentially a complete overlap between molecular diagnostics and pharmacogenomics, at least as far as the technologies are concerned, so attention will now be focused on pharmacogenetics, a discipline that promises to turn the vision of personalized medicine into reality. Pharmacogenetics predates DNA sequencing by many years, since the first recorded case of different individual responses to an ingested chemical was reported in 1932 by Arthur Fox (Fox 1932). The first paragraph of his paper sums up the situation:

> Some time ago the author had occasion to prepare a quantity of phenyl thio carbamide, and while placing it in a bottle, the dust flew around in the air. Another occupant of the laboratory, Dr C.R. Noller, complained of the bitter taste of the dust, but the author, who was much closer, observed no taste and so stated. He even tasted some of the crystals and assured Dr Noller that they were tasteless but Dr. Noller was equally certain it was the dust he tasted. He tried some of the crystals and found them extremely bitter. With these two diverse observations as a starting point, a large number of people were investigated, and it was established that this peculiarity was not connected with age, race or sex. Men, women, elderly persons, Children, Negroes, Chinese, Germans and Italians were all shown to have in their ranks both tasters and non-tasters.

This paper was followed by one written by A.F. Blakeslee, who showed that this variation had a genetic explanation. Many years later, the gene that differs between tasters and nontasters was shown to encode a taste receptor protein (a GPCR, Chap. 6).

In the intervening years between this first description of a pharmacogenetic response and the advent of whole genome sequencing, clinicians noticed examples of inherited responses to medicines that related to their effectiveness and safety.[7] Now that it is possible to compare genes at the level of DNA sequence, the stage is

---

[7] Arno Motulsky pioneered the field of pharmacogenetics (Motulsky 1957).

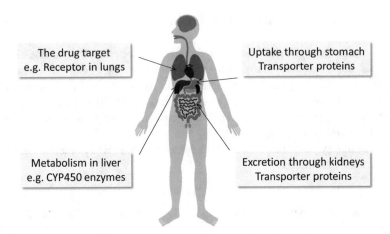

The drug target
e.g. Receptor in lungs

Uptake through stomach
Transporter proteins

Metabolism in liver
e.g. CYP450 enzymes

Excretion through kidneys
Transporter proteins

Fig. 14.8 Anatomical sites with specific functions that may contribute to individual responses to medicines

set for a new era of personalized medicine, based on the use of genetic information to guide drug prescription.

The Targets of Pharmacogenetic Mutations

Since mutations in DNA lead to altered levels or functions of proteins, the targets of these mutations in drug responses are to be found in the proteins that affect both the pharmacodynamics and pharmacokinetics of the drug. Figure 14.8 shows the different points where DNA changes could influence drug efficacy or toxicity.

Examples of SNPs Associated with Drug Responses

This area of drug development is of critical importance for understanding drug interactions and avoiding serious adverse drug reactions (iatrogenic diseases) in individual patients. The reviews from Yiannakopoulou (2013) and Krebs and Milani (2019) are taken from an extensive scientific and clinical literature; they highlight the main pharmacogenomic targets of interest, as well as areas where the results are less clear-cut. However, pharmacogenomic changes may also determine the efficacy of the drug on its target as discussed in the section below.

- Drug targets

Changes in genes for drug target proteins could affect the actual binding site for the drug, or the amount of protein produced by the relevant cell. Although most of the variation in drug responses observed clinically appear to relate to changes in pharmacokinetics (see below), there are some published cases of genetic variation in drug pharmacodynamics. One example is the variation in responses in patients with non-small cell lung cancer (NSCLC) to gefitinib (Iressa®). This drug targets the tyrosine kinase domain for the epidermal growth factor receptor (EGFR) responsible for tumor growth but works in only 10% of patients due to activating mutations that confer sensitivity (Lynch et al. 2004). In clinical practice it is therefore possible to use the mutational status of patients to determine who should receive Iressa®.

A more complex situation exists with cystic fibrosis (CF) in which mutations in the CF transmembrane conductance regulator (CFTR) lead to mucus hypersecretion in the lungs and elsewhere, followed by considerable morbidity and mortality. The problem for a personalized medicine approach is the large number (over 1700) of pathogenic CFTR gene variants. Nonetheless, some small molecule drugs (ivacaftor, lumacaftor, tezacaftor, and elexacaftor) are available that target specific mutation types and are clinically effective. This is summarized in Table 14.2 and shows the relative proportion of CF sufferers in each category (adapted from Manfredi et al. 2019).

The situation is complex, however, as the drugs have different effects across the five categories making it difficult to make the therapy truly a precision medicine.

- Drug metabolism enzymes

So far, the most useful and informative pharmacogenetic targets have been the enzymes which are involved in both phase I and phase II drug metabolism (Chap. 11). The FDA is now sufficiently confident in the genetic association between certain drug responses and the rate of metabolism that it is allowing genetic data to be included in the package insert provided with the drug. The following example concerns the metabolism of the drug warfarin, marketed as Coumadin®, by Bristol Myers Squibb. Although originally used as a rat poison, warfarin is an effective anticoagulant (blood thinning) drug but one in which the effective dose may vary as much as tenfold between patients. If too much drug is given, the patient may hemorrhage uncontrollably with fatal consequences. The large dose variation is due to differences in warfarin metabolism, which are associated with SNPs in two enzymes CYP2C9 and VKORC1. These variants have been used to guide warfarin dosing in clinical trial settings to see whether pharmacogenetics offers any advantage over conventional dose estimation. The results of a 2010 trial in the USA indicate that prior knowledge of the CYP2C9 and VKORC1 genetic variants (or genotypes) reduces admissions of patients to hospital for adverse bleeding events by over 30%. This type of clinical analysis provides the evidence needed to support the introduction of this pharmacogenetic test into routine clinical practice (Epstein et al. 2010).

Table 14.2 Summary of mutation types and drug interventions in CF

Class	I	II	III	IV	V
Mutation type	Protein synthesis	Maturation processing	Ion channel gating	Ion channel conductance	Reduced protein
% CF patients	22	88	6	6	5
% patients with approved drug	<0.5	39.2	4.6	2.6	3.5

Other findings include mutations in UDP-glucuronosyltransferase (UGT), an enzyme responsible for transforming lipophilic metabolites into water-soluble metabolites that can be excreted from the body. The reduced enzyme activity associated with the UGT1A1*28 variant increases toxicity to metabolites of the anticancer drug irinotecan. Lastly, adverse reactions to clopidogrel (another anticoagulant) are linked to CYP2C19, adding cardiovascular medicine to real-world pharmacogenetic testing.

- Drug transporters

Transporter proteins act as gatekeepers to control the entry of drug compounds into different tissues as well as their exit during excretion. Genetic variation, in the form of SNPs and other mutations, could limit the efficiency of uptake or clearance of a drug and therefore affect its efficacy or safety. Members of the ATP-binding cassette (ABC) and solute carrier (SLC) transporters have been implicated in various studies, but their genetics are variable in the general population, thus complicating the selection of robust SNP markers that could be used clinically (Yiannakopoulou 2013).

One transporter that has been fully validated has nothing to do with drug responses, however, but controls whether earwax is wet or dry. The ABCC11 transporter has a SNP that affects the rate of transport of fluids in the ear and hence the state of the wax, which is dry in East Asians and wet in other populations who have a G instead of A in the transporter DNA.

Summary of Key Points

Diagnostics are fundamental to clinical practice and are used to support drug development.

The technology is wide ranging and includes imaging, biochemical assays for small molecules, immunoassays for proteins, and nucleic acid detection.

DNA sequence information is being used for personalized medicine through the identification of SNPs and other mutations which associate with responsiveness to drugs, as well as certain diseases.

The integration of biomarkers and companion diagnostics into personalized medicine is dependent on clinicians establishing the clinical validity and utility of the tests.

References

ASCEND Trial (2020) https://clinicaltrials.gov/ct2/show/NCT04213326. Accessed 29 Apr 2020

Cell Signaling Technology (2020) https://www.cellsignal.com/common/content/content.jsp?id=types-of-elisas. Accessed 28 Apr 2020

cancer.gov (2020) https://www.cancer.gov/about-cancer/diagnosis-staging/diagnosis/tumor-markers-list. Accessed 4 May 2020

Eisenstein M (2020) Taking cancer out of circulation. Nature 579:S6–S8

EMA Medical devices (2020) https://www.ema.europa.eu/en/human-regulatory/overview/medical-devices. Accessed 29 Apr 2020

Epstein RS et al (2010) Warfarin genotyping reduces hospitalization rates results from the MM-WES (Medco-Mayo Warfarin Effectiveness study). J Am Coll Cardiol 55:2804–2812

FDA CDRH (2020) https://www.fda.gov/about-fda/center-devices-and-radiological-health/reorganization-center-devices-and-radiological-health. Accessed 29 Apr 2020

FDA Companion diagnostics (2020). https://www.fda.gov/medical-devices/vitro-diagnostics/list-cleared-or-approved-companion-diagnostic-devices-vitro-and-imaging-tools. Accessed 4 May 2020

Fox AL (1932) The relationship between chemical constitution and taste. Proc Natl Acad Sci U S A 18:115–120

Gaster GS et al (2009) Matrix-insensitive protein assays push the limits of biosensors in medicine. Nat Med 15:1327–1332

Hornung S et al (2020) CNS-derived blood exosomes as a promising source of biomarkers: opportunities and challenges. Front Mol Neurosci 13:38. https://doi.org/10.3389/fnmol.2020.00038

ICH E15 guideline (2020) https://www.ich.org/page/efficacy-guidelines Accessed 1 May 2020

Introduction to the study of experimental medicine (2017 edition). Claude Bernard. Routledge, Abingdon, Oxford

Komura D, Ishikawa S (2018) Machine learning methods for histopathological image analysis. Comput Struct Biotechnol J 16:34–42

Krebs K, Milani L (2019) Translating pharmacogenomics into clinical decisions: do not let the perfect be the enemy of the good. Human Genomics. https://doi.org/10.1186/s40246-019-0229-z. Accessed 4 May 2020

Langbein T et al (2019) Future of theranostics: an outlook on precision oncology in nuclear medicine. J Nucl Med 60:13S–19S

Lee HH et al (2010) Simple amplification-based assay: a nucleic acid-based point-of-care platform for HIV-1 testing. J Infect Dis. 201 Suppl 1:S65–72

Ledford H (2016) AstraZeneca launches project to sequence 2 million genomes. Nature 532:427

23andMe Therapeutics (2020) https://therapeutics.23andme.com/. Accessed 30 Apr 2020

Lynch TJ et al (2004) Activating mutations in the epidermal growth factor receptor underlying responsiveness of non–small-cell lung cancer to gefitinib. N Engl J Med 350:2129–2139

Lyon E, Wittwer CT (2009) LightCycler technology in molecular diagnostics. J Mol Diagn 11:93–101

Manfredi C et al (2019) Making precision medicine personal for cystic fibrosis. Science 365:220–221

Mansfield E et al (2005) Food and drug administration regulation of in vitro diagnostic devices. J Mol Diagn 7:2–7

Microfluidic Reviews (2020) https://www.elveflow.com/microfluidic-reviews/general-microfluidics/introduction-to-lab-on-a-chip-2015-review-history-and-future/. Accessed 29 Apr 2020

Micronit Microfluidics (2020). https://www.micronit.com/microfluidics/lab-on-a-chip.html. Accessed 29 Apr 2020

Motulsky AG (1957) Drug reactions, enzymes, and biochemical genetics. JAMA 165:835–837

Genomics England (2020) https://www.genomicsengland.co.uk/about-genomics-england/the-100000-genomes-project/. Accessed 30 Apr 2020

Poore GD et al (2020) Microbiome analyses of blood and tissues suggest cancer diagnostic approach. Nature 579:567–557

Qdot probes technology overview (2020) https://www.thermofisher.com/uk/en/home/brands/molecular-probes/key-molecular-probes-products/qdot/technology-overview.html#applications. Accessed 29 Apr 2020

Spitzer MH, Nolan GP (2016) Mass cytometry: single cells, many features. Cell 65:780–791

The 1000 Genomes Project Consortium (2015) A global reference for human genetic variation. Nature 526:68–74

Suntharalingam G et al (2006) Cytokine storm in a phase 1 trial of the anti-CD28 monoclonal antibody TGN1412. N Engl J Med 355:1018–1028

ThermoFisher (2020) https://www.thermofisher.com/uk/en/home/life-science/cell-analysis/cell-analysis-learning-center/molecular-probes-school-of-fluorescence/flow-cytometry-basics/flow-cytometry-fundamentals/how-flow-cytometer-works.html#applications. Accessed 5 May 2020

ThermoFisher Overview of western blotting (2020) https://www.thermofisher.com/uk/en/home/life-science/protein-biology/protein-biology-learning-center/protein-biology-resource-library/pierce-protein-methods/overview-western-blotting.html. Accessed 5 May 2020

UK Biobank (2020) https://www.ukbiobank.ac.uk/. Accessed 1 May 2020

Wu T-Y et al (2010) Ten-year trends in hospital admissions for adverse drug reactions in England 1999–2009. J R Soc Med 103:239–250

Yiannakopoulou EC (2013) Pharmacogenomics of phase II metabolizing enzymes and drug transporters: clinical implications. Pharmacogenomics J 13:105–109

Chapter 15
Pulling It All Together: A Drug Development Case History

Abstract This chapter presents a drug development case history that aims to consolidate the information previously given on drug target discovery, selection of compounds or biologicals, clinical development, and personalized medicine. It takes the form of a hypothetical program to discover and develop a medicine to treat Alzheimer's disease. It attempts to be as realistic as possible, mixing current research findings with comments about the strategic thinking that would occur in a biopharmaceutical company that undertakes a project of this type.

15.1 Introduction

The purpose of this chapter is to consolidate into a single case history the various aspects of drug discovery and development that have been described previously. The example chosen is a medicine for treating Alzheimer's disease, a major cause of suffering for patients and their families and a financial burden on health and welfare systems. While a clinically effective drug for stopping disease progression sadly does not yet exist, the background to how success might be achieved is intended to be as realistic as possible, based on current literature and the author's own experiences in the biopharmaceutical industry. Of course, the logical progression of activity laid out in this example is an idealized situation that rarely occurs in real life. The former British Prime Minister Harold MacMillan, when asked to name factors that make a government go off course, replied "Events, dear boy, events." This is as true for drug discovery as it is for politics, and of course human behavior always plays its part. Nevertheless, the objective here is to show how the key scientific areas covered in the preceding chapters are integrated with the commercial decision-making that is central to the drug discovery industry.

E. D. Zanders, *The Science and Business of Drug Discovery*,
https://doi.org/10.1007/978-3-030-57814-5_15

15.1.1   *The Discovery and Development Strategy*

Once the idea of developing a novel drug to treat Alzheimer's disease has been accepted in principle, several actions must be taken to turn this into practice. The main headings for these are summarized in Fig. 15.1.

Before the project is approved by the company management, it must be evaluated on its scientific and commercial merits; if either one of these is lacking, it is unlikely that the necessary human and financial resources will be forthcoming. These deliberations are covered in the following section.

15.1.2   *Scientific and Commercial Analysis:*
The Product Profile

Since the discovery and development of prescription medicines is market driven, it is important to take this into account at the outset of a drug discovery program. A decision on whether to invest years of time and millions of dollars in a drug for a disease area will depend upon several factors, including the likelihood of success and the commercial return. The decision-making process starts with the drawing up of a product profile that should answer the following questions:

- Is there an unmet medical need?
- Will a new drug show significant advantages over existing medicines?
- How does it fit into the company's research and commercial portfolio?

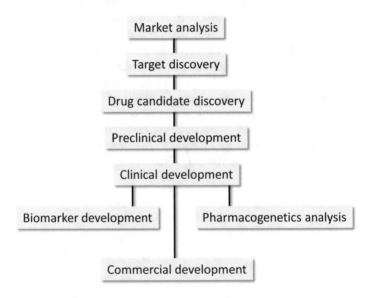

Fig. 15.1 Main components of a drug discovery and development program

- Will a new drug make a return on investment?
- How will the drug be administered?

15.1.2.1 Unmet Medical Need

Most people in the developed world are aware of the impact that dementia is having on an aging population (WHO 2020). The topic has become the subject of a great deal of media attention, mostly focused on the disease described by Alois Alzheimer at the beginning of the twentieth century. The fact that the prevalence of all dementias is increasing suggests that medical intervention is not working well enough. Current medicines for Alzheimer's disease treat the symptoms of memory loss but do not arrest the breakdown of nerve cells in the brain (neurodegeneration), so progression of the disease is inevitable. There is therefore a considerable unmet medical need for effective treatments.

15.1.2.2 Comparison with Existing Medicines

Given the nature of Alzheimer's disease and the commercial opportunities for effective drugs (see later), it is no surprise that there is a great incentive for the biopharmaceutical industry to introduce new treatments as quickly as possible. The mainstay of current treatment is the range of drugs such as Aricept® and Namenda® that improve symptoms (i.e., loss of memory and cognition) by increasing neurotransmitter levels in the brain. This could be compared with trying to keep a car with a faulty engine from stopping completely by pumping the accelerator to increase the level of fuel; unfortunately, this only serves to postpone the inevitable failure of the engine. The two available classes of memory-enhancing drugs are enzyme inhibitors (of acetylcholinesterase) and receptor antagonists (NMDA). Acetylcholinesterase inhibitors work by increasing levels of the neurotransmitter acetylcholine in the brain, and the NMDA antagonist may regulate levels of the neurotransmitter glutamate. While these drugs provide some beneficial effect, they do not work in all patients, and they do not arrest the underlying neurodegeneration found in Alzheimer's disease, i.e., they are not disease modifying. This means that there is plenty of opportunity to improve upon current medicines, particularly if a novel drug slows down or even reverses disease progression.

15.1.2.3 Company Research and Commercial Portfolio

The major biopharmaceutical companies have programs in different therapeutic areas but dominate in one or two commercial franchises depending on which drugs generate the highest revenues. Alzheimer's disease comes under the CNS (central nervous system) diseases therapeutic area, where companies like Eli Lilly have historically been highly active. As a result of historical and current activities, these companies will have no problem in marshalling the resources (in-house or contracted out) that are required for a CNS drug discovery program. Internal lobbying by project champions who support the proposal will be necessary to make these resources available from the research and commercial management. Although Alzheimer's disease is chronic and difficult to treat, there is unlikely to be much opposition to attempting at least one drug discovery program because of the potential to produce a commercial blockbuster. Quite often, a target or even a drug candidate will be produced by a small company or a spinout from a university where the academics wish to exploit a laboratory discovery. Since it should be clear from reading this book that no small organization can possibly resource a full drug development program of this type on its own, some partnering or licensing to larger biopharmaceutical companies is inevitable. The stage at which this happens is partly based on estimates of risk and the escalating cost of each clinical trial stage. Assuming the human and financial capital is available to pursue the program, the next stage is to see whether this can be recovered through drug sales.

15.1.2.4 Return on Investment

The amount of money recovered over the commercial lifetime of the drug will obviously depend on the number of patients and the amount that healthcare providers are prepared to pay for each course of treatment. In the case of Alzheimer's, the number of patients worldwide is increasing inexorably as the average life span of the general population increases. Although there is an early onset form of the disease affecting those younger than 65, most cases occur in later years. The risk of contracting the disease rises from about 10% between the ages of 65 and 85 to about 45% thereafter[1]. The number of cases in the USA for the over 65s is 5.8 million projected to rise to 13.8 million by 2050 (2020 Alzheimer's disease facts and figures). In the (unlikely) event that every one of those patients was to be given the proposed new drug at $10,000 per year, this would produce revenues of over $50Bn. Given that the best-selling blockbuster drugs are creating over $10Bn in annual sales, even a fraction of this market would be worth the investment. Another commercial aspect is the willingness of healthcare providers (governments and managed care organizations) to pay for new medicines or to support research and development efforts. There is a strong incentive for this in the case of Alzheimer's disease as the costs are truly terrifying; the Alzheimer's Association quotes $305Bn in projected annual costs to the US economy in 2020 (2020 Alzheimer's disease facts and figures). Cost/benefit considerations are evaluated by the commercial departments of biopharmaceutical companies and government and academic departments wishing to assess the value for money provided by medicines of this type (see pharmacoeconomics, Chap. 16).

---

[1] I recall hearing a talk in which a graph of prevalence compared with age was extrapolated to 150 years, in which case everyone would have the disease, something that those who want to dramatically extend the human life span might wish to ponder.

Lastly, the intellectual property (IP) position is important (see Chap. 16), as the time taken from patent filing to marketing the new medicine may significantly erode its patent life and therefore commercial attractiveness. This could be an important issue if the drug candidate is in-licensed from a third party, as is often the case.

15.1.2.5   What Type of Drug and How Will It Be Administered?

The next item on the product profile list concerns the type of drug that should be produced and how it should be delivered.

The ideal medicine is an orally bioavailable small molecule that can be administered once or twice a day and be safe for chronic use over many years; furthermore, because Alzheimer's is a CNS disease, in many cases the drug must cross the blood-brain barrier to get to its target. It is possible that there will be too many obstacles to be overcome in achieving this idealized situation, so alternatives to small molecules may be considered instead. Given the clinical success with monoclonal antibodies in various chronic diseases, these biological agents will be used for this hypothetical example. The route of administration of the antibody will have to be considered, as well as frequency of dosing as these factors have practical implications for patients and carers alike.

15.1.3   Discovering and Testing the Drug

The search for drug targets and molecules interfering with their action has been extensively covered in earlier chapters, and the basic principles described for bio-therapeutics are applied to this case history.

15.1.3.1   Identifying a Suitable Drug Target

Although important target discoveries are made in company laboratories, many ideas that stimulate drug development come from university or hospital research laboratories as part of investigations into basic cell biology and mechanisms of disease. Experimental findings are published in scientific journals and presented at conferences and seminars, often with the conclusion that "such and such may make a good drug target." Systematic surveys of the biomedical literature using machine learning algorithms are being applied to target discovery to reveal potential targets that may have been missed by traditional means. However, company scientists will still have to make the judgment as to the practicality of the idea and, if sufficiently enthused by it, may later develop a formal project proposal based on the target. Ideas may even come from non-scientific sources; the example of compounds from Chinese herbal medicines (Chap. 4) came from a newspaper cutting passed on by the author's mother. Alternatively, the target may be the same as that being worked on by a competitor, so that the objective of the drug development program will be to produce a therapeutic that is safer and more effective.

A suitable drug target for Alzheimer's is only discovered through having a basic understanding of the disease. Alois Alzheimer himself was a Munich physician who followed the decline in mental function of a patient over 5 years and then (in 1904) examined her brain postmortem to look for abnormalities. He identified a series of tangles and plaques (translucent areas) in the brain (using histochemical staining, Chap. 6) that appeared to be a hallmark of Alzheimer's as opposed to other degenerative conditions in the brain. In later years, the chemical nature of the molecules in the tangles and plaques was established, along with an analysis of their effects on the viability of nerve cells. The results of these analyses showed that neurofibrillary tangles consist of a protein (called tau) that kills nerve cells from within. The production of tau, however, is stimulated by β-amyloid protein (also called Aβ) which is the main component of amyloid plaques. Opinion is divided as to whether tangles or plaques are more important for disease progression; it depends on whether one subscribes to the tau hypothesis (a tauist) or the amyloid hypothesis. In this example, the amyloid hypothesis will be used as a guide to the drug target.

Different techniques are employed to provide evidence for the role of Aβ in Alzheimer's. For example, transgenic mice have been engineered to produce human Aβ peptides in their brains, the result being a pathology with similarities to the human disease. Other evidence comes from epidemiological studies and from genetic evidence in families with early-onset Alzheimer's disease, where mutations in APP (see below) have been detected by DNA sequencing. Although these are rare mutations, they at least provide some support for the role of amyloid peptides in the disease.

The next logical step is to ask where the Aβ peptide comes from and how it is formed. Peptides found in biological tissues are often derived from the breakdown of proteins by the action of protease enzymes; this proteolysis is used in the digestion of food as well as in a series of important biological control mechanisms such as blood clotting and the regulation of blood pressure. In the case of amyloid peptide, the parent protein was identified as a molecule called amyloid precursor protein (APP) that is expressed on the surface of cells in many parts of the body, including the brain. This identification was achieved by comparing the amino acid sequence of the peptide with the sequences of proteins held in computer databases. The sequence of the Aβ peptide was duly found to be contained within the sequence of the APP protein. With this information, the search for a drug target can then be narrowed down to either removing the toxic Aβ peptide from the brain or else preventing their formation from APP. Figure 15.2 is a schematic diagram of APP and aggregation of the Aβ peptide from monomers to the fibrils that are believed to contribute to the neuronal damage and cell death found in Alzheimer's disease. For details see Parsons and Rammes (2017).

The scenario outlined in this chapter will focus on the removal of Aβ, but it should be pointed out that considerable effort has been made by biopharmaceutical companies to inhibit its production by targeting the protease enzymes responsible, namely, β and γ secretases (Kumar et al. 2018).

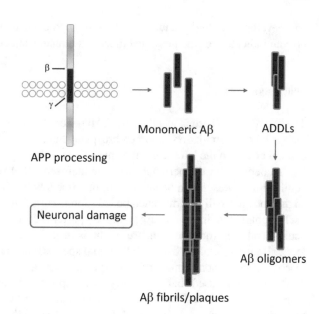

Fig. 15.2 Formation of Aβ peptide from APP protein embedded in the cell membrane. β and γ represent cleavage sites for β and γ secretase enzymes. The monomeric Aβ peptides go through a series of aggregation steps leading to the fibrils/plaques that cause the neuronal damage. *ADDLs* Aβ-derived diffusible ligands. (Adapted from Parsons and Rammes 2017)

15.1.3.2 Generating Monoclonal Antibodies to the Aβ Peptide

On the face of it, it should be straightforward to create a monoclonal antibody (mAb) to a peptide of 42 amino acids (the most prominent Aβ peptide in Alzheimer's); the problem lies in the heterogeneity of peptide structures in the brain, such as oligomers of amyloid fibrils and the free peptide (see Van Dyck 2018). Decisions must be made about whether to target amyloid aggregation and what part of the peptide (N-terminus, middle, or C-terminus) that the antibodies should be directed at. The resulting antibody candidate will have its own potential as an effective therapeutic depending on the fundamental biology of Alzheimer's disease and other issues such as dosing (see later).

The mAb will normally be produced rodents by immunization with the desired form of Aβ peptide and the resulting murine antibody humanized by genetic engineering (a rodent form can be used for testing in rodent models). Alternatively, a human antibody can be generated directly from human immune cells (B lymphocytes) taken from patients.

15.1.3.3 Preclinical Development

Pharmacodynamics

It will be important to establish that the candidate mAb will reduce Aβ peptide (and possibly amyloid fibrils) in a suitable animal model such as the transgenic mouse Tg2576. This expresses a human APP variant in the brain, has elevated Aβ peptides, and develops amyloid plaques. Antibody-related changes in these parameters will give an indication of the possible efficacy in human subjects.

Pharmacokinetics

The key question here is whether the mAb penetrates the brain or whether action in the peripheral circulation (and cerebrospinal fluid (CSF)) will be sufficient for a clinical effect. In the latter case, the "peripheral sink" hypothesis is based on lowering Aβ peptides in the periphery to drive their removal from the brain and restore equilibrium levels. Brain penetration of mAbs is low (0.1%, with some exceptions) so the question will be how much penetration is sufficient for a therapeutic effect.

The pharmacokinetics of large antibody proteins differs from those of small molecules, and therefore the standard methods of estimating the first in human dose (e.g., allometric scaling from small animal species) may not apply. For mAb pharmacokinetics, a non-human primate species such as the cynomolgus monkey is most commonly used, but these may still respond differently to human subjects. During safety studies, there will be a concern that the catastrophic response encountered during the phase I study of TeGenero's mAb TGN1412 (Chaps. 11 and 12) will not be repeated. The overall aim is to determine a minimal anticipated biological effect level (MABEL) to guide phase I dosing.

Drug Interactions

Data on drug interactions from therapeutic mAbs are nowhere nearly as comprehensive as with small molecule drugs, but there is clearly a need to determine potential problems, particularly with coadministration of a second antibody for another condition. It might also be anticipated that the Alzheimer's medicine may have to be administered over many years; this is even more likely in the event that the drug is used prophylactically for high-risk individuals before symptoms of dementia become apparent.

Formulation

mAbs, being proteins, require careful formulation to ensure stability. Normally, the optimum pH of the buffer solution will be established followed by an evaluation of different stabilizing agents and excipients. The material may be kept as a liquid or as a lyophilized (freeze dried) powder for storage.

15.1.3.4 Clinical Trials

At this stage, a starting dose of antibody will have been established for first time in human (FTIH) trials. As with other serious diseases like cancer, the initial phase I trial will be conducted in patients to determine the safety and tolerability of the drug, plus its pharmacodynamics and pharmacokinetics. Having obtained approval from the regulatory authorities in the form of an IND or CTA (Chap. 5), the sponsoring company must identify the clinical centers that will conduct all the safety, efficacy, and confirmatory studies as detailed in Chaps. 12 and 13. Clinical trials for drugs that slow down the progression of a neurodegenerative disease are going to be particularly challenging, however. Firstly, patients enrolled in efficacy trials must be correctly diagnosed as having Alzheimer's disease in the first place. Although this condition may be present in most enrolled patients, it is quite possible that some people with unrelated dementias will be included in the trials and confound the results. Standardized tests have been devised, such as the Alzheimer's Disease Assessment Scale (ADAS) and Mini-Mental State Examination (MMSE) which contain a series of evaluations of memory, speech, and other functions considered central to the disease. Although useful, these tests are not definitive, so brain imaging (Chap. 14) is required to exclude vascular dementia (MRI scans) and to observe amyloid plaques directly by PET scanning using florbetapir, an amyloid-binding diagnostic compound labeled with ^{18}F. More complexity in trial design is introduced when different levels of cognitive impairment are considered, since these may respond differently to the drug in question.

The primary outcome of a placebo-controlled phase I trial may be safety and tolerability of the mAb, with secondary outcomes being reduction in amyloid plaques as observed using PET scans. Later efficacy trials will be concerned with measuring improvement in mental cognition in the drug-treated group. If the drug can be shown to have achieved a statistically significant reduction in disease progression in pivotal phase III trials, marketing approval will be sought from the regulators. If granted, the drug discovery and development program will be near end of its primary phase. With a complex chronic disease such as Alzheimer's, it would be anticipated that more clinical trials would be undertaken in selected patient groups or with combination therapies.

In practice the whole process from inception of the project to final approval may have taken 12–15 years. It is therefore quite possible that the people who started the project will have long departed the company, rather like the scientists and engineers who design a spacecraft and then have to wait years before it reaches its destination in space. In fact, the drug discovery business tends to recycle staff through companies of varying size, so that experience gained in one is transferred to another. This can have the effect of increasing efficiency through not repeating mistakes made in other organizations, but it can also result in a certain uniformity of approach to discovery projects.

15.1.3.5   Biomarkers and Pharmacogenetics

The hypothetical drug discovery program described here would certainly include a search for biomarkers to guide clinical trials and even screen individuals at risk before disease symptoms have become apparent; otherwise, it may be too late to reverse the neuronal damage in the brain. The term "biomarkers," as well as encompassing the proteins found in blood or CSF (e.g., Aβ peptide), includes imaging, of which MRI and PET scanning have already been described.

The pharmacogenetics of responses to the experimental drug would also be investigated based on known genetic factors that influence Alzheimer's disease, notably the APOE4ε4 mutation that confers a significant risk[2]. The APOE gene encodes a lipid carrier protein in the brain with the E4 variant disrupting the blood-brain barrier near centers of cognition (Montagne et al. 2020).

The Situation with Real Anti-amyloid Therapies

The above case history is based on scenarios that are being played out by companies and academic institutions. The amyloid hypothesis is still considered sufficiently compelling that companies are prepared to invest millions of dollars in clinical trials, but their commitment has been sorely tested by the multiple failures to demonstrate clinical efficacy. Different approaches to removing amyloid plaques have been tried, including the following: using small molecules and antibodies to disrupt the Aβ peptide and remove it from the brain, producing a vaccine based on the APP protein, and creating inhibitors of both the β and γ secretase enzymes. None of these approaches have been successful so far, either because of lack of efficacy or because of serious adverse events (as occurred with an otherwise promising vaccine trial and a BACE inhibitor). The design of the clinical trial may be part of the problem; some patients may not have been correctly diagnosed with Alzheimer's (through a lack of suitable biomarkers or imaging techniques), or the disease may have progressed too far to be reversible. The physical form of Aβ may also be critical, so that mAbs targeting the "wrong" type may never be clinically effective. In the case of β-secretase inhibitors, the difficulty lies in producing small molecules that satisfy all the criteria of potency, selectivity, oral bioavailability, and ability to penetrate the brain. However, despite this dispiriting summary, there is a glimmer of hope that the amyloid hypothesis may still work out in the clinic. The aducanumab antibody to Aβ from Biogen showed great promise in phase I trials as it clearly removed amyloid plaques from the brain (as imaged by PET scans) but appeared to be unable to improve cognition in two pivotal phase III trials that were terminated on the basis of futility. However, in late 2019, Biogen reported that a subset of patients in both trials showed measurable improvement in cognition; as a result, the company requested

---

[2]This high risk has meant that some people (like James Watson) who have had their genomes sequenced have preferred to remain ignorant about their APOE4 status.

FDA approval for aducanumab (Schneider 2019), but no more information is available at the time of writing.

The case history presented in this chapter should underline the fact that uncertainties and complexities are an inevitable part of modern drug discovery. Sometimes it may be necessary to walk away from a hypothesis, even after years of effort and vast financial (and emotional) cost. This may not be the case for amyloid and Alzheimer's, but it will be for something else.

The end of this chapter is also the end of the description of drug discovery from a predominantly scientific and technical perspective. The next two chapters cover the commercial side of the drug discovery industry and the roles played by companies, academia, and charities.

References

Alzheimer's disease facts and figures (2020) Alzheimer's Dement 16:391–460

Kumar D et al (2018) Secretase inhibitors for the treatment of Alzheimer's disease: long road ahead. Eur J Med Chem 148:436–452

Montagne A et al (2020) *APOE4* leads to blood-brain barrier dysfunction predicting cognitive decline. Nature 581:71–76

Parsons CG, Rammes G (2017) Preclinical to phase II amyloid beta (Aβ) peptide modulators under investigation for Alzheimer's disease. Expert Opin Investig Drugs 26:579–591

Schneider (2019) A resurrection of aducanumab for Alzheimer's disease. Lancet Neurol. https://doi.org/10.1016/S1474-4422(19)30480-6. Accessed 12 May 2020

Van Dyck CH (2018) Anti-amyloid-β monoclonal antibodies for Alzheimer's disease: pitfalls and promise. Biol Psychiatry 1583:311–319

WHO (2020) https://www.who.int/news-room/fact-sheets/detail/dementia. Accessed 8 May 2020

Part IV
The Global Pharmaceuticals Business

Chapter 16
Commercial Aspects of Drug Development

Abstract This chapter moves away from the technical aspects of drug discovery to cover the commercial operations of biopharmaceutical companies. Starting with the global market for prescription medicines, the chapter moves on to discuss the different types of organization that conduct pharmaceutical R&D and then covers the customer base, pharmacoeconomics, and portfolio management. Lastly, the all-important subject of intellectual property and generic competition is reviewed in some detail.

16.1 Introduction

The pharmaceuticals market is truly global, as the need for medicines has no geographical boundaries; it is, however, strongly biased towards the affluent western-style economies, since the enormous costs of drug development have to be covered by those who have the resources to pay. Of course, this is a rapidly changing business like many others, so specific details about companies and legislation may become outdated relatively quickly. Fortunately, there are numerous sources of market information, including private business intelligence companies such as the IQVIA Institute, Evaluate Pharma, and academic groups such as the Tufts Center for the Study of Drug Development (TCSDD). In addition, there are printed and online journals and magazines that provide up-to-date information on commercial, scientific, clinical, and regulatory developments in the biopharmaceutical industry. Some of these publications are listed in Appendix 1, Further Reading.

16.1.1 The Pharmaceuticals Marketplace

Before describing the current marketplace for drug sales, it is useful to describe the actual costs required to bring a single medicine through from discovery to the marketplace. A much quoted $2.6Bn figure from the Tufts Center for the Study of Drug Development was based on the cost of failed compounds, as well the cost of capital required over the many years before the drug can be marketed (DiMasi et al. 2016).

© The Editor(s) (if applicable) and The Author(s), under exclusive license to
Springer Nature Switzerland AG 2020
E. D. Zanders, *The Science and Business of Drug Discovery*,
https://doi.org/10.1007/978-3-030-57814-5_16

A later publication (Wouters et al. 2020) gives a median cost of $985 M, whose values varied according to the therapeutic areas served by the drugs under study; nervous system therapeutics accounted for $765.9 M and cancer and immune diseases $2.7716 Bn. Inevitably the multimillion dollar figures have been used as a political football between the companies that have to recoup this expenditure and those who think that the industry is profiteering from medicines (see Chap. 17).

16.1.1.1 Global Pharmaceutical Sales

A figure of $1245Bn (i.e., over one trillion dollars) for the total market size in 2019 was estimated by the IQVIA Institute (IQVIA Institute Report 2019). This is broken down by region in Table 16.1.

The USA continues to dominate the market, but the situation with the rest of the world has changed markedly since the end of the twentieth century. In the 1990s, for example, Japan was the second largest market, with other Asian countries much further down the list. Since then, the European Union has expanded, and Russia has increased its share of the pharmaceuticals market. The most dramatic change was the emergence of the BRIC economies, namely, Brazil, Russia, India, and China, as major consumers of healthcare. These countries, and others like Turkey and South Korea, have been called pharmerging markets (further proof of abuses to the English language) which have greater current growth than the mature markets. Figures for individual countries relative to the USA are shown in Table 16.2.

The Emergence of China

China is now the second largest biopharmaceuticals market after the USA as shown in Table 16.2. This has occurred for several reasons, including a reform of the state regulatory authority in 2015 (see Chap. 13) and changes to the internal marketplace and increased cooperation with external drug makers. One of the goals of the national "Made in China 2025" objectives is the development of novel products from the rapidly growing homegrown biopharmaceuticals sector. Several Chinese companies are developing their own drugs against novel targets, initially with the local market in mind (an aging population of over one billion people) but with a

Table 16.1 The global pharmaceuticals market. Figures represent estimated 2019 sales in $Bn

Territory	2019 revenues $Bn
USA	507
Pharmerging	293
Top 5 Europe	182
Japan	89
ROW	174

IQVIA Institute Report (2019)

Table 16.2 2018 spending on pharmaceuticals by countries relative to the USA

	2018	% USA
1	USA	100
2	China	28
3	Japan	18
4	Germany	11
5	France	7
6	Italy	7
7	Brazil	6
8	UK	6
9	Spain	5
10	Canada	5
11	India	4
12	South Korea	3
13	Russia	3
14	Australia	3
15	Mexico	2
16	Poland	2
17	Turkey	2
18	Saudi Arabia	2
19	Argentina	1
20	Belgium	1

IQVIA Institute Report (2019)

view to selling to the USA. Examples include Zensun Science and Technology's Neucardin, a recombinant human neuregulin-1 fragment for heart failure and Roxadustat from FibroGen China for anemia. More details can be found in: The next biotech superpower (2019) and Innovation nation (2019).

16.1.1.2   The World's Top Selling Medicines

The state of the pharmaceuticals market at any given time is partly reflected by the top selling medicines that bring in the major revenues. The top ten best-selling branded prescription medicines of 2019 are shown in Table 16.3.

Several things stand out from the above table: firstly, the size of the revenues, secondly the proportion of oncology products, and lastly the preponderance of biologics. The situation has therefore changed dramatically since the time when the top selling branded medicines included anti-ulcerants and cholesterol-lowering drugs.

Table 16.3 Global sales revenues for top ten selling medicines in 2019. The majority are biotherapeutics (highlighted in bold), all monoclonal antibodies except Prevnar 13 a vaccine. Revlimid and Eliquis are small molecules

Brand name	Company	Indication	Sales ($Bn)
Humira	AbbVie	Rheumatoid arthritis	19.2
Keytruda	Merck & Co.	Oncology	11.1
Revlimid	Celgene	Oncology	9.7
Eliquis	BMS	Antithrombotic	7.9
Opdivo	BMS	Oncology	7.2
Avastin	Roche	Oncology	7.1
Rituxan	Roche	Oncology	6.5
Stelara	Johnson & Johnson	Inflammation	6.4
Herceptin	Roche	Oncology	6.1
Prevnar 13	Pfizer	Infectious disease	5.8

Data from EvaluatePharma in Urquhart (2020)

16.1.2 Drug Discovery Organizations

While drug discovery and development are dominated by the long-established pharmaceutical companies, this is certainly not the whole picture, since major contributions to this field are made by biotechnology companies, academia, charities, and contract research organizations (CROs). Some brief comments about the main drug discovery players are listed below:

Large (Big) Pharma
International sites.
　Multiple therapeutic areas.
　Large sales and marketing activity.
　Dependent upon blockbuster sales.

Small Pharma
Similar structure to large pharma, but scaled down.
　Can originate from a specialist biotechnology company.
　Proportion of products may be in-licensed from other companies.

Small Biotech
Develops technology and out-licenses discoveries for other companies to develop.
　Small number of employees, often ex university scientists.

Not-for-Profit Organizations
Academia, funded by governments, industrial grants, and charities.
Medical research charities.
Philanthropic research foundations.

Contract Research Organizations (CROs)
Used by companies to outsource drug development operations.

16.1.2.1 Large Pharmaceutical Companies

The drug discovery industry is dominated by the multinational pharmaceutical corporations such as Pfizer and GlaxoSmithKline, with 90,200 and 98,500 employees, respectively, in 2019 (Francisco 2019). Many of these companies have a long and tortuous history, having grown through mergers and acquisitions. GlaxoSmithKline, for example, was formed in 2001 after a merger between GlaxoWellcome and SmithKline Beecham[1]. These in turn were formed by mergers between Glaxo and Wellcome and Smith, Kline, & French and Beechams, so that two once-famous names disappeared. In my experience, the name Glaxo is so catchy that people instinctively use it to refer to the Company, despite that fact that it has been subsumed into a new organization. The Glaxo name was created in 1906 as a trademark for the powdered milk that was exported from New Zealand to England by the company's founder Joseph Nathan. Meanwhile, Smith Kline and French had developed as a drug store, founded by John Smith in Philadelphia in 1830. Beechams originated as a laxative pill invented by Thomas Beecham in 1842, and the Burroughs Wellcome Company was formed in London in 1880. Each of these companies produced innovative medicines that have had a major impact on the treatment of infectious diseases, asthma, cancer, and metabolic diseases.

Although major pharmaceutical companies may be headquartered in one country, they have research, development, and manufacturing sites in many parts of the world. While in the past, the research sites may have been mostly confined to Europe and the USA, centers have been established in countries such as Singapore, China, and Japan, with a significant local market and a good supply of qualified personnel. Large companies have the resources to work in multiple therapeutic areas, thereby spreading risk and improving the chances of finding the blockbuster products they require for survival. As well as having to support clinical trials and regulatory affairs, big pharma must also employ large marketing departments and a trained sales force to maximize the return on their investments. Most large pharmaceutical companies are publicly listed on international stock exchanges and therefore have a

---

[1] A more detailed historical timeline can be found in GSK History 2020.

large shareholder base. The financial performance of these companies is followed carefully by pension fund managers, as the pharmaceuticals sector is conventionally seen as a defensive "safe haven," in contrast to the more volatile biotechnology sector.

16.1.2.2 Smaller Pharmaceutical Companies

These organizations may be based on the traditional pharmaceutical company model or else have developed from a biotechnology company such as Amgen Inc. (20,800 employees) or Biogen Inc. (7300 employees) after achieving commercial success. Although these companies develop fewer products for a smaller number of diseases than the major pharmas, they can be multinational organizations with a portfolio of small molecule and biologic drugs resulting from in-house research or acquisitions from third-party companies.

16.1.2.3 Small Biotechnology Companies

These companies are often formed as spinouts from university departments and may only have a handful of employees. They are strong on innovative technology, but short on capital, and must form collaborations to be viable. Funding for small biotech companies often takes the following route: initial venture capital funding, further rounds of finance and hope that the investment can be recovered at an initial public offering (IPO) on the stock exchanges, or a sale to a pharmaceutical company. When things go wrong, the company may collapse and have its assets sold off in a fire sale. The biotechnology sector has in the past suffered a similar fate to the dotcom companies in that valuations were unrealistically high for businesses that were nowhere near profitability. However, this has not deterred those biotech entrepreneurs who have weathered the investment storm and managed to create small companies; these may not always be profitable, but many do make a significant contribution to drug discovery. This is often the case when their drug candidates are licensed to partners who can afford to move them into the clinic.

16.1.2.4 Contract Research and Manufacturing Organizations

In principle, contract research organizations (CROs) can undertake many of the drug development and commercialization activities that are normally undertaken by employees of major pharmaceutical companies. Contract manufacturing organizations (CMOs) of course do the same for drug manufacture. The financial advantages are obvious since the company does not have to pay salaries to people who may not be needed at a point in the development cycle that has been outsourced. On the other hand, projects outsourced to CROs must be carefully managed by each party to avoid costly mistakes. CROs are particularly useful for biotech companies that wish to bring a drug candidate up to phase I clinical trials or even beyond. The value of the product increases significantly as evidence is gathered to show that it has genuine commercial potential; if this is the case, it can be licensed to larger companies with the expertise and facilities to progress it through to the marketplace. Large CROs offer a range of preclinical, clinical, and regulatory activities in different pharmaceutical markets. As an example, see PPD Inc. 2020.

16.1.2.5 Academia

The relationship between academia and industry is not always straightforward, but successful drug discovery and development is dependent upon good interactions between the two. Much of the science and technology behind target identification, small and large molecule discovery, and clinical research arises from university and hospital laboratories that have been funded by the taxpayer. The National Institutes of Health (NIH) in the USA, for example, receives nearly $42Bn in taxpayer funding for basic and applied medical research; it is a strong advocate of applying laboratory findings to clinical practice in as short a time as possible (see translational research, Chap. 6).

16.1.2.6 Charities and Philanthropic Foundations

Medical research charities have been around for a long time. Often founded by individuals who have a personal reason for wanting to find cures for a specific disease, these charities depend upon private donations and government tax breaks. Some, like the Wellcome Foundation in London, have large financial assets (created through the profits gained by the Burroughs Wellcome pharmaceutical company) and can influence the course of medical research through the disposition of their research grants. Britain's large investment in the Human Genome Project, for example, was made possible with Wellcome Trust funding. Similarly, charitable foundations like the Howard Hughes Medical Institute and the Broad Institute contribute indirectly to the business of drug discovery by supporting some of the best biomedical research to be found anywhere in the world.

The application of venture capital funding to charitable (including drug discovery) activity is known as venture philanthropy. It operates using the same business principles of due diligence and accountability that is found with other commercial investments. For example, the UK charity Cancer Research UK has partnered with Boston-based SV Health Investors to gain access to management talent to support its scientific expertise. The largest funding organization of this type, Bill & Melinda Gates foundation, tends to focus on developing world diseases which are generally underserved by the major biopharmaceutical companies.

Table 16.4 lists the top ten pharmaceutical companies of 2019 by sales revenue along with selected organizations involved in drug development. It also includes examples from other research-intensive industries to place these revenues in a larger

Table 16.4 2019 sales for major pharmaceutical companies, CROs, IT companies, and aerospace manufacturers

Company	2019 sales ($Bn)
Major pharma	
Roche	48.2
Novartis	46.0
Pfizer	43.9
Johnson & Johnson	40.0
Sanofi	35.0
AbbVie	32.4
GlaxoSmithKline	31.3
Takeda	29.1
Bristol Myers Squibb	25.2
Contract research organizations	
LabCorp (Covance)	11.6
IQVIA	11.08
PPD	4.03
IT companies	
Apple	260.17
Alphabet	161.8
Facebook	70.7
Aerospace	
Boeing	76.6
Airbus	70.5

Data for pharma companies from Urquhart 2020, remainder from web searches

context. While these sales figures are impressive, they are dwarfed by IT companies and aerospace. Furthermore, the market share of the largest companies is around 5%, which is low compared with similar sized industries as illustrated by the approximately 50% division of market share between Boeing and Airbus.

16.1.3 Commercial Operations

Prescription medicines are, in many ways, like any other branded product: they are marketed to customers with specific needs and interests. The main difference between promoting drugs or, say, consumer electronics lies in the fact that major restrictions have been placed on the biopharmaceutical industry regarding advertising and inducements. This section will cover some of the main areas covered by the commercial arm of a biopharmaceutical company.

16.1.3.1   Customers

While the patient is clearly the end user of the marketed drug, he or she is rarely the direct customer. The primary customers will be the healthcare systems and managed care organizations that buy medicines in bulk and negotiate discounts from the manufacturers. The customer categories are listed below:

Individual Patients

The medical awareness of the public is being enhanced through responsible medical websites and support groups on social media for patients suffering from a given disease. As a result, there is a more informed knowledge of the types of drugs that are available to treat specific conditions. This, along with direct-to-consumer advertising, can stimulate demand for specific products that may not be provided by the primary care physician. Countries, such as the USA, which permit such advertising, employ all of the communication media to advertise medicines; this can have a downside, such as excruciating revelations of embarrassing personal medical problems being broadcast on prime time television, or legal disclaimers that take as long to read out as the description of the product itself. Of far more concern is the high cost of innovative drugs means that patients may not be able to get them through their healthcare system; if they are made aware of these products through direct advertising, they may feel compelled to pay large amounts of money for them out of their own pockets. This situation is not restricted to the USA but is coming about because of increasing pressure on healthcare budgets (see next chapter).

Physicians and Pharmacists

Physicians, whether primary care or hospital-based, represent the most important customer group, since they choose which drugs to prescribe to their patients. As a result, biopharmaceutical companies spend a great deal of money on targeting doctors in their offices; they do this by employing sales forces to make face-to-face visits (detailing), through running advertisements in medical publications, or sponsoring conferences and training courses. Some doctors are identified as opinion leaders in each therapeutic area; their views can have a profound influence on the prescribing habits of their colleagues within a medical specialty. It is well-known that the relationship between doctors and the drug discovery industry is not without controversy, and rules are in place to limit abuses. However, as I have stated in previous chapters, I can only say that the vast majority of my collaborations with physicians have been highly professional, focusing on the research and its potential benefit to patients.

Pharmacists are also major customers since they have responsibility for dispensing medicines through the primary care practice, hospital, private retail pharmacies, or wholesalers.

Government-Financed Healthcare Services

The relative proportion of social healthcare provision and private/insurance funding varies according to country. At one extreme, Britain's National Health Service (NHS) provides most of the country's healthcare from taxation with a total expenditure of £152.9Bn in 2018/2019 (NHS Expenditure 2020). Of this figure, £18.9Bn was spent on drugs (just over 12%), while hospital use accounted for 53.7% of the total.

In the USA, Medicaid and Medicare Part D are administered by a government department, namely, the Centers for Medicare & Medicaid Services (CMS). Medicaid is a social welfare system that provides healthcare support for those on low incomes, while Medicare is a social insurance system available to older people through both direct taxation and employee contributions. The total cost of Medicare in 2019 was $765Bn with the prescription drugs component (part D) accounting for $83Bn, representing just over 10% of the total budget (Medicare 2020). Not surprisingly, the above government agencies, and similar organizations worldwide, are major customers of the pharmaceutical industry.

Private Healthcare Organizations

Marketplace health insurance plans operate alongside state-funded healthcare systems, the relative proportion varying according to the country (e.g., higher in the USA than in Europe). The Health Maintenance Organization Act was passed in the USA in 1973 to regulate the private health management organizations (HMOs) that have arisen over the years, including Cigna and Kaiser Permanente. There are different HMO models, from companies with dedicated staff and premises to looser affiliations of physicians and hospitals (Healthcare.gov 2020). Since individual HMOs control access to their customers' healthcare, they can negotiate with pharmaceutical companies for favorable terms on price and exclusivity. Employers use market-based insurance to manage the healthcare provision for their workforce, but this is expensive, costing the Ford Motor Company, for example, roughly $1 billion for its 56,000 employees.

16.1.3.2 Pricing and Reimbursement

Setting the price of a branded medicine is a difficult process; healthcare markets are under great cost pressures and yet the industry needs to be able to cover its costs and invest in new products. The gross or list price of a drug offered by manufacturers can be reduced to a net price by as much as 60%. This will be through discounts and rebates in the chain of payer organizations (Edison 2019).

Reimbursement is the money that the holders of healthcare funding will pay the front-line suppliers of that care. In the UK, for example, the pharmacies dealing with primary care are reimbursed by the NHS for the cost of drugs dispensed, but hospitals will pay the manufacturers and wholesalers directly. Governments or managed care organizations will naturally try to keep prices as low as possible, often through using generic products that are substantially cheaper than patented medicines (see later). This enables healthcare providers to treat a larger number of people with cheaper products while at same time reserving money for more expensive innovative products such as biologicals. Comparative effectiveness research has been undertaken by government and private research foundations for many years, but with increased pressures on healthcare funding, the field is being reinvigorated. Organizations which are set up to be independent of industry and government have, or are being, established to objectively evaluate the cost and benefits of new medicines. Their remit is to recommend, or otherwise, the provision of a medicine by state-funded healthcare systems. In Britain, the National Institute for Health and Care Excellence (NICE) was established in 1999 to provide these recommendations (NICE 2020). The situation is different in the USA because of the insurance-based model, but the 2010 Patient Protection and Affordable Care Act mandated the establishment of a Patient-Centered Outcomes Research Institute (PCORI) aimed at addressing healthcare costs and issues (PCORI 2020).

Pharmacoeconomics

The discipline of pharmacoeconomics exists to apply a cost-effectiveness analysis (CEA) to drug pricing. Information can be found through the professional body, International Society for Pharmacoeconomics and Outcomes Research, Inc. (ISPOR 2020). Health outcomes are often quantified as quality-adjusted life years (QALYs), the CEA in this case being a cost-utility analysis (CUA). Analyses of this type are used by the UK's NICE (see above) to make decisions about prescribing expensive drugs through the state healthcare system.

The cost-effectiveness of medicines has already been alluded to in Chap. 15 where a successful drug to treat Alzheimer's disease would save billions of dollars in care costs. Further aspects of pharmacoeconomics will be covered in the next chapter (Chap. 17).

16.1.3.3 Portfolio Management

Portfolio management is central to pharmaceutical business strategy; this is because a biopharmaceutical company must maximize its financial returns to cover the soaring costs of drug development and to return value to its shareholders. In looking at a company's portfolio of products, managers will have to make decisions about the balance between novel targets and those for which there are several competitor drugs already available. There is also the specter of patent expiration hanging over any drug development program, given the long times between initial patent filing and commercial launch (see later). Pharmaceutical portfolio management is designed to make a business case for embarking upon a given drug development program or investing in new manufacturing processes (see Jekunen 2014 for review). To make decisions about projects, there must be a set of value-added indicators (VAIs) that will help managers to decide whether it is worth investing in a drug development project. One of the most used VAIs is the net present value (NPV) calculation. NPV is the current value of all cash receipts and outgoings during the lifetime of a project. It takes account of the changing value of money over time (discount rate) that may occur, for example, through inflation. Details of the simple formula used for the NPV calculation can be found on financial or business websites. The NPV figure is a monetary value which, if positive, means that the project is likely to be worth pursuing. If it is negative, however, then the company will gain nothing by taking the program any further. The standard NPV formula does not consider factors such as the risk inherent in a stage of drug development, so the risk-adjusted NPV (rNPV) can be calculated instead.

Other value-added indicators include internal rate of return (IRR) and return on investment (ROI) calculations. Positive values for each of these, as with net present values, give portfolio managers the confidence to initiate a development program. Financial tools such as Real Options Analysis, Decision Tree Analysis, and Monte Carlo simulation are also used to make decision-making more scientific (based on statistical principles) without having to rely upon "gut instinct." It is understandable that this instinct is frowned upon by business schools, particularly with such a complex area as modern drug development. Nevertheless, it is tempting to speculate that few of the great medicines of the past would have seen the light of day if companies had relied totally upon economic forecasting.

An example of how a risk-adjusted NPV may be calculated for each stage of drug development is shown in Table 16.5 (taken from Alactrita White paper (2019) with kind permission). The values for the different criteria figures fall within the average range of real-life examples. The discount rate accounts for the time-dependent value of money and commercial risk.

Table 16.5 Hypothetical example of rNPV (risk-adjusted net present value) calculation

	Phase I	Phase II	Phase III	Registration	Approval
POS (%)	63.2	30.7	58.1	85.3	100
Duration (Yrs)	1	2	4	1	10
Net cash flow ($M)	2	14	30	1	1000
rNPV calculation					
Year	1	2	4	8	9
Discount rate (%)	11	11	11	11	11
Valuation ($M)	11.9	24.1	114.9	451.7	588.9

Taken from Alactrita White paper (2019). POS is probability of success. Cash flow post-approval based on 10 years market presence at $100 M per year

16.1.4 Intellectual Property

The statutory protection of inventions by legislation is the foundation stone of commercial research and development. The main elements of IP are patents, copyright, trade secrets, trademarks, and bailments. These are summarized, except for patents, which will be covered in more depth.

- Copyright

  This often relates to written material or images where the originator has an automatic right to ownership. From a pharmaceutical R&D perspective, this will most likely apply to material in technical or commercial publications.

- Trade secrets

  These only last so long as they remain secret, i.e., are not disclosed to others without a prior legal contract, such as a confidentiality or non-disclosure agreement (CDA or NDA).

- Trademarks

  These can be registered in most countries and are used for company logos and related branding material.

- Bailments

  This is a legal term used to describe the property rights of a donor who gives physical materials to a third party. These materials may be biotechnology products, for example, in which there is a great deal of associated know-how that must be protected. A common bailment is the Material Transfer Agreement that occurs between laboratories in academia and industry.

16.1.4.1 Patents

Patents (or letters patent) are legal instruments which are designed to prevent the exploitation of an invention by competitors. The modern system used by the USA and UK originated in the eighteenth-century England, in the reign of Queen Anne. From this point in history onwards, the patent had to be submitted to the relevant office as a written document detailing the invention. Each country has its own patent law, but harmonization through the World Trade Organization encourages individual countries to comply with global regulations. The Agreement on Trade-Related Aspects of Intellectual Property Rights (TRIPS) has helped to bring the trade in pharmaceuticals under a global umbrella. In the past, IP protection has been insufficient to prevent the manufacture of generic western medicines in developing countries, but the situation has changed through changes to their patenting laws and desire to sell their own products in the international marketplace.

Types of Pharmaceutical Patents

Composition of Matter Patents

These protect a novel compound or biological based on it having a unique chemical structure. For example, cimetidine and ranitidine are two compounds which both bind to the histamine H2 receptor but are protected by composition of matter patents. This is because the patent examiners in these cases felt that the chemical structures of the two compounds and their derivatives were sufficiently different from each other. Patent applications are written in such a way that makes it difficult for a rival company to mount a legal challenge. This means that many different analogues of the compound must be laboriously documented in large, often heavy-going, documents. The skills required to prepare these applications and to maintain them over the years of the patent mean that many biopharmaceutical companies run dedicated IP departments.

A hint of the level of detail required for pharmaceutical composition of matter patents is given in Fig. 16.1, which is taken from the Espacenet online patents resource (Espacenet 2020).

This figure shows a basic backbone structure consisting of rings joined by single bonds. The letters (X, Y, Z, R and variants) refer to the possible substitutions to the basic structure, and these are listed in the patent. This is a so-called Markush claim, named after the US chemist who devised a system of notation for chemical structures in patents. An example of this is shown with the benzene ring at the far right of the figure. The line joining $(R^3)_y$ to the benzene is directed at the center of the structure. This means that the $(R^3)_y$ group can substitute at any point in the ring.

Patenting Genes

One of the most contentious issues in the commercialization of life sciences has been the patenting of DNA sequences. In the early days of the Human Genome Project, companies such Celera and Incyte Genomics filed patents for many of the novel gene sequences they identified in their laboratories. The feeling, in academic and other circles, was that this amounted to the patenting of life and should therefore be discouraged. Large pharmaceutical companies were relaxed about using sequences for drug targets that were patented by others, but the situation with diagnostics is more problematical. This is exemplified by the patenting of the BRCA1 gene by Myriad Genetics in the USA. Mutations in this, and the BRCA2 gene, provide strong diagnostic evidence for increased risk of breast or ovarian cancer. Myriad were granted patents on these genes by the US Patent and Trademark Office (PTO) and had monopoly on this genetic test, which they offer for sale. This monopoly position changed in the USA in 2013 after a court ruling in the case between the *Association for Molecular Pathology* v. *Myriad Genetics, Inc.* stated that "human genes can't be patented because they are a "product of nature." This has not had an adverse effect on patenting individual genes, as the Human Genome Project has placed much of this information in the public domain; however, there are concerns

WO 02/28433

PCT/GB01/04370

6

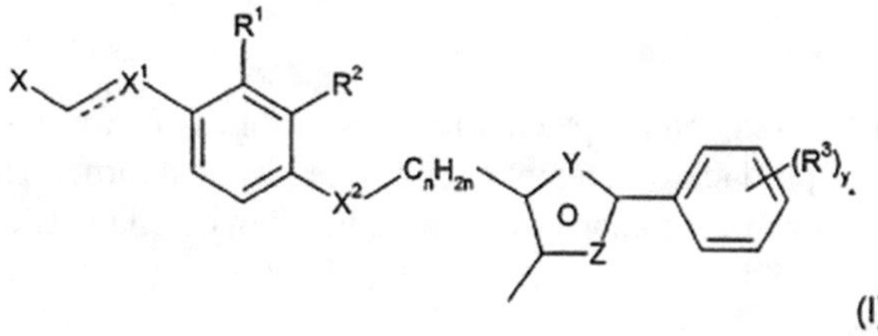

(I)

wherein:

5   X represents a COOH (or a hydrolysable ester thereof) or tetrazole group;
     X^1 represents NH, NCH_3, O, S, a bond (i.e. is absent), CH_2, or CH where the
     dashed line indicates that when X1 is CH the depicted bond is a double bond;
     X^2 represents O or S;
     R^1 and R^2 independently represent H, CH_3, OCH_3, or halogen;
10   n is 1 or 2;
     one of Y and Z is N and the other is S or O:
     y is 0, 1, 2, 3, 4 or 5;
     Each R^3 independently represents CF_3 or halogen.

Fig. 16.1 Portion of typical composition of matter patent taken directly from the Espacenet online patents resource, reproduced by kind permission

that lack of IP protection in this area may create problems with, for example, diagnostic tests that use a defined panel of genes which may not fall into the category of "obviousness." In 2019 a proposal was made by two US congressmen to modify the law to create clarity and reduce the influence of individual court decisions, thus allowing some flexibility for biopharmaceutical companies in patenting genetic processes (Servick 2019). This issue is unresolved at the time of writing in 2020.

Process Patents

The chemical and biological processes that lead to patented small molecule and biotech drugs are often an important part of a company's patent estate. Sometimes companies will even attempt to patent manufacturing processes for out-of-patent drugs. Patents for these so-called analogy processes can be granted if genuine inventiveness can be proved. Process patents are important for biotechnology; the humanization of monoclonal antibodies is a good example, where a highly competitive technology was developed by Gregory Winter in Cambridge and patented by the UK's Medical Research Council (MRC). Their technology transfer arm, LifeArc, has been able to recoup large royalties from companies wishing to develop humanized therapeutic antibodies (see Chap. 8). For example, they have received a lump sum of $1.29Bn by trading in some rights for their work on the checkpoint inhibitor antibody Keytruda, marketed by Merck & Co. (LifeArc 2019)

Secondary Patents

These are used to extend the patent life of drugs through reformulation, finding new indications, or producing different chemical or physical forms. These topics will be covered at the end of the chapter under the heading of Life Cycle Management.

16.1.4.2   Patent Timelines

The patenting formalities begin with an application (filing) to a national office such as the US Patent and Trademark Office or the UK Intellectual Property Office. In the USA, it is possible to file a provisional patent that lasts for 1 year before making the decision to file a non-provisional patent (which is published after 18 months). This is a useful way of keeping the costs down while assembling the data required to proceed to a full patent or to abandon the process at an early stage. As there is no such thing as an international patent, it is possible to use the Patent Cooperation Treaty (PCT) to file patents in multiple countries that are valid for about 30 months before having to be submitted to individual countries.

Once a patent application is filed in a specified territory, it has a life span of 20 years before the patent expires. This might seem like a long time, but given the years that it takes to develop the drug, the amount of patent protection left once it goes on the market is significantly reduced. In a commentary on this situation, S. Knowles from GSK wryly notes that a novelty umbrella that can be attached to beer bottle has 18 years of patent life compared with 13.5 years for a breast cancer therapy (Knowles 2010, reference 14).

A question that is often raised during my courses is "why don't companies wait as long as possible before patenting?" This is tempting, but given the fluid nature of employment in the industry and the general leakiness of information, this would be tantamount to commercial suicide. Sometimes risks are taken, however; I recall a situation in which a novel compound was isolated from a natural product that had been screened against a disease target. It was felt that delaying the filing of a patent would not be a problem, since the chance of a competitor finding the same drug was small. Although it later turned out that the compound could not be turned into a drug, a rival company did in fact come up with the same compound from a similar source, completely by chance.

16.1.5 Generic Drugs

Once the patent on a branded prescription medicine expires, the field is open to the generic drug manufacturers who will then sell the product at a lower price. Not surprisingly, consumers of healthcare encourage the use of generics wherever possible to save money. Equally unsurprisingly, the developers of branded medicines wish to recoup the costs of drug discovery and development and to invest in future programs. There is an inevitable tension between these different parties, assisted by various pressure groups, including those who accuse the biopharmaceutical industry of profiteering from disease. Whatever the pros and cons of the argument, the existence of generic competition is one of the major challenges for the drug discovery industry.

Attempts have been made to encourage the introduction of more generic medicines while at the same time providing a "soft landing" for the patent holder once the term has expired. The key legislation in the USA was introduced in 1984 as the Drug Price Competition and Patent Term Restoration Act, otherwise known as the Hatch-Waxman Act. Approval of a patented medicine requires a New Drug Application, supported by large amounts of data, as outlined in Chap. 13. It was considered that expecting generics manufacturers to go through the entire process themselves for a given compound (biologicals excepted) would be too great a burden. The Hatch-Waxman Act therefore specifies an Accelerated New Drug Application (ANDA) that just requires the generics company to demonstrate bioequivalence with the patented drug. In practice, this means testing absorption pharmacokinetics in patients and accepting a difference of 20% or less. In return, the patent holders can receive an extension of patent life up to a maximum of 5 years; this covers the time that the drug was under FDA review prior to marketing. The act allows patent holders to create legal objections that can delay the introduction of a generic product by a period of 30 months.

A summary of pharmaceutical patent timelines is shown in Fig. 16.2.

16.1.5.1 Generics and Biologicals

Generic small molecule drugs are identical atom for atom with the branded equivalent, but this is not the case with biologicals. This is due to the chemical nature of proteins and particularly to their glycosylation (Chap. 8), where variations occurring in different batches of recombinant protein can have a profound influence on the pharmacokinetics of the drug. Although a highly detailed analysis of glycoprotein structure and purity is possible with mass spectrometry, it is not possible to generate these molecules in a totally uniform way using standard manufacturing processes. For this reason, generic biologicals, or follow-on biologicals, are commonly known as biosimilars. Generic biologicals with improved pharmacokinetics, or other properties, are called biobetters or biosuperiors by some companies. Where will this end: Are bioawesomes on the way?

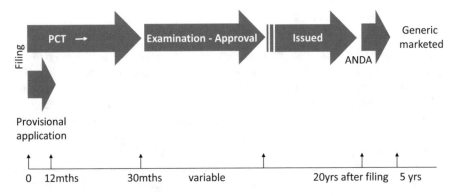

Fig. 16.2 Patent timelines for branded and generic medicines. The original drug has 20 years of patent protection from date of filing (a non-provisional patent). A PCT application is for application to multiple territories. Patent examination and approval times are variable. A generics manufacturer can submit an Accelerated New Drug Application (ANDA) and wait 5 years before marketing approval

The Hatch-Waxman Act did not consider generic biological products, since insulin and growth hormones were the only recombinant proteins available at the time the legislation was drafted, a situation addressed by the Biologics Price Competition and Innovation Act (BPCI Act) passed in 2010[2]. This requires the biosimilar's manufacturer to show interchangeability between their product and the FDA-approved medicine while giving a 12-year market exclusivity for branded biologics prior to generic competition (FDA Biosimilarity Guidance 2020).

European legislation for biosimilars has been in place since 2004 (EMA Biosimilars Overview 2020) with manufacturers of the branded product having 8 years of market exclusivity.

The development of generic proteins is inevitably more costly than with their small molecule equivalents, but the rewards are significant given the prominence of biologics in the list of top ten best-selling drugs (Table 16.3). Further comments on the commercial aspects of generics and biosimilars will be made in Chap. 17.

Orphan Drug

This is a term for a branded medicine that has been developed for rare diseases. In the USA, the Orphan Drug Act of 1983 provides 7 years of market exclusivity for drugs that treat fewer than 200,000 people (Office of Orphan Products Development 2020). The EMA's Orphan Regulation (for 5 in 10,000 people) provides 10 years of exclusivity, plus fast access to regulators, unless the drug becomes too profitable, in

---

[2] The FDA has also introduced a "Biosimilars Action Plan" to deal with the key issues arising from these complex generic products.

which case the period is decreased (Orphan Designation Overview 2020). For some examples of orphan drug approvals, see Miller and Lanthier 2018.

Pediatric Exclusivity

Under Section 505A of the Food, Drug, and Cosmetic Act passed in 2002, the FDA can grant "Pediatric Exclusivity," i.e., an extra 6 months of patent protection if a drug can produce health benefits in children. This can be very lucrative to companies; Pfizer, for example, was granted pediatric exclusivity for its Lyrica medicine after a successful phase III trial to treat epileptic seizures in children. Since US sales were around $3.5Bn before patent expiry, the company should recoup at least half of this figure in sales (Sagonowsky 2018).

16.1.6 Life Cycle Management

Product life cycle management (PLM) is a key component of the pharmaceutical development process and can be defined in the following stages: product development, introduction, growth, maturity, and decline. This section focuses on managing the decline of a branded medicine due to generic competition. It is in a biopharmaceutical company's interest to manage the life cycle in such a way as to maximize profits over the time available. This can involve a process of "evergreening" to maximize the commercial returns on a product through secondary patents. This can lead to a battle of wits between the company and the patent examiners who must decide whether the relevant patents are admissible. Patent life extension can be achieved through line extensions and secondary medical uses as follows:

16.1.6.1 Line Extensions

Also known as reformulations, or follow-on products, line extensions may be reformulations of the original product (patented NCEs) or else different chemical or physical forms that may affect formulation or delivery (see Fowler 2017 for summary). For example, ranitidine, the active ingredient of the antiulcer drug Zantac®, was reformulated as ranitidine bismuth citrate and marketed in 1996 as Tritec®. This line extension took advantage of new findings that implicated the bacterium *Helicobacter pylori* in peptic ulcer disease. This bacterium can be killed by bismuth, so the combination of an H2 antagonist and this metal helped Glaxo to maintain sales in the gastrointestinal area for a while longer. The patent for ranitidine was also extended by producing it in a different crystalline (polymorphic) form (Chap. 10). Form 2 ranitidine hydrochloride was patented in 1985 based on it having "favorable filtration and drying characteristics." As the original (form 1) 1978 patent was due to expire in 1995, the form 2 patent gave Zantac® an extra lease of life up to 2002. Although highly lucrative in terms of sales, this period also saw extensive litigation between Glaxo and Novopharm, a generics drug maker that attempted to market ranitidine itself. Nevertheless, Glaxo's strategy of holding off generic competition by patenting different polymorphs has been eagerly adopted by other companies. Another strategy is to identify different salt forms, or hydrates, of active compounds which may have superior properties to the original drug. These properties may include increased stability in the stomach or ease of formulation. The approach adopted by AstraZeneca involved the separation of chiral (mirror image forms) of its drug to extend patent life. This was achieved with its anti-ulcer drug omeprazole (Losec®) that replaced Zantac® as the treatment of choice for peptic ulcers. Losec® is a racemic mixture (Chap.7) of two mirror image forms, R-omeprazole and S-omeprazole, the latter named esomeprazole (Fig. 16.3). Both enantiomers are prodrugs that are metabolized to an active molecule lacking a chiral center. However, esomeprazole has a greater metabolic stability than the R form, thereby achieving a higher serum concentration and requiring a lower dose for the therapeutic effect. This strategy allowed the company to market esomeprazole as Nexium® and to maintain its position in the anti-ulcerants marketplace until 2014 when generic competition took over.

Finally, it is possible to patent based on new formulations so long as the product is demonstrably better than the original, that is, it shows an inventive step. An example of this is the anti-inflammatory drug diclofenac produced by Novartis. The original patent on the oral formulation of this drug has expired, but Novartis has produced a line extension by formulating the drug as Voltaren® gel for topical application.

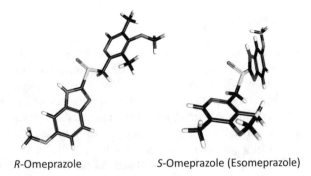

R-Omeprazole   S-Omeprazole (Esomeprazole)

Fig. 16.3 Three-dimensional structures of chiral molecules that comprise the anti-ulcerant Losec®. The esomeprazole prodrug (S-omeprazole) has greater metabolic stability than the R-enantiomer and was patented by AstraZeneca as Nexium® thereby extending the patent life of this molecule. The * indicates the sulfur atom that is the center of symmetry where the two mirror image forms overlap

16.1.6.2 Second Medical Use Patents

These are sometimes referred to as use patents; they may be used by companies who want to extend the patent life of their own products or else by companies that want to patent the use of an unpatented drug for a new indication. There appears to be more resistance to this from the patenting authorities than other line extension strategies, but examples do exist, such as thalidomide. This compound, discussed earlier in the book, gained notoriety through the birth defects it caused after being prescribed to pregnant women for morning sickness. Academic investigators subsequently noticed that the drug was useful in treating leprosy, some AIDS symptoms, and multiple myeloma. Later still, the FDA authorized the use of thalidomide in leprosy, and an analog lenalidomide was developed by Celgene and marketed under the trade name Revlimid®. Other examples include sildenafil (Viagra®), which was developed for angina but successfully marked for erectile dysfunction by Pfizer. For biologics, Roche's Rituximab® was developed for B-cell cancers (leukemia, lymphomas) but is also indicated for rheumatoid arthritis.

> **Summary of Key Points**
> The cost of developing a new drug can be over one billion dollars.
> The USA is the largest market for prescription medicines, followed by China.
> Global sales of the top ten medicines in 2019 include eight biological products.
> Patent life for branded medicines is compromised by the time taken for them to be developed and approved.
> The approval process for generic medicines is underpinned by the US Hatch-Waxman Act that provides opportunities for generic manufacturers while protecting patent holders.
> Generic products based on biologics (biosimilars) are challenging to manufacture but are very important commercially.
> Portfolio and Life Cycle management are key commercial functions within a biopharmaceutical company.

References

Alactrita White paper (2019) https://www.alacrita.com/whitepapers/valuing-pharmaceutical-assets-when-to-use-npv-vs-rnpv. Accessed 18 May 2020

DiMasi JA et al (2016) Innovation in the pharmaceutical industry: new estimates of R&D costs. J Health Econ 47:20–33

Edison (2019) https://www.edisongroup.com/investment-themes/drug-pricing-market-access-and-reimbursement/23551/. Accessed 16 May 2020

Espacenet (2020) https://worldwide.espacenet.com/patent/. Accessed 18 May 2020

FDA Biosimilarity Guidance (2020) https://www.fda.gov/regulatory-information/search-fda-guidance-documents/quality-considerations-demonstrating-biosimilarity-therapeutic-protein-product-reference-product. Accessed 19 May 2020

Fowler AC (2017) Pharmaceutical line extensions in the United States. nber.org. Accessed 20 May 2020

Francisco M (2019) Third-quarter biotech job picture. Nat Biotechnol 37:1381

GSK History (2020) https://www.gsk.com/en-gb/about-us/our-history/. Accessed 14 May 2020

Healthcare.gov (2020) https://www.healthcare.gov/choose-a-plan/plan-types/. Accessed 15 May 2020

ISPOR (2020) https://www.ispor.org. Accessed 16 May 2020

IQVIA Institute Report (2019) https://www.iqvia.com/insights/the-iqvia-institute/reports/the-global-use-of-medicine-in-2019-and-outlook-to-2023. Accessed 13 May 2020

Jekunen A (2014) Decision-making in product portfolios of pharmaceutical research and development – managing streams of innovation in highly regulated markets. Drug Des Devel Ther 8:2009–2016

Knowles SM (2010) Fixing the Legal Framework for Pharmaceutical Research. Science 327:1083–1084

LifeArc (2019) https://www.lifearc.org/news/2019/lifearc-monetises-keytruda-royalty-interests-20052019/. Accessed 19 May 2020

Medicare (2020) https://www.kff.org/medicare/fact-sheet/an-overview-of-the-medicare-part-d-prescription-drug-benefit/. Accessed 15 May 2020

Miller KL, Lanthier M (2018) Investigating the landscape of US orphan product approvals. Orphanet J Rare Dis 13:183. https://doi.org/10.1186/s13023-018-0930-3. Accessed 19 May 2020

NHS Expenditure (2020) https://commonslibrary.parliament.uk/research-briefings/sn00724/. Accessed 15 May 2020

NICE (2020) https://www.nice.org.uk/. Accessed 16 May 2020

Office of Orphan Products Development (2020) https://www.fda.gov/industry/developing-products-rare-diseases-conditions#About%20OOPD. Accessed 19 May 2020

Orphan Designation Overview (2020) https://www.ema.europa.eu/en/human-regulatory/overview/orphan-designation-overview. Accessed 19 May 2020

PCORI (2020) https://www.pcori.org/. Accessed 16 May 2020

PPD Inc (2020) https://www.ppd.com/. Accessed 14 May 2020

Sagonowsky (2018) https://www.fiercepharma.com/pharma/pfizer-wins-blockbuster-patent-extension-for-lyrica-exclusivity-now-stretches-until-june. Accessed 19 May 2020

Servick K (2019) Controversial U.S. bill would lift Supreme Court ban on patenting human genes. Science. https://doi.org/10.1126/science.aay2710. Accessed 19 May 2020

Urquhart L (2020) Top companies and drugs by sales in 2019. Nat Rev Drug Disc 19:228

The next biotech superpower (2019) Nat Biotechnol 37:1243

Innovation nation (2019) Nat Biotechnol 37:1264–1276

Wouters OJ et al (2020) Estimated research and development investment needed to bring a new medicine to market, 2009–2018. JAMA 323:844–853

Chapter 17
Challenges and Responses

Abstract This chapter examines the state of the biopharmaceutical industry at a point where the genomics revolution has begun to bear fruit through the introduction of biotherapeutics including advanced therapy medicinal products such as gene and cell therapy. As well as highlighting these successes, this chapter also lists the challenges facing the industry, which include falling productivity despite increased investment in R&D, pressures from healthcare providers and patients, and loss of patent protection for several major drugs. The second half of the chapter ends on a more positive note, with a description of the responses that companies are making by adopting new technologies such as artificial intelligence/machine learning, acquiring real-world evidence from clinical data, and restructuring through mergers and acquisitions. Finally, some comments are made about the future of drug discovery, including speculations about science, industry structures, and new opportunities provided by online social and professional networking.

17.1 Introduction

This chapter concludes the long journey from drug target to marketed product by examining the state of the drug discovery business one fifth of the way through the twenty-first century. It describes some of the challenges facing the biopharmaceutical industry and the different responses that are being made to meet those challenges. There are many difficulties to be overcome, as those who have lost their jobs in biopharmaceutical companies will attest. Also, to be frank, there can be a sense of doom and gloom about the future of the biopharmaceutical industry in its present form. This is partly due to the nature of drug discovery itself, coupled with external pressures from governments, regulators, and patients; the global economic downturns and political pressures do not help matters either[1]. On a positive note, it is unthinkable that the industry will collapse since it has immense human and financial resources and human disease will never disappear completely. There will always be

---

[1] This was written during the COVID-19 pandemic and attendant economic catastrophe. In contrast to other financial crashes during the last 100 years, the biopharmaceutical industry is a key part of the solution to this problem.

the need for a viable drug discovery industry, even one that may have been through a painful period of restructuring.

17.1.1   Pressures on the Industry

17.1.1.1   Productivity Slowdown

In 2012 Scannell and colleagues pointed out that the number of new drug approvals per billion US dollars R&D spend had approximately halved every 9 years since 1950, an 80-fold fall when allowing for inflation (Scannell et al. 2012). The authors named this phenomenon Eroom's law, a mirror of Moore's law where productivity (semiconductor density) increases at a fixed rate over time. However, more recent studies have shown that Eroom's law has been broken[2], with 0.7 launches per billion $US of R&D spending per year by 2018 where 59 drugs were approved by the FDA (48 in 2019). Some of the reasons for this improvement have been discussed in (Ringel et al. 2020) and will also be highlighted in this chapter.

Despite this promising recovery in industry fortunes, there are still plenty of expensive phase III failures (e.g., Alzheimer's drugs, Chap. 15). The following section outlines the different parts of the development process where these failures occur.

Rates of Attrition by Development Phase

There is clearly a need for manufacturers and their investors to understand the reasons for failure and use them to devise countermeasures. These will include "failing fast" to avoid wasting time and money on projects with a poor chance of clinical or commercial success.

One analysis was undertaken by AstraZeneca in an analysis of their drug development programs up to 2016. They demonstrated that the percentage of drug candidates completing phase III clinical trials improved from 4% in 2005–2010 to 19% in 2012–2016 (Morgan et al. 2018), the latter shown in Table 17.1:

An analysis by Informa Pharma Intelligence (2019) of all trials terminated by different companies in 2018 gives a more detailed breakdown of trial phase and therapeutic areas as summarized in Table 17.2.

Of note is the highest proportion of trial failures being in phase II and in oncology. Apart from lack of efficacy, some terminations are due to business decisions related to company strategy.

---

[2] Moore's law no longer holds either as the problem of heat dispersal limits the density of circuits that can be fabricated on a computer chip.

Table 17.1 AstraZeneca development failures 2012–2016 as percentages of 47 (Preclinical), 43 (Phase I), and 22 (Phase II)

	Preclinical	Phase I	Phase II
Safety	50	38.5	8
Efficacy	17	38.5	84
PK/PD	0	8	0
Strategy	33	15	8

From Morgan et al. (2018)

Table 17.2 Termination of clinical trials by phase in 2018 (Informa Pharma Intelligence 2019 with kind permission)

Therapeutic area	I	I/II	II	II/II	III	III/IV	Total
Oncology	101	72	150	2	42	1	368
CNS	19	6	40	7	40		112
Autoimmune/inflammation	13	8	50	4	26		101
Infectious disease	13	3	26	4	8		54
Metabolic/endocrinology	13	3	18	2	11		47
Cardiovascular	8	2	16	2	17		45
Ophthalmology	1	1	13		8		23
Genitourinary	2		5	1	4		12
Vaccines			6		1		7

17.1.1.2 Healthcare Cost Pressures

Biopharmaceutical companies attempt to maximize their revenues as soon as possible after the launch of a drug; there are several reasons for this: for example, patent life is ticking away so there is a relatively short period of time available before generic competition sets in. There is also the need to recoup the considerable investment required to produce the drug in the first place; finally, the business must be profitable to be able to invest in future research and development. For the reasons just outlined, newly introduced medicines are sold at the highest price that the healthcare provider will bear. Healthcare costs have soared in the developed world and will continue to do so as the average age of the population increases. The situation is, of course, far worse in the developing world, where branded medicines are mostly unaffordable at Western prices. This has led to considerable political pressure on companies who have been forced to adopt new pricing models in different markets, particularly for drugs to treat infectious diseases. In the UK, medicine prices are constrained by negotiation with government using the Voluntary Scheme for Branded Medicines Pricing and Access which places a cap on the maximum profits that a company can make.

Drug profits may be reduced because of the parallel trade that occurs between different countries, for example, the different countries of the EU or the USA and Canada. This is because the same drugs are priced differently in each country; they can therefore be bought in bulk from countries where they are cheaper to be sold on in other markets at a higher price. Parallel trading is not illegal, and the actual impact on profits is quite small, but counterfeit drugs certainly are illegal. This is becoming more of an issue with the developed world as "lifestyle drugs" such as Viagra® and weight loss treatments are sold over the Internet. Many counterfeit drugs contain reduced levels of API or none. This is not only potentially dangerous for the patient, but it also reduces confidence in the biopharmaceutical industry itself.

The last point about lack of confidence in the biopharmaceutical industry is exemplified in a dramatic way by exchanges in the US Senate between congressmen and drug company executives in 2018: Emotionally charged comments such as "Drugmakers behave as if patients and taxpayers are unlocked ATMs full of cash to be extracted, and their shareholders are the customers they value above all else" and drug pricing practices are "morally repugnant" sadly reflect the image in many people's minds about the biopharmaceutical industry in general (Reuters 2019).

17.1.1.3 Product Withdrawals and Litigation

Clinical trials can only go so far when it comes to assessing drug safety. Even large phase III studies do not continue for long enough to be able to reveal longer-term safety issues. The medicine will usually have been taken by millions of patients before any problems are brought to the attention of doctors, companies, and regulators (see post-marketing surveillance Chap. 12).

Two examples of product withdrawal come from the glitazone class of anti-diabetic medicines. Troglitazone (Rezulin®) was marketed in the late 1990s by Parke-Davis/Warner-Lambert (now Pfizer) and rosiglitazone (Avandia®) from GSK in the early 2000s. Rezulin® was sold in the UK by GSK who withdrew it from sale in 1997 after 135 reported cases of serious liver toxicity and 6 deaths in the USA. The drug remained in sale in the USA until 2000 and was the subject of extensive litigation against the manufacturer. Avandia® was a commercially successful drug for its maker GSK until it was linked to a 43% increased risk of heart attacks by a meta-analysis of 42 clinical trials conducted by the chief cardiologist at the Cleveland Clinic. The product was withdrawn from the EU but is still approved by the FDA, albeit with restrictions on its use.

Another reason for a product withdrawal is a problem with the chemical stability of the drug product. For example, the H2 antagonist ranitidine (Zantac®) has been withdrawn from sale in 2020 after concerns about its contamination with the probable human carcinogen, N-nitrosodimethylamine.

The above examples highlight the negative impact of drug safety concerns on the business and reputation of individual companies. If a medicine causes death and disability in a proportion of patients who take them, the company is generally considered responsible. There are situations, however, where the company has withdrawn a drug from market because of safety concerns, but patient groups then demand that it be reinstated. This has occurred with the therapeutic antibody natalizumab, marketed as Tysabri® for the treatment of multiple sclerosis. It works by blocking an adhesion molecule in blood vessels that allows the migration of inflammatory leukocytes into the brain. In 2005, the drug was voluntarily withdrawn from sale by its manufacturer Biogen Idec after three patients developed a severe brain inflammation called progressive multifocal leukoencephalopathy (PML). Tysabri® was relaunched in 2006 after the FDA had taken evidence from interested parties, including patients with MS who were desperate for the medicine. The relaunch was authorized on the condition that the drug was to be used in the clinic under strict guidelines. This example illustrates the dilemma faced by drug makers and patients; if few treatments exist for a chronic illness other than a drug with potentially serious side effects, should the patient be allowed to make a choice, or will this be denied because of regulation and probable litigation? The Tysabri® case showed that it is possible to follow a path between the two; most companies, however, will become more risk averse, particularly if bad publicity about their drugs is rapidly broadcast through online resources, with little chance of them being able to redress the balance.

17.1.1.4 Falling Off the Patent Cliff

Perhaps the most immediate and dramatic problem for individual biopharmaceutical companies is the loss of patent protection on the best-selling medicines that bring them blockbuster profits. This loss of protection, called the patent cliff, is very costly for the manufacturers of branded medicines since almost $90Bn of revenues will be lost in the period 2020 to 2023, as shown in Table 17.3.

In Europe, the top selling branded product adalimumab (Humira®) is facing biosimilar competition, with the same happening in the USA in 2023. Only Opdivo and Keytruda from the list (presented in Chap. 16) will not be facing generic competition by the end of 2023.

17.1.2 Meeting the Challenges

Having described the scientific and commercial pressures felt by the biopharmaceutical industry, it is now time to conclude the chapter (and book) on a positive note. The problems facing the industry are immense but essentially solvable, given enough time, resources, and political will. Assuming that most people who are involved in the drug discovery industry agree that there is a problem, solutions may

Table 17.3 Sales loss in developed markets ($Bn) due to loss of exclusivity

	2020	2021	2022	2023
Small molecules	21.7	14.9	21.8	19.6
Biologics	5.1	3	1.7	1.9

Data taken from IQVIA Institute (2019)

emerge quite rapidly over the next few years. Alternatively, with apologies to US readers, the spirit of Winston Churchill's quotation: "one can always trust the United States to do the right thing, once every possible alternative has been exhausted" might be applied to the biopharmaceutical industry. Whatever the future holds, there are some emerging trends which will shape it, as discussed below with reference to previous sections on productivity, healthcare costs, and patent protection.

17.1.2.1 Company Structures and the Pharmaceuticals Marketplace

Company Spinoffs

Large multinational biopharmaceutical companies have the same problems of scale as other major corporations. Although large organizations have strong financial and human resources, the culture of management hierarchies, meetings, committees, and fighting for internal resources does not compare well with that of the leaner biotech companies that are attractive to creative people. The business world is changing, however, and companies such as Alphabet (Google parent) or Apple demonstrate that it is possible to sustain a more youthful business culture within a multinational enterprise (although not without controversy).

Some large pharmaceutical companies have created revenues from drug development alongside "safer" consumer products or less risky technologies. The financial and organizational pressures of keeping these different businesses together have prompted companies like Abbott to split into two companies. Abbott Laboratories was founded in 1888 and sold diagnostics, nutritional products, and medical devices alongside research-based branded pharmaceuticals. A separate company, AbbVie, was spun off in 2013 to focus on high-value pharmaceuticals, including the Humira® monoclonal antibody for rheumatoid arthritis and Crohn's disease. GSK is following this trend through a consumer healthcare joint venture (JV) with Pfizer to create a stand-alone company that allows GSK in its CEO's words "to become a biopharma company with an R&D approach focused on science related to the immune system, use of genetics and new technologies."

Therapeutic Areas

The choice of therapeutic area for drug development is determined by the prevalence of the disease (i.e., number of patients), the technical challenges, and the required financial investment and returns. Large companies focus on the chronic diseases of an aging population, including cancer, Alzheimer's, and inflammatory diseases. However, the once dominant central nervous system (CNS) therapeutic area has fallen out of favor in large companies as the complexities of depression and schizophrenia plus failures in Alzheimer's and Parkinson's disease made drug development less attractive. Of course, this will inevitably change as new science refreshes the field and smaller companies introduce their own products (for a commentary, see Bell 2020).

Although large companies will continue to develop drugs for major diseases affecting large numbers of patients, there is a noticeable increase in their portfolio of orphan drugs[3]. It is estimated that these drugs will make up one fifth of branded medicine sales in 2024, a total of $242Bn (EvaluatePharma Orphan Drug Report 2019). Inevitably, the small number of patients requiring the orphan medicine leads to a high price tag; for example, the list price of the Luxturna® gene therapy for a rare eye disorder is $425,000 per eye. Apart from price negotiations to bring these prices down (see later), there is the concept of "one and done" in which a single therapy (perhaps gene or cell-based) may provide a permanent solution, unlike cheaper treatments that must be administered over many years.

Mergers and Acquisitions

Merger and acquisition (M&A) activity with the biopharmaceutical industry has increased in response to falling revenues resulting from poor drug pipelines and patent expirations. The purpose of M&A is either to acquire another company's products and marketing infrastructure or else to buy in technical expertise that is lacking in the parent company. Some examples of M&A activity are shown in Table 17.4.

The value of different drug classes acquired by M&A in 2019 is shown in Table 17.5.

The mergers of large independent companies can, of course, lead to considerable disruption of science and business activities and hence productivity. It also means site closures and job losses, which, of course, is of great concern to employees, their families, and local communities. Some, but by no means all, can find new employment in the industry through bringing their skills and experience to new drug discovery companies. Ironically, managers of these startup organizations often form business alliances with their former companies. There has always been a strong need large pharma expertise in the biotechnology industry, and there is no longer a

---

[3] See Chap. 16 for definitions.

Table 17.4 M&A activity in 2019

Acquirer	Target	Value $Bn
BMS	Celgene	74
AbbVie	Allergan	63
Pfizer	Array BioPharma	11.4
Novartis	The Medicines Company	9.7
Eli Lilly	Loxo oncology	8
Roche	Spark therapeutics	4.8

Source Micklus and Giglio (2020)

Table 17.5 Asset classes acquired in 2019 M&A and their values. (BioPharma Dealmakers 2020)

Asset	Value $Bn
Small molecule	90.3
Biologics	96.4
Medtech	23.3
Diagnostics	3.3
Immunotherapy/cell therapy	4.8
Gene therapy	11

Table 17.6 Licensing deals for FDA approved NMEs 2014–2018 according to development stage

Stage	Percentage
Discovery	4
Preclinical	8
Phase I	10
Phase II	23
Phase III	19
Pre-registration	5
Marketed	31

Data from GlobalData PharmSource (2020)

shortage of qualified personnel to guide smaller companies through the minefield of drug development.

In-Licensing

Apart from M&A activity to refresh or increase product pipelines, the (less expensive) option is to in-license drugs of all types from external organizations (who, of course, out-license them). A typical scenario will be a large pharma's acquisition of a promising early stage drug from a smaller biotech company. An analysis of FDA approvals from 2014 to 2019 showed that 41% of drugs from the top ten biopharmaceutical companies originated in-house, while 30% were in-licensed and 29% acquired through M&A (HBM New Drugs Approval Report 2019).

Licensing deals occur at different stages of the drug development cycle with attendant risk tending to decrease as the product gets closer to marketing. Table 17.6 shows the percentage of NMEs licensed by stage between 2014 and 2018.

Academia-Industry Liaison

The profit-driven culture of industry has always sat uneasily with academia, despite many successful collaborations between the two, including the creation of numerous startup companies by university scientists (see Thomas and McKew 2014 for a review). The current dearth of new drugs and the need for innovation is prompting the large biopharmaceutical companies to fine-tune their approach to academic collaborations. One way of doing this is to move into purpose-built facilities near a university, an example being the Genomics Institute of the Novartis Research Foundation (GNF) with access to major Californian research centers such as the Scripps Institute. The reverse can occur, with academic institutions establishing translational research centers to turn their laboratory discoveries into products. Examples include the Institute for Translational Medicine and Therapeutics (ITMAT) at the University of Pennsylvania and the Academic Drug Discovery Consortium (ADDC).

Commercial organizations must protect their intellectual property, often a point of contention in academic collaborations where technology transfer offices are tasked to negotiate equitable licensing deals. Despite this, however, biopharmaceutical companies are offering open-source collaborations, particularly around small molecule screening. An early example is the GSK's open access to structures and data for over 13,000 compounds for antimalarial screening (Gamo et al. 2010).

This initiative is a major departure from the conventional practice of keeping proprietary data securely within the walls of a company, unless publicized in a patent or scientific publication. Of course, this way of sharing data is ultimately to the benefit of patients and clinicians if the development of new antimalarial drugs is accelerated, but does it benefit the pharma company? It certainly does from a public relations point of view and possibly in tangible commercial benefits; however, it remains to be seen how this will work with highly competitive disease areas like cancer and neurodegeneration.

Drug Discovery for the Developing World

One of the major criticisms of the biopharmaceutical industry has been its neglect of diseases that are more prevalent in the developing world. Although these illnesses affect huge numbers of individuals, the healthcare systems of the most affected countries do not have the resources to pay Western prices for medicines. From the perspective of the discovery stage for diseases such as malaria, mention has already been made of "open-source" drug discovery where compound screening data are made freely available to the global scientific community. So, there is no shortage of basic research into diseases like malaria and tuberculosis, but funding for the clinical development of promising medicines is not easy to obtain. Therefore, product development partnerships (PDPs) have been established between biopharmaceutical companies, governments, and non-government organizations (NGOs), to direct money towards focused drug development programs. These are run on commercial lines, with input from drug discovery experts, and include the Global Alliance for TB Drug Development, Medicines for Malaria, and the Drugs for Neglected Diseases Initiative (DNDi).

The situation with HIV in developing countries is particularly controversial, as the drugs with proven efficacy are freely available to Western populations but are priced at an unaffordable level for other markets. The creation of ViiV Healthcare in

2009 as a joint venture between GSK and Pfizer (later joined by Shionogi) has helped to redress the balance by sponsoring R&D into new HIV treatments (ViiV 2020). The company also sells branded anti-HIV drugs at reduced prices to developing countries, for example, through agreements with organizations such as the Medicines Patent Pool (MPP) based in Geneva (Medicines Patent Pool 2020).

Outsourcing

One way of dealing with the increasing costs of pharmaceutical R&D is to cut operational costs by outsourcing key operations to other companies. These operations cover much of the activity required for preclinical and clinical development, with an emphasis on the last two functions. For example, a drug candidate discovered in a small biotech or academic lab will often be evaluated further by a contract research organization (CRO, see Chap.16) prior to negotiating a licensing deal with a larger pharmaceutical company. CROs can also be involved in discovery science, such as medicinal chemistry.

The sector has a projected value of around $90Bn by 2026 (Fortune Business Insights 2020) and, while being fragmented into thousands of companies worldwide, is dominated by a small number of companies such as IQVIA. The main market is the USA, along with a mature market in Europe. The Asian market is developing rapidly through changes in clinical trial regulations, large patient populations, a skilled workforce, and a strong interest in biosimilars.

Contract manufacturing organizations (CMOs) must not only meet the challenges of producing small molecules at scale but also complex biological products, including gene and cell therapies. The products may be the complete medicine (finished dosage form, FDF) or the active pharmaceutical ingredient (API) alone. Contracting out to CMOs is normally undertaken by small- to medium-sized companies. Examples of large CMOs include the Lonza Group in Switzerland and Patheon (Thermo Fisher).

17.1.2.2 Healthcare Costs

The area of drug pricing is a highly contentious issue which has partly been covered in Chap. 16. This section covers some potential solutions to the tension between manufacturers who need to recoup the costs of drug development and invest in future research and healthcare providers. The latter, whether government sponsored or private, have limited resources and must provide for an aging population of patients who need innovative medicines to treat chronic diseases[4].

---

[4] And of course, an increasing number of younger patients with rare diseases requiring expensive therapies

Value-Based Arrangements

One way of ensuring that drug prices reflect benefit to patients is to use value-based arrangements (VBAs) (otherwise known as outcomes/performance-based contracts or risk-sharing agreements). This could involve a rebate from the manufacturer to the user if the drug or device does not perform as expected (see Mahendraratnam et al. 2019). This area is in an early stage of development and includes possible variations such as indication-based pricing, which is self-explanatory.

Precision Financing

The MIT NEWDIGS (New Drug Development ParadIGMS) Consortium was established to develop strategies for healthcare financing (MIT NEWDIGS 2020). One of their areas of interest is treatments, such as gene and cell therapies, with high upfront prices. Their FoCUS Project was launched in 2016 to consider alternative financing approaches that included 1-year milestone-based contracts and multi-year performance-based annuities. They also considered the idea of Orphan Reinsurer and Benefit Manager (ORBM), which assigns gene therapies to an intermediary which takes responsibility for all aspects of care management (Barlow et al. 2019).

Netflix Model

Subscription-based pricing for drugs has been termed a "Netflix Model" like that company's payment system for streaming films and TV. A prominent example is the price for the hepatitis C antiviral drug sofosbuvir negotiated by the Australian government with the manufacturer Gilead. An upfront payment (subscription) of $776 M was made for 5 years' worth of the drug whose list price was $73,000. The resulting price per patient was between $7 and 12,000 thus representing a substantial saving (Moon and Erickson 2019).

17.1.2.3   Intellectual Property

Mention has already been made of the patent cliff affecting the sales of lucrative branded medicines. One survival strategy for large companies which are exposed to generic competition is to aggressively protect the patent on the drug product or its manufacturing process. The other is for companies to adopt the "if you can't beat them, join them" approach and establish or acquire a generic drug business themselves.

Patent Thickets

Biologics such as monoclonal antibodies are lucrative products that are vulnerable to replacement with cheaper biosimilars. The top selling antibody Humira®, marketed by AbbVie, has lost patent protection in the EU, resulting in the availability of much cheaper biosimilars. In the USA, however, AbbVie has established a "patent thicket" of over 100 patents to maintain control over the US market, potentially to 2034. This approach has resulted in legal action from rival companies and patient groups and may eventually be made untenable through US government legislation (National Law Review 2020).

Spinoff Generics Companies

The spinout of generics companies from established branded pharmaceutical manufacturers is not new: The Swiss companies Ciba-Geigy and Sandoz merged in 1996 to form Novartis which has retained Sandoz as a generics and biosimilars subsidiary. More recently in 2015, the Upjohn division of Pfizer produced a range of generic products, and as of 2020, it is planned to merge it with Mylan to form Viatris, potentially a top global player.

On a smaller scale, the antivirals manufacturer Gilead Sciences will sell its own branded hepatitis C drugs in the USA as cheaper "authorized generics" through a subsidiary Asegua Therapeutics. In essence the pricing will be discounted from the list price in much the same as other medicines, but this will be a more transparent process.

17.1.2.4   Improving Productivity: Drug Discovery and Development

The previous sections have covered the commercial structures that are put in place to improve productivity in the biopharmaceutical industry, but central to this is the clinical effectiveness of the drugs and devices. This of course is a major challenge given the comments about human biology and predictability given earlier in this book, the example of failed Alzheimer's trials, and the amyloid hypothesis being given several times. Every part of the drug discovery pipeline from early research to clinical development is being examined in detail to identify areas where technical improvements could be made. These are summarized in Fig. 17.1 below:

Data Science

Efforts to improve the processes in the above figure large datasets and the means to extract information from them. Large databases of DNA sequences, metabolites, and detailed clinical/lifestyle information are being built to gain a more detailed understanding of diseases and their causes. At the same time, many novel drug

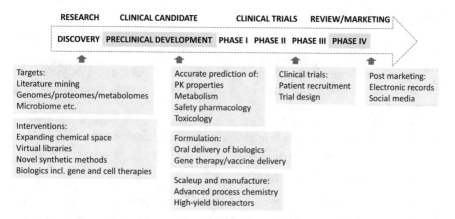

Fig. 17.1 Improving the efficiency of drug discovery and development. Some areas of interest for R&D

targets may be revealed that could otherwise have been missed using more traditional low-throughput methods. Some of these data gathering initiatives like the UK Biobank and 100,000 genomes program have been described already in Chap. 14. Others, such as the China Kadoorie Biobank with data from over half a million people, help to shift the emphasis away from studying people of European descent only; this has obvious attractions for drug discovery efforts directed towards different markets (China Kadoorie Biobank 2020).

Artificial Intelligence

To extract meaning from the types of data mentioned above, artificial intelligence (AI) is being deployed to identify actionable points at each part of the pipeline[5]. AI is a general term used to describe computer-aided learning using mathematical algorithms. Two important areas are listed here:

- Machine learning.

    Finding patterns in data without prior instruction, for example, associating images with a specified object (tumor cell, etc.) Unsupervised learning requires no prior training with the "correct" outputs unlike supervised learning.

- Deep learning.

    Uses neural networks to identify features that become progressively more abstract as it develops. An example would be the AI required for self-driving cars.

---

[5] To quote Leo Anthony Celi "Health data is like crude oil. It is useless unless it is refined."

Applications

- Literature mining.

The millions of articles that comprise the biomedical literature are far too numerous for human investigators to read and analyze, so interesting material related to drug discovery may be "hidden in plain sight." Literature or text mining algorithms are designed to extract textual and graphic data and to make associations across a wide span of literature. Example applications include Semantic Scholar and Iris.AI (Semantic Scholar 2020, Iris.AI 2020).

- Optimizing compound selection.

Screening for compounds that bind to a drug target protein and optimizing the hits using medicinal chemistry is a time-consuming process. Mention has already been of generating novel hits by screening billions of virtual compounds (Chap. 9), but AI can be applied to other steps. For example, a single compound binding to a defined target will also bind to many other proteins, albeit at lower affinities, so these should be evaluated before the compound enters development, as is being undertaken by companies like Cyclica in Canada (Cyclica 2020).

The opposite approach to the above involves looking at a single target protein in fine detail and using AI to identify interacting molecules that will also have the right drug-like properties and safety profile. This is the approach of the Scottish company Exscientia (Exscientia 2020) who have reduced discovery times for compounds affecting targets of interest to them and their collaborators (which include Celgene and GSK).

Preclinical Evaluation

Apart from problems with efficacy, some drug development programs fail through problems with pharmacokinetics and safety in humans. It follows that in an ideal world, highly predictive information should be available before a drug candidate enters clinical trials.

In some ways this can be an extension of the drug-protein target analysis since pharmacokinetic problems can arise through the action of metabolizing enzymes like the CYP450s, drug transporters, and issues like protein binding. AI approaches for predicting these interactions should help to reduce clinical failures and facilitate the move away from using animals in preclinical development.

Clinical Trials

Some of the problems with clinical trial design and execution as well as potential solutions like adaptive clinical trials have been described previously (Chap. 12). This section follows this up, by commenting about real-world data and evidence

(RWD and RWE) and the ways in which companies are using these to improve clinical trial design and execution.

The relevant clinical data must be in a form that can be analyzed computationally (e.g., using AI). Electronic clinical data records and remote sensing devices provide such information.

Electronic Health Records

In 2009 the USA began the process of digitizing medical records so that in 2017, 96% of hospitals and 86% of doctors' offices could access electronic records (Hecht 2019). This has not been without controversy, particularly with privacy concerns, the online theft of data, and the fact that doctors perhaps too much of their time on a computer (like the rest of us?). The system will mature with improved human-computer interfacing such as "natural language processing" for patients and clinicians alike.

Despite these caveats, biopharmaceutical companies like Roche, through their 2018 acquisition of Flatiron Health, are using electronic records to conduct prospective clinical studies in lung cancer. The Roche subsidiary Genentech has launched the Prospective Clinico-Genomic (PCG) Study of 1000 patients with the aim of correlating genomic changes in a patient's tumor with drug response or resistance. The study will draw upon data from imaging, liquid biopsies, and outcomes in a coordinated way and provide valuable information for Roche's ongoing drug trials for lung cancer (Flatiron 2020).

Wearable Devices

The ubiquity of smartphones and wearable electronic devices has allowed clinicians to gain data remotely to assess drug safety, efficacy, and compliance. Some examples are shown in Table 17.7.

Table 17.7 Applications of remote sensing devices and smartphone apps

Application	Examples
Patient phenotype, safety monitoring	Blood pressure, heart rate, sleep, respiration, skin temperature
Novel endpoints	Tremor in Parkinson's disease, cognition in MS
Adherence to medication	Reminders, surveys, smart caps on bottles
Patient enrollment and retention	Remote enrollment and consent forms, communication about trial progress

Adapted from Izmailova et al. (2018)

AI and Clinical Trials

Two main issues with clinical trials are patient recruitment and trial design. AI and natural language processing can help patients navigate the complexities of sites such as clinicaltrials.gov and allow clinicians to identify patients from healthcare providers (Woo 2019). This can produce impressive results, such as the recruitment of 16 patients for a heart study in 1 hour compared with a conventional process which recruited 2 in 6 months (Deep6 AI 2020). This area is in its infancy and perhaps subject to exaggerated expectations, but there is huge potential in using AI to select patients from a truly global population. As mentioned earlier, most trial participants are of European descent: 79% of genomic data come from this group even though they represent only 16% of the world's population (Martin et al. 2019).

AI is being employed for clinical trial design by accessing different types of data that can guide the patient selection criteria; see, for example, Trials.AI 2020.

Finally, could AI replace clinical trials altogether? If enough workable data are available through health records for millions of people, it is at least conceivable that trials could be simulated to identify problems and avoid expensive failures if conducted in real trials.

Social Media

In 2008 I published an article with David Bailey entitled "Drug Discovery in the Era of Facebook-new Tools for Scientific Networking" (Bailey and Zanders 2008). This was a tentative effort to review the role of (the then new) social media in drug discovery, a world which has moved on dramatically in the intervening years.

Facebook itself has of course expanded considerably, and Twitter has become a significant part of human communication, with both positive and negative consequences. Pharma companies keep a close eye on these platforms as would any commercial organization, but there is a specific need for them to be alerted to adverse drug reactions (ADRs) caused by their products. The WEB-RADR project was established in 2017 to apply new technology to pharmacovigilance (WEB-RADR 2020). Their surveys of social media indicated that the number of actionable drug alerts was insignificant; however, GSK has reported a search of Facebook and Twitter in which they found 21 million mentions of its products, resulted in the recall of one. The company used Epidemico (acquired by Booz Allen Hamilton) to extract useable information that would be almost impossible to acquire through more formal channels.

17.1.3   The Future of Drug Discovery

Predicting the future for anything is fraught with hazards; to use Benjamin Franklin's observation: "In this world nothing can be said to be certain, except death and taxes." These words were included in the first edition of this book and are still valid now, except the COVID-19 pandemic has intervened to create more uncertainty about the future in general and not just drug development. Of course, the race to find drugs and vaccines against the SARS-CoV-2 virus has raised the profile of the biopharmaceutical industry in a mostly positive way and will (hopefully) ensure that investment and technical preparedness will be maintained for new viruses that may arise in the future. Perhaps the huge challenges of manufacturing synthetically complex small molecules, as well as therapies based on viral vectors, will stimulate further R&D as well as changed economic models. The subject of infectious diseases inevitably reminds us about antibiotic resistance, again something that can only be addressed through novel compounds/biologicals and financial incentives.

When it comes to the technicalities of drug discovery, it is possible to make an informed guess about what might happen over the next decade or so because the results of actions taken now may only be felt years later. For example, new technologies such as AI are still in their infancy as far as drug discovery and development are concerned, but it is reasonable to assume that it will have a positive impact on productivity[6].

One of the most talked about aspects of the biopharmaceutical industry is how declining sales through patent expiries and lean pipelines will affect the major companies. These conversations are along the lines of: "end of the blockbuster model which will lead to the fragmentation of large companies, who may give up discovery research altogether to concentrate on clinical trials and marketing." I can recall these same conversations years ago, but while this fragmentation has occurred to an extent (e.g., Abbott and Abbvie), other factors have come into play, such as increasing Merger and Acquisition (M&A) activity. I have always been struck by the fact that certain companies were admired for their research or commercial organization yet somehow were never emulated by others; changing the internal culture of a company was extremely difficult and tended to happen only as a result of the mergers and acquisitions mentioned above. Despite this conservatism, large pharma companies are now making significant changes to the way they operate internally and present themselves externally to the world. The futures of all organizations involved in drug discovery and development depend upon advances in science and technology. It is reasonable to expect that many technical obstacles will eventually be overcome or sidelined, and we can expect to see a broadening of the range of advanced therapy medicinal products (i.e., gene and cell therapies) to treat more than just orphan diseases.

A possible future scenario for drug discovery and development by the biopharmaceutical industry is summarized below:

---

[6] Despite the inevitable hype that accompanies these things

17.1.3.1 Industry Structures

The major biopharmaceutical companies still have large financial resources despite the decline in productivity. This will ensure that they will remain at the top of the "drug discovery tree," since there is no other organization (private or public) that has the ability or willingness to support major drug development programs on their own unless they serve specialized markets. It is unlikely that large pharma will abandon its internal discovery research completely, but it will continue to augment it by forming partnerships and acquiring technology from small biotech companies. The demographics of the industry will continue to change as China and other non-Western countries contribute more to innovative R&D and increase their share of the market for prescription medicines. Biotech companies, charities, and new entrants into drug development, such as software companies, will contribute innovative ideas but still be hampered by lack of investment. This can only improve if the huge costs of drug discovery and development can be brought down by introducing more efficient and flexible processes at every stage of the pipeline. Since governments, regulators, industry scientists, and patient advocacy groups are all aware of this, changes will surely be forthcoming, albeit slowly and perhaps painfully.

Social Attitudes

The drug discovery industry is making efforts to repair its image problem by becoming more transparent in its dealings with the medical profession and the public. A long running issue has been the perceived lack of interest by Western companies in diseases of the developing world, as well as controversies over the pricing of drugs for these markets. These problems may not disappear completely, but new collaborations and business models will be implemented that allow the costs of drug development to be shared between companies and allow cheaper pricing for certain markets. A more fundamental problem relates to how medicines are perceived by an increasingly risk averse society. The remit of the biopharmaceutical industry is to provide safe and effective medicines, but these can never be 100% safe for every patient. The challenge for drug makers will be to ensure that as much as possible is done to minimize risk, through technology and post-marketing surveillance, while at the same time trying to educate the public about the risk versus benefits in a completely open way. Another problem is the public's perception of the so-called lifestyle drugs being introduced to treat diseases that supposedly do not really exist. It could be argued that drugs are being developed for conditions that could be prevented by simply maintaining a healthy lifestyle.

The future will see the drug discovery industry attempting to overcome its image problems by being much more transparent about its products and procedures through dedicated websites and social media. In addition, the companies will increasingly form alliances with organizations that promote individual health and well-being. This is, in a sense, similar to the situation with electricity companies that promote energy savings measures; it seems counterintuitive to encourage customers to use less of your product, but this is how the business world is changing in response to its highly vocal customers.

17.1.3.2 Science and Technology

Drug Products

An examination of the early part of the drug discovery pipeline should provide clues about the sort of products that might enter the clinic several years later. Small molecules will probably dominate overall sales due to their relative ease of manufacture and administration compared to biologicals. Furthermore, advances in medicinal chemistry and proteomics will increase the number of small molecule drugs that can affect hitherto "undruggable" targets, such as KRAS (Chap. 7). New synthetic methods, computer-aided design, and AI will contribute to all of this.

There is no doubt that the branded biological products such as fully humanized monoclonal antibodies will continue to occupy the lists of best-selling medicines, and variations such as bi- and trispecific antibodies, immunotoxins, and nanobodies will enter the clinic in greater numbers.

The development of RNA-based drugs for interfering with gene expression had stalled until recently but has now gained a new lease of life with approvals of siRNAs and antisense therapies for rare diseases. This gives confidence that some of the delivery and toxicity issues have been overcome to allow expansion of these technologies into other disease areas. The same principles apply to gene and cell therapy where some of the difficulties that arise in their production and deployment are being overcome through practical experience.

Drug Delivery

This area has already been alluded to in the discussion of nucleic acid-based drugs. An exciting possibility is the oral delivery of peptides and larger proteins, including even monoclonal antibodies (see Chap. 8). Research is still ongoing, but this is an area with great potential for delivering a wide range of therapeutics that can be taken in a tablet or capsule form.

Concluding Remarks

There are many more scientific issues in drug discovery, including personalized medicine and the role of the microbiome in health and disease. Biological mechanisms like cellular waste disposal and epigenetics are currently hot areas of research that will lead to novel drug products. Over the much longer term, the elusive problem of how human (and other) biology works at the finest level of detail may be solved using a combination of experiment and mathematical theory. Maybe one day, it will be possible to predict biological behavior with the same degree of accuracy as physical behavior in the natural world and the universe at large. Until then, we must use our imperfect tools the best we can, building on the impressive amount of knowledge that has been gained over the past couple of hundred years and advancing drug discovery through the rest of the twenty first century.

References

Bailey DS, Zanders ED (2008) Drug discovery in the era of Facebook—new tools for scientific networking. Drug Discov Today 13:863–868

Barlow et al (2019) http://www.pharmexec.com/precision-financing-durable-potentially-curative-therapies. Accessed 26 May 2020

Bell J (2020) https://www.biopharmadive.com/news/pharma-neuroscience-retreat-return-brain-drugs/570250/. Accessed 23 May 2020

Biopharma Deal Makers (2020) biopharmadealmakers.nature.com. Accessed 23 May 2020

China Kadoorie Biobank (2020) https://www.ckbiobank.org/site/. Accessed 27 May 2020

Cyclica (2020) https://cyclicarx.com/. Accessed 27 May 2020

Deep6 AI (2020) https://deep6.ai/. Accessed 28 May 2020

EvaluatePharma Orphan Drug Report (2019) www.evaluate.com/OrphanDrug2019. Accessed 23 May 2020

Exscientia (2020) https://www.exscientia.ai/. Accessed 27 May 2020

Flatiron (2020) https://flatiron.com/press/press-release/pcg-study/. Accessed 28 May 2020

Fortune Business Insights (2020) https://www.fortunebusinessinsights.com/industry-reports/contract-research-organization-cro-services-market-100864. Accessed 26 May 2020

Gamo F-J et al (2010) Thousands of chemical starting points for antimalarial lead identification. Nature 465:305–310

GlobalData PharmSource: Contract Manufacturing of Novel In-Licensed Drugs (2020) https://store.globaldata.com/report/gdps0030mar%2D%2Dpharmsource-contract-manufacturing-of-novel-in-licensed-drugs-2020-edition/. Accessed 25 May 2020

HBM New Drugs Approval Report (2019) http://www.hbmpartners.com/en/industryreports. Accessed 25 May 2020

Hecht J (2019) Fixing a broken record. Nature 573:S114–S116

Informa Pharma Intelligence (2019) 2018 Completed trials: state of industry – sponsored clinical trials pharmaintelligence.informa.com › files › whitepapers. Accessed 21 May 2020

Scannell JW et al (2012) Diagnosing the decline in pharmaceutical R&D efficiency. Nat Rev Drug Disc 11:191–200

IQVIA Institute (2019) The global use of medicine in 2019 and outlook to 2023. www.iqvia.com. Accessed 22 May 2020

Iris.AI (2020) https://iris.ai/. Accessed 27 May 2020

Izmailova ES et al (2018) Wearable devices in clinical trials: hype and hypothesis. Clin Pharmacol Ther 104:42–52

Mahendraratnam N et al (2019) Value-based arrangements may be more prevalent than assumed. Am J Manag Care 25:70–76

Martin AR et al (2019) Clinical use of current polygenic risk scores may exacerbate health disparities. Nat Genet 51:584–591

Medicines Patent Pool (2020) https://medicinespatentpool.org/. Accessed 25 May 2020

Micklus A, Giglio P (2020) Biopharma dealmaking in 2019. Nat Rev Drug Disc 19:87–88

MIT NEWDIGS (2020) https://newdigs.mit.edu/. Accessed 26 May 2020

Moon S, Erickson E (2019) Universal medicine access through lump-sum remuneration – Australia's approach to hepatitis C. N Engl J Med 380:607–610

Morgan P et al (2018) Impact of a five-dimensional framework on R&D productivity at AstraZeneca. Nat Rev Drug Disc 17:167–181

National Law Review (2020) https://www.natlawreview.com/article/humira-how-far-can-drug-makers-go-to-protect-their-branded-market. Accessed 26 May 2020

Reuters   (2019)   https://www.reuters.com/article/us-usa-congress-drugpricing-idUSKC-N1QF1WC. Accessed 22 May 2020

Ringel EM et al (2020) Breaking Eroom's Law. Nat Rev Drug Disc. https://www.nature.com/articles/d41573-020-00059-3. Accessed 21 May 2020

Semantic Scholar (2020) https://www.semanticscholar.org/. Accessed 27 May 2020

Thomas CJ, McKew JC (2014) Playing well with others! Initiating and sustaining successful collaborations between industry, academia and government. Curr Top Med Chem 14:291–293

Trials.AI (2020) https://www.trials.ai/. Accessed 28 May 2020

ViiV (2020) https://viivhealthcare.com/en-gb/. Accessed 25 May 2020

WEB-RADR (2020) https://web-radr.eu/. Accessed 28 May 2020

Woo M (2019) Trial by artificial intelligence. Nature 573:S100–S102

Part V
Professional Interactions with the Drug Discovery Industry

Chapter 18
Technology Transfer Executives

Abstract The next three chapters are written as supplements to the main book, each being written for a defined group of professionals. This chapter focuses on university technology transfer managers and business development executives who offer their products and services to biopharmaceutical companies. Starting with a brief background to technology transfer and pharmaceuticals, the chapter provides a series of suggestions on how to prepare for a meeting with pharmaceutical executives and how to conduct the meeting itself. These suggestions are based on the author's own experiences in dealing with these meetings from both sides of the negotiating table.

18.1   Introduction

The biopharmaceutical industry directly employs hundreds of thousands of people worldwide, but there are also large numbers of people who deal with the industry from the outside. This book has been written as a guide for anyone who is professionally or personally interested in the complex business of drug discovery. The next three chapters are, however, dedicated to aspects of the biopharmaceutical industry that are relevant to specific groups of professionals, namely, technology transfer managers, recruitment consultants, and technical translators or interpreters.

18.1.1   Background to Technology Transfer

The modern era of government-supported science (at least in the USA) was heralded by Vannevar Bush's 1945 report to the US government entitled "Science - The Endless Frontier" (Science The Endless Frontier 1945)[1]. His recognition of the

---

[1] The title may have been inspired by a 1944 speech by President Roosevelt: "New frontiers of the mind are before us, and if they are pioneered with the same vision, boldness, and drive with which we have waged this war we can create a fuller and more fruitful employment and a fuller and more fruitful life."

© The Editor(s) (if applicable) and The Author(s), under exclusive license to   365
Springer Nature Switzerland AG 2020
E. D. Zanders, *The Science and Business of Drug Discovery*,
https://doi.org/10.1007/978-3-030-57814-5_18

importance of science in post-war society encouraged the idea of technology transfer and even spawned the term "basic research."

The passing of the University and Small Business Patent Procedures Act (Bayh–Dole Act) in the USA in 1980 meant that public research organizations (PROs) could benefit financially from the patenting of government-funded research[2]. This Act, and the later Federal Technology Transfer Act of 1986 (FTTA) and other improvements, has made technology transfer a key factor in the development of a knowledge-based economy in the USA (Lamm 2009). The benchmarks for success in technology transfer (for all products) include the number of licensing deals with companies and the revenues generated for the PRO. Table 18.1 gives some figures for technology transfer activity in the USA in 2018 taken from the AUTM 2018 Licensing Activity Survey (AUTM 2020).

The UK has arguably always operated under a Bayh-Dole-like system without feeling the need to enshrine this in statute. Universities own IP generated from grant funding because funded scientists are employees of the host institution. However, the UK government is aware that it is necessary to encourage commercial innovation from research (through UK Research and Innovation (UKRI), even though a relatively small proportion of inventions arise from academic institutions (House of Commons Science and Technology Committee 2017).

A reference point for technology transfer expertise within the EU was created in 2018 with the launch of the Competence Centre on Technology Transfer (CCTT) by the Joint Research Centre (JRC) of the European Commission.

The commercialization of university research is generally managed by dedicated offices within the university or by companies set up for the purpose. In practice, technology transfer groups broker deals with companies to license IP generated by a university or spinout company. One example of successful technology transfer in biopharmaceuticals is the humanization of monoclonal antibodies by the UK's Medical Research Council (see LifeArc in Chap. 16). This has generated royalties

Table 18.1 IP and technology transfer data in USA 2018

Category	Number
Licenses and options executed	9350
Startups formed	1080
Research expenditure	$71.7Bn
New products created	828
US patent applications	17,087
US patents issued	7625
Invention disclosures	26,217

From AUTM 2020 with kind permission

[2] Whether the entrepreneurial success of US universities compared with the rest of the world is a consequence of the legislation is a matter of debate. It is amusing to note, however, that identical complaints about "having plenty of home-grown Nobel Prize winners but none of the commercial benefits" can be heard on both sides of the Atlantic.

of over £580 M although, ironically, the original discovery of monoclonal antibodies by Kohler and Milstein in Cambridge was not patented.

18.1.2   Some Practical Considerations

Some practical tips for dealing with pharmaceutical companies are laid out in the following sections. These are not intended as a guide to the negotiation of licensing terms or to any of the financial and legal activities associated with technology transfer. The only exception to this is to remind readers that when negotiating terms for a pharmaceutical invention, the length of patent protection may be considerably shorter when compared with other high technology products; this is a consequence of the long development times before a drug reaches the market (Chap.16).

18.1.2.1   The View from the Client Company

Because large pharmaceutical companies have significant financial resources, there is no shortage of outside visitors who have something to sell to them. These visitors range from representatives of laboratory supply companies to managers from biotech companies and university technology transfer offices. Sometimes a visitor may be a private individual trying to sell their "discovery" to the company. Some of these visitors can turn out to be quite bizarre. I recall having to test a sample delivered by the son of a Nigerian witch doctor who claimed that it would cure a whole range of unrelated diseases. Quite how this got through the door I do not know, but a herbal mixture dissolved in gin was duly delivered to us in a used bleach bottle for testing in various in vitro assays. Not surprisingly this led nowhere, but it does demonstrate that pharmaceutical company managers have a sense of humor.

Identifying the needs of the client/customer and then satisfying it with products or services are, of course, central to sales and marketing. It is therefore logical when approaching a pharmaceutical company to ensure that their way of thinking is understood from the outset. Some companies make this very straightforward by proactively seeking out new products at clearly defined stages of development. These products may be compounds/biologicals for disease areas or novel technologies in chemistry, biology, or informatics. This approach ensures that time is not wasted in trying to sell anything outside the core wish list (which might take the form of a printed brochure). In most cases, however, the university or other inventor will have developed something that they feel will interest a drug company, and therefore approaches will be made to the company on that basis. This is entirely reasonable, but it is important that the questions that will be raised by company scientists are fully anticipated in advance. Many of the key topics are related to material covered in this book and could include the following:

New Compounds or Biologicals

It is assumed that the basic criteria of potency and selectivity will have been established, at least in a relevant in vitro test. Has activity been demonstrated in an animal model of the disease? A convincing demonstration of such activity will attract immediate interest, but sometimes this is not possible if the target is different in animals and humans; extra effort therefore has to be made to show that the drug has a reasonable chance of working in the clinic. Assuming that the target and disease is of interest to the company (as established by prior research by the visitor), the compounds themselves will come under great scrutiny. A medicinal chemist will identify functional groups within the molecule that are likely to cause problems during development. These problems may include difficulty of synthesis, scale up of the compound itself, poor pharmacokinetics, or potential toxicity. Certain classes of compounds are known to cause specific side effects, so it may need to be avoided. If the university or spinout does not have access to medicinal chemistry expertise, it should employ a consultant chemist to check the compounds before attempting to contact a company. If the compound looks really promising, it may be submitted to a contract research organization for some basic metabolism studies and even some preliminary toxicology such as an Ames test. The problem facing the inventor is that money must be spent up front to provide these data and improve the chances that the compound will find a buyer. The academic community understandably does not see itself as a drug development company, and in any case, it has limited resources. Unfortunately, companies must apply very stringent criteria for compound selection because of the high failure rate in small molecule drug development. This creates a high hurdle for the university to overcome, so whatever the outcome, some serious money will have been spent. The irony is that the pharmaceutical company will spend the money all over again as it repeats each test prior to committing a drug candidate to full clinical development.

The main issues when dealing with biologicals relate to manufacturing and drug delivery. Serious consideration must be made of the costs involved in producing novel drug types such as gene and cell therapies. Because these technologies are so new, it will be some time before companies feel able to commit the resources necessary to bring them to market. The adoption of monoclonal antibodies as drugs is a good example of how biological products were viewed by pharmaceutical companies. Once a small number of pioneer companies demonstrated the commercial viability of this technology, momentum built up among the other companies to get involved as well. This was the "herd mentality" in action because nobody wanted to miss out on some potential blockbuster revenues[3]. Nucleic acid drugs and stem cells are in the next phase of acceptance with some of these products approved and marketed by major pharma. Perhaps the next phase will be the introduction of microbiome-based drugs with quite different pharmaceutical properties compared to other medicines.

---

[3] Or FOMO: fear of missing out

New Processes

Fashions in pharmaceutical technology transfer come and go. One year it may be for clinical development candidates and the next for platform technologies; sometimes it is about 50:50. A few points are made in his section about technologies as opposed to drug molecules.

Technical advances, for example, in drug delivery systems or stem cell creation from adult tissues will be of great interest to companies, but the technology will have limited up-front value. If they become standard techniques with full IP protection, however, their value will be greatly increased because everyone who wants to use them will only be able to do so under license.

When it comes to the process of drug development, the biopharmaceutical industry must reduce the number of clinical trials that fail because the drug candidate lacks efficacy or has problems with ADMET. There may also be problems with manufacturing and formulation, so there are plenty of opportunities for selling improved processes to industry; these could range from chemical synthesis to streamlining clinical development. It is worth noting that the later stages of clinical development are highly regulated and therefore less flexible when it comes to the introduction of new technology. Those selling technology may have more success with products that affect earlier stages of drug development, for example, preclinical toxicology, where there is a great need to replace the animal models in current use. The key message for anyone who wishes to provide products and services to the drug discovery industry is that research moves fast and regulation much more slowly.

18.1.2.2   Getting a Foot in the Door

The following observations on arranging and conducting a meeting with pharmaceutical companies are based on my personal experience from being on "both sides of the table." These should be considered as highly subjective and a statement of the obvious, but it does no harm to set them out as a reminder. As in all business, personal contacts within the client organization are invaluable. Senior executives often prove to be the best conduits to the part of the company that might be interested in a product or technology. I have often received communications that were sent from external organizations to a very senior research director and then passed to me, accompanied by a request to "look into the proposal." Given the provenance of the request, this is a guaranteed way to ensure that the work will be thoroughly assessed and critically reviewed.

Major collaborations with academic groups often emerge from conversations between the relevant professor and a senior company executive. Although it helps to have the buy-in of more junior staff, the final decision to proceed is generally made at the top.

Senior company executives are highly visible even if their personal contact details are not made public (the use of animals in research being part of the reason). There may be situations where it is more appropriate to contact other people within the organization. Many companies have university liaison managers who should circulate proposals around the relevant departments. This generally works well, but there is a danger that the documents will be passed to someone who has neither the time nor the inclination to look at it within a short time frame. Sometimes negotiations with large companies can appear to move at the speed of the glaciers. The ideal situation is to target an individual or group who is most likely to be interested in the proposal, but this is more easily said than done; where do you start? Large pharmaceutical companies are not homogeneous entities but are made up of hundreds of different groups, each with a different culture relating to the opinions and personalities of the people who work in them. Sometimes the groups work together harmoniously and at other times are in competition. Many company scientists publish in scientific journals that are freely searchable on the Internet; failing that, company patents contain the names of individual scientists who have contributed to the work, so online patent searches for companies or topics could pay dividends. Finally, general Internet searches using names or companies may pick up useful leads from online conference agendas or social/business networking sites.

18.1.2.3   Conducting the Meeting

Busy pharmaceutical executives or scientists generally do not like being taken away from their work to listen to outsiders pitching for business. They are also engrained with a large dose of cynicism, brought on by listening to claims that the products are "the greatest thing since sliced bread." Long experience with the realities of drug development has made those who evaluate the proposals more probing with their questioning than might perhaps be the case elsewhere. I had direct experience of this the first time I presented my research to a company research management committee. I had to keep justifying the use of certain biological tests that I used in my experiments, even though these were considered standard procedures in academia[4]. After my initial feeling of annoyance at the relentless questioning, I began to accept that this was not going to go away, so for future presentations, I thought carefully about every experiment and interpretation in advance to prepare myself for any awkward questions. If this can happen with a research manager in a large pharma company, it is also likely to occur with a visitor trying to sell/license a product or service, so rigorous planning is essential. Other things to bear in mind are:

- Getting the company name right

  With all the mergers and acquisitions, the name of the company may have changed since your last visit. Having visitors talk about "Glaxo" when the company is GlaxoWellcome or GlaxoSmithKline is potentially annoying if you used to work for Wellcome or SmithKline Beecham.

---

[4] Comparisons of the academic and business worlds sometimes feature the stiletto knife that goes silently into the back of the victim, so it is not noticed until it is too late. In industry the attack is made from the front, so at least you know exactly who delivered the blow.

- Do not tell the company they are doing everything the wrong way

I have had to deal with combative visitors who are openly aggressive in their criticism of how the company operates. Even if this were true, it does not bode well for any future collaboration. Similarly, getting angry and defensive if your idea is not received enthusiastically does not help. I have been close to a fight after one particularly irate visitor tried to grab the notes of the meeting out of my hands because he was worried about what I might have written down.

- Keep to time

For better or for worse, scientists and managers in large pharma companies spend a great deal of time in meetings. Most of the scientists who are evaluating a proposal will also have experiments running in their labs and could therefore be keen to leave the meeting at the earliest opportunity. This means that the presentation must be planned to convey the maximum amount of relevant information in the minimum amount of time. This is all part of developing good presentation skills through meticulous preparation and keeping the number of PowerPoint slides (if used) to a minimum. There are also danger points to be noted, for example, when personal reminiscences start to take up too much of the allotted time.

18.1.2.4   The "Not Invented Here" Syndrome

The reluctance of companies to bring in inventions from the outside because of the perceived competition with internal programs can be a real problem. However, the "not invented here" syndrome is less likely to occur with drug candidates than with technical processes such as drug design or chemical synthesis. A significant percentage of in-licensed drugs now feed the pipelines of major companies because of their more open attitude towards external providers (see Chap. 17). Where there is a problem, the solution is not always easy to find; it will depend on the technology and the personalities involved in the discussions. Since the company will probably be reluctant to sideline its own research groups, the ideal solution is to establish formal collaborations between external and internal groups to allow a share of the credit for any future success.

References

AUTM (2020) https://autm.net/surveys-and-tools/surveys/licensing-survey/2018-licensing-activity-survey. Accessed 29 May 2020

House of Commons Science and Technology Committee (2017) managing_intellectual_property_and_tt_by_house_of_commons.pdf. Accessed 29 May 2019

Lamm M (2009) Technology transfer: United States policy and laws. http://www.unece.org/. Accessed 29 May 2020

Science The Endless Frontier (1945) https://www.nsf.gov/od/lpa/nsf50/vbush1945.htm. Accessed 29 May 2020

Chapter 19
Recruitment Executives

Abstract This short chapter is written for recruitment executives who work with pharmaceutical industry clients. Most of the job descriptions and backgrounds to different areas of work will have been covered in the main part of the book. The intention here is to summarize this information as a series of figures which list the job categories and titles as they relate to the drug discovery pipeline.

19.1 Introduction

The sheer number of employees working for the global drug discovery industry means that there will always be a need for specialist pharmaceutical recruiting agencies. Many of the large pharmaceutical companies are undergoing major structural changes due, in part, to a significant reduction of revenues generated by their best-selling products as their patent protection comes to an end. The result is a spate of reorganizations and efficiency drives that inevitably lead to job losses; these fuel the labor market with scientists and managers who are moving from job to job (voluntarily or otherwise). The increased globalization of the industry also means that there are large numbers of qualified personnel in China, India, and other growing economies; this will clearly have an impact upon recruitment in Western nations (Figs. 19.1 and 19.2).

Social and business networking sites are being enthusiastically embraced by recruiters who can advertise their presence to thousands of candidates (and check out their Facebook activities). For example, a major recruitment consultants' group on the business networking site LinkedIn has over one million members. The intense competition for clients and candidates, along with the global reach of online communication, means that more than ever before, the recruitment agencies need knowledgeable personnel to ensure the success of their business. It is obviously important that their consultants and managers should have a reasonable understanding of the work required for each job specification and the type of individual who would be suitable for the role. The purpose of this chapter is simply to list the job titles and functions that are encountered by pharmaceutical recruiters since it is hoped that this book has already provided enough background information. The titles and job functions are listed next to the various sections of the drug discovery pipeline in the

E. D. Zanders, *The Science and Business of Drug Discovery*,
https://doi.org/10.1007/978-3-030-57814-5_19

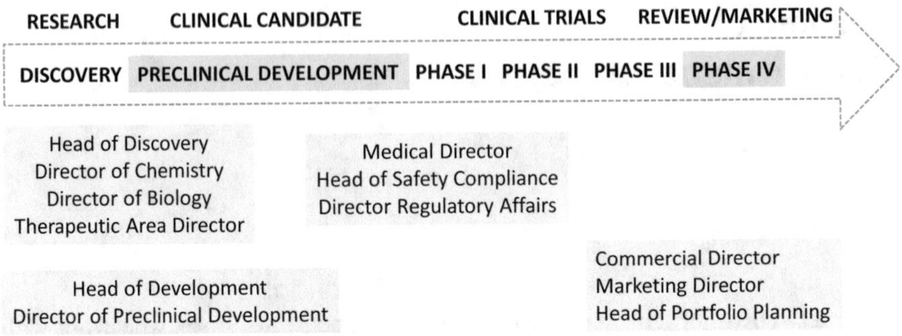

RESEARCH	CLINICAL CANDIDATE		CLINICAL TRIALS			REVIEW/MARKETING
DISCOVERY	PRECLINICAL DEVELOPMENT	PHASE I	PHASE II	PHASE III	PHASE IV	

Head of Discovery
Director of Chemistry
Director of Biology
Therapeutic Area Director

Medical Director
Head of Safety Compliance
Director Regulatory Affairs

Commercial Director
Marketing Director
Head of Portfolio Planning

Head of Development
Director of Preclinical Development

Fig. 19.1 Some representative examples of senior management roles associated with each part of the drug discovery pipeline. The titles will vary according to the terminology followed by a company or country, for example, Head of, Director of, VP, etc.

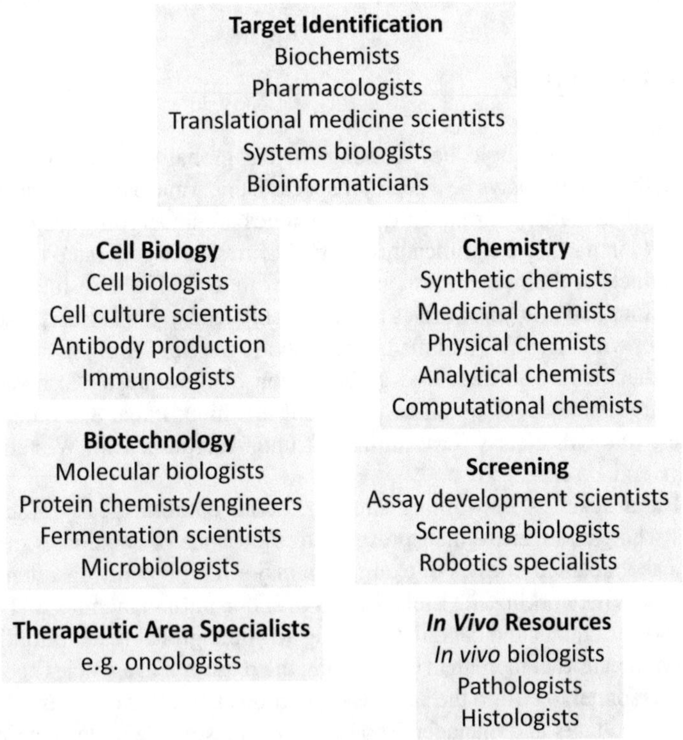

Target Identification
Biochemists
Pharmacologists
Translational medicine scientists
Systems biologists
Bioinformaticians

Cell Biology
Cell biologists
Cell culture scientists
Antibody production
Immunologists

Chemistry
Synthetic chemists
Medicinal chemists
Physical chemists
Analytical chemists
Computational chemists

Biotechnology
Molecular biologists
Protein chemists/engineers
Fermentation scientists
Microbiologists

Screening
Assay development scientists
Screening biologists
Robotics specialists

Therapeutic Area Specialists
e.g. oncologists

In Vivo **Resources**
In vivo biologists
Pathologists
Histologists

Fig. 19.2 Job roles associated with early drug discovery. This list is not exhaustive because of the wide variety of titles that scientists give themselves (e.g., genomics and proteomics are not included). Some of the roles also apply to the later stages of drug development, particularly in vivo work

following figures. There is, however, no mention of computer scientists and related personnel who are involved in artificial intelligence/machine learning. This is because the specified job role could be applicable (in principle at least) to any aspect of the discovery and development pipeline (Figs. 19.3, 19.4 and 19.5).

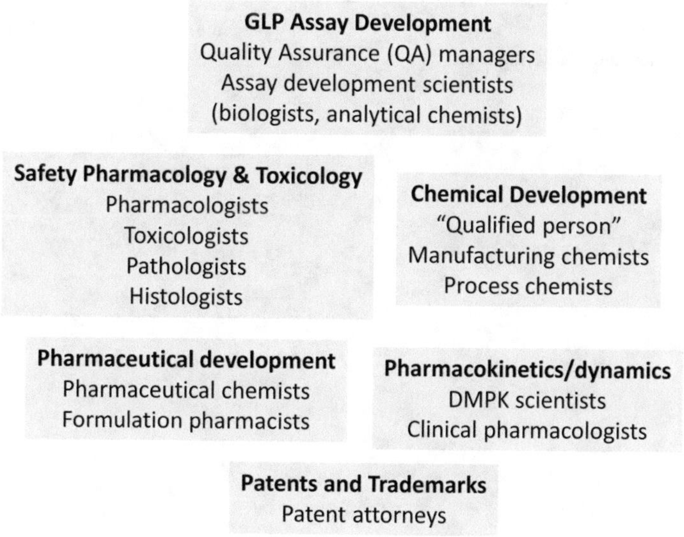

GLP Assay Development
Quality Assurance (QA) managers
Assay development scientists
(biologists, analytical chemists)

Safety Pharmacology & Toxicology
Pharmacologists
Toxicologists
Pathologists
Histologists

Chemical Development
"Qualified person"
Manufacturing chemists
Process chemists

Pharmaceutical development
Pharmaceutical chemists
Formulation pharmacists

Pharmacokinetics/dynamics
DMPK scientists
Clinical pharmacologists

Patents and Trademarks
Patent attorneys

Fig. 19.3 Preclinical development job titles. The abbreviation DMPK is "drug metabolism pharmacokinetics"

Clinical Research & Development
Clinical research (project) managers
Clinical research associates (CRAs)
Clinical pharmacologists

Regulatory Affairs
Regulatory affairs managers
Compliance managers

Biostatistics & Data Management
Statisticians
Data analysts
SAS programmers

Pharmacogenetics
Molecular biologists
Genomics specialists

Pharmacovigilance
Pharmacovigilance managers
Drug safety scientists

Fig. 19.4 Clinical development job titles. SAS programmers (in the Biostatistics section) use the industry standard statistical software package from the SAS Institute in the USA

Marketing Communications
Medical affairs manager
Medical writers

Healthcare Economics
Strategic pricing executive
Health economist
Health outcomes executive

Sales Force
Territory sales manager
Sales representative

Fig. 19.5   Sales and marketing roles

Chapter 20
Pharmaceutical Translators and Interpreters

Abstract The last of the supplementary chapters covers the specific requirements of pharmaceutical translators and interpreters. There are many aspects of drug discovery and development that may require the services of these professionals, but the majority will concern regulatory and other documents produced during clinical trials and marketing. After briefly reviewing the types of documents which translators may encounter during their work, the remainder of the chapter deals with the electronic documentation required for the EMA's Summary of Product Characteristics and Patient Information Leaflets. Finally, some links to online resources on terminology are provided to supplement those already given in previous chapters (e.g., on drug nomenclature in Chap. 3).

20.1 Introduction

The global nature of the biopharmaceutical industry, combined with the mountain of paperwork involved at almost every level of drug development, ensures that there will always be work for pharmaceutical translators. However, translation of medical and regulatory material carries with it a great responsibility, since any misunderstanding arising from inaccurate translation of a key document could have serious consequences.

Someone once said that the language of science is broken English. Since the USA is the largest market for prescription medicines, it is no surprise that the language of pharmaceuticals is intact US English. Even though English is still being spoken by a clear majority, the demographics of the USA are changing. Data collected by the US Census Bureau's American Community Survey for 2018 reveal that in a population of about 307 million aged 5 years and over, 78.1% spoke only English. The remaining 21.9% was broken down into Spanish (13.5%), other Indo-European languages (3.7%), Asian and Pacific Island languages (3.6%), and "other languages" (0.2%). This last figure may be a small proportion of the total, but it still represents around half a million people (American Community Survey 2018).

Europe has over 60 minority or regional languages, its largest market being the European Union, which recognizes 24 official languages (Table 20.1).

E. D. Zanders, *The Science and Business of Drug Discovery*,
https://doi.org/10.1007/978-3-030-57814-5_20

Table 20.1 List of official EU languages

EU language	
Bulgarian	Irish
Croatian	Italian
Czech	Latvian
Danish	Lithuanian
Dutch	Maltese
English	Polish
Estonian	Portuguese
Finnish	Romanian
French	Slovak
German	Slovenian
Greek	Spanish
Hungarian	Swedish

Lastly, in the Asia Pacific region, China, India, and South Korea have become rivals to Japan in the pharmaceuticals marketplace. This means that the demand for translations from these native languages can only increase.

20.1.1 Translation Challenges

Technical translators are expected to understand the correct terminology displayed in English and to preserve the exact meaning in translation. Unfortunately, there are difficulties from the outset since there are clear differences between US English and British English, as well as English usage by the UK regulators (MHRA) and the EMA. Some pharmaceutical and medical phraseology is specific to individual languages, but this book covers American or British usage only. Readers who seek to understand how particular terms are used during the drug development process should by now have obtained much of the information they need from the main chapters of this book. The problem for translators lies, however, in the fact that the borderline between correct and colloquial usage of some technical words and phrases is quite blurred. Questions along these lines get raised frequently during the courses that I run for freelance translators and they can give me a few headaches. For example, "what is a side effect, or an adverse event or drug reaction?" The term "side effect" is a catch-all phrase used to describe any unexpected reaction to a drug, whether beneficial or harmful. Adverse drug reactions (ADRs) are any noxious and unintended response associated with the use of a drug in humans. Adverse events are any events that occur during the clinical evaluation of the drug, *whether caused by the treatment or not*. If this were taken too literally, it could mean an adverse reaction caused by, say, the hospital building collapsing. In practice, the term "adverse event" is most likely to be found in a clinical trial report or scientific publication where clinical symptoms are noted for patients treated in a placebo-controlled drug trial. While context is important, there may, however, be particular usage guidelines laid down by the originator of the translation, so these obviously have to be followed even if at first sight they might appear to be counterintuitive.

20.1.2 Types of Documentation

It is impossible to cover all the types of document that a translator is likely to encounter in commissions from the pharmaceutical and biotechnology industries. The nature of the material will also depend on the product being manufactured; this could be a branded medicine, a medical device, or a diagnostic test. Like drugs, devices and tests are highly regulated consumer products where the instructions for their use have to be translated into different languages following European directives.

The following illustration outlines the main types of material that pharmaceutical translators are likely to encounter. The abbreviations are as follows: GLP, GCP, GMP, good laboratory, clinical, manufacturing practice; SOP, standard operating procedure; CMC, chemistry manufacturing and controls; PROs, patient reported outcomes; PIL, Patient Information Leaflet; SmPC, Summary of Product Characteristics; EPAR, European Public Assessment Report (Fig. 20.1).

Research & IP
Biotechnology
Chemical nomenclature
Patents
Scientific publications

Preclinical Study Reports
Pharmacology & toxicology
GLP documents
SOPs
CMC documents

Clinical Trial Documents
Clinical study protocols
Investigator brochures
Ethics Committee papers
Informed consent forms
Case Report forms
GCP documents
Clinical trial reports
Patient records
PROs

Regulatory Documents
Marketing applications
Clinical trial authorizations
Package labels
Package inserts
PIL
SmPC
EPAR

Marketing Documents
Promotional materials

Fig. 20.1 Examples of different types of documentation required throughout the drug development process. Although not exhaustive, the list shows items commonly encountered by technical translators. The research and intellectual property (IP) section covers the general themes of biotechnology and chemical nomenclature, both of which can occur in many different documents. Patent translation of course is a major activity in its own right

20.1.2.1 Electronic Documents

The key documents required for marketing drugs in the European Union are the Summary of Product Characteristics (SmPC or SPC), labeling information, and the Patient Information Leaflet (PIL). These are contained within the European Public Assessment Report (EPAR). For more information, see EPAR Background & Context 2020.

How to Find SmPCs and PILs for a Particular Medicine

Taking the biotherapeutic Herceptin® as an example, accessing the EMA website, and following the Medicines link (https://www.ema.europa.eu/en/medicines), a search on Herceptin reveals the European Public Assessment Report page (https://www.ema.europa.eu/en/medicines/human/EPAR/herceptin). Here, the SPC and PIL are available, in multiple European languages, as sections within the following five annexes from the downloadable pdf document:

Annex I – Summary of Product Characteristics.
Annex IIA – Manufacturing Authorization Holder Responsible for Batch Release.
Annex IIB – Conditions of the Marketing Authorization.
Annex IIIA – Labelling.
Annex IIIB – Package Leaflet.

These documents for specific drugs can be particularly useful when translating an SmPC or PIL for a similar drug or drug type or more generally to find the correct wording or terminology for a given drug or procedure.

Notes about Product Details

The Product Details section lists the INN as trastuzumab. INNs have already been referred to in Chap. 3 in connection with drug nomenclature. A downloadable pdf file from the WHO containing detailed guidelines for using INNs is available from reference WHO Guidance on INNs 2020. The MeSH field is the Medical Subject Headings system curated by the US National Library of Medicine (MeSH 2020).

The Anatomical Therapeutic Chemical (ATC) classification is a WHO-sponsored system that groups drugs in a hierarchy, beginning with the target organ system and then continuing with chemical class, etc. (ATC Classification system 2020). It is used for the statistical analysis of drug use across different territories and for the reporting of adverse drug reactions.

Terminology to be Used in Translating SPCs

For formal pharmaceutical texts such as the Summary of Product Characteristics and the Patient Information Leaflet (PIL), standard product information templates have been prepared by the QRD (Quality Review of Documents) Working Group of the EMA. These set out the required structure of the documents such as section numbers, headings, fonts, and font sizes. They also suggest some standard phrases, such as "Hypersensitivity to the active substance(s) or to any of the excipients <or {residues}>"; extreme care should be taken to follow these templates; even the style or terminology used does not seem to be correct. The only exception would be if the source text deviates significantly from the template in the relevant language. The templates are available in 24 languages and can be downloaded in both English and the working language(s). To access this from the EMA website (http://www.ema.europa.eu), follow the links (top or side navigation bar as appropriate) from the Home Page as follows:

Home Page.
Human Regulatory.
Marketing Authorization.
Product Information.
Product Information Templates.

The QRD Human Product Information Template v10.1 can then be downloaded in the appropriate language. In addition to downloading the "clean templates," translators can download and use the annotated template and guidance documents that are available in English only.

Standard Terms

It is possible that translators will be required to use the "Standard Terms" prescribed in the European Directorate for the Quality of Medicines (EDQM), published by the Council of Europe (Standard Terms Database 2020). This multilingual glossary currently covers over 900 terms in 34 world languages.

EU Versus UK Terminology

UK terminology and style differ somewhat from the EU standards. In general, the British SPCs and PILs follow the same format, but slight differences in terminology and usage may exist. In this case, it is a good idea to cross-check the information available in the EMA templates with any existing SPCs or PILs found at the electronic Medicines Compendium (EMC 2020). The eMC provides electronic Summaries of Product Characteristics (SmPCs) and Patient Information Leaflets (PILs), as well as information on thousands of licensed medicines available in the UK.

References

American Community Survey (2018) https://data.census.gov/cedsci/table?q=languages&hidePrev
iew=false&tid=ACSST1Y2018.S1601&vintage=2018. Accessed 1 June 2020

ATC Classification system (2020) http://www.whocc.no/atc/structure_and_principles/. Accessed
1 June 2020

EPAR Background & Context (2020) https://www.ema.europa.eu/en/medicines/what-we-publish-
when/european-public-assessment-reports-background-context. Accessed 1 June 2020

EMC (2020) http://www.medicines.org.uk/emc/. Accessed 1 June 2020

MeSH (2020) https://www.nlm.nih.gov/mesh/meshhome.html. Accessed 1 June 2020

Standard Terms Database (2020) https://www.edqm.eu/en/standard-terms-database. Accessed 1
June 2020

WHO Guidance on INNs (2020) https://www.who.int/medicines/services/inn/innguidance/en/.
Accessed 1 June 2020

Appendices

Appendix 1: Further Reading

Most of the references in the short list at the end of each chapter support a specific item mentioned in the text, rather than a more general theme. This appendix contains a more comprehensive list of printed and online material that will hopefully be of interest to readers, whether they are scientifically trained and want further information about the biopharmaceutical industry, or whether they would like to learn a bit more about the basic science behind drug discovery. It is, of course, possible to access vast amounts of information about drug discovery and development from searching the Internet; there is a great deal of useful material out there, but the reader has to be able to identify credible work. Unfortunately when dealing with pharmaceuticals, the incredible can get in the way, particularly with online sources. Authoritative work has generally been scientifically peer reviewed or at least published by an organization with a good scientific reputation. This does not necessarily mean that every finding is correct, however; science is always moving ahead by overturning preexisting concepts. A review system that uses the informed opinions of international experts helps to identify errors and to correct them, because no one voice (at least in theory) is supposed to dominate the scientific debate.

Accessing the Literature

Literature on the scientific aspects of drug discovery is most easily located by searching PubMed, a freely available database hosted by the US National Library of Medicine (http://www.ncbi.nlm.nih.gov/sites/entrez?db=pubmed).

A search for subjects, authors, journals, etc. brings up a list of abstracts from articles published in scientific and medical journals in printed and/or online formats. The articles can then be downloaded from the website of each journal, but

© The Editor(s) (if applicable) and The Author(s), under exclusive license to
Springer Nature Switzerland AG 2020
E. D. Zanders, *The Science and Business of Drug Discovery*,
https://doi.org/10.1007/978-3-030-57814-5

most will require a subscription or charge per article. This is obviously not satisfactory for those who are not working for an institution which will pay for the journals, but it is possible to find free full-length articles in PubMed. The latter are either made freely available by the journal publisher or are available from Open Access Journals. PubMed Central (https://www.ncbi.nlm.nih.gov/pmc/) is a collection of scientific and medical articles which are all freely available. Although the PubMed/PMC articles are technical and therefore challenging for non-scientists, it is worth examining how they are structured and how English is used within a scientific context; this is something that could be particularly useful for translators.

There is no shortage of textbooks covering scientific subjects relevant to drug discovery for school or college students; unless the reader is a student of these subjects and therefore issued with appropriate books, it is of course possible to obtain many of them secondhand at bargain prices or to borrow them from a library.

Finally, online video sites such as YouTube and many educational sites from universities are extremely helpful in explaining complex science. A search for "polymerase chain reaction," for example, will highlight links to animations and even videos of the laboratory procedure, giving non-scientist readers an idea of how these experiments are actually performed in the lab.

References

A.1.2.1. General Drug Discovery Books

This is a series of books that cover (with varying levels of detail for each topic) the main scientific and business aspects of drug discovery and development. Some are aimed at the general reader and others at science students and those studying pharmacy. It is now quite straightforward to browse the contents online to locate information on a topic which may have been highlighted in this book but not explored in great detail.

Blass BE (2015) Basic principles of drug discovery and development. Elsevier Academic, Burlington

Evens R (ed) 2007 Drug and biological development. From molecule to product and beyond. Springer, New York

Chorghade MS (ed) (2006) Drug discovery and development: drug discovery. Wiley-Blackwell, Hoboken

Chorghade MS (ed) (2007) Drug discovery and development: drug development. Wiley-Blackwell, Hoboken

Rang HP (2005) Drug discovery and development: technology in transition. Churchill Livingstone, Edinburgh

Bartfal T, Lees G (2006) Drug discovery: from bedside to wall street. Elsevier Academic, Burlington

Ng R (2008) Drugs: from discovery to approval. Wiley-Blackwell, Hoboken

Giordanetto F (2018) Early drug development. Bringing a preclinical candidate to the clinic. Fabrizio methods and principles in medicinal chemistry. Wiley VCH, Weinheim

Jacobsen TM, Wertheimer AI (2010) Modern pharmaceutical industry: a primer. Jones & Bartlett Learning, New Dehli

Bartfal T, Lees G (2013) The future of drug discovery: who decides which diseases to treat? Academic Press, Cambridge MA

Schacter B (2005) The new medicines: how drugs are created, approved, marketed and sold. Praeger, Westport

Smith CG, O'Donnell JT (2006) The process of new drug discovery and development, 2nd edn. Informa Healthcare, New York

Campbell JJ (2005) Understanding pharma: a primer on how pharmaceutical companies really work. Pharmaceutical Institute, Raleigh

A.1.2.2. Case Histories of Drug Discovery

Sneader W (2005) Drug discovery: a history. Wiley, West Sussex

Corey EJ, Czako B, Kurti L (2007) Molecules and medicine. Wiley, Hoboken

Lednicer D (2007) New drug discovery and development. Wiley, Hoboken

Li JJ (2015) Top drugs: their history, pharmacology, and syntheses. Oxford University Press, Oxford

A.1.2.3. Journals and Magazines

The scientific and medical journals with content relating to drug discovery cannot all be listed here since they are far too numerous. The same applies to pharmaceutical industry magazines and of course the numerous blogs and wikis found online. The following list contains titles which provide news features and updates on a regular basis.

A.1.2.4. Peer-Reviewed Journals

Cell
Drug Discovery Today
Nature magazine
Nature Biotechnology
Nature Reviews Drug Discovery
Proceedings of the National Academy of Sciences (USA)
Science Magazine
Science Translational Medicine

A.1.2.5. Online and Printed Magazines

BioPharm International
http://www.biopharminternational.com/

Derek Lowe's pharma blog.
https://blogs.sciencemag.org/pipeline/
Drug Discovery and Development Magazine.
http://www.dddmag.com

Drug Target Review.
https://www.drugtargetreview.com

Fierce Biotech.
https://www.fiercebiotech.com/

Genetic Engineering & Biotechnology News.
http://www.genengnews.com/

in-Pharma Technologist.com

Pharmafile.
http://www.pharmafile.com/

Pharmaceutical Technology
https://www.pharmaceutical-technology.com/

Pharmtech.com.
http://www.pharmtech.com/

A.1.2.6. Organizations

Some organizations and regulators, such as the EMA, FDA, and ICH, have already
been referenced, and there are of course many more throughout the world. The following are repeated here, as they are leading industry sites containing useful facts
and figures about the drug discovery business:

The Association of the British Pharmaceutical Industry (ABPI).
http://www.abpi.org.uk/

The Pharmaceutical Research and Manufacturers of America (PhRMA).
http://www.phrma.org/

A.1.2.7. Chemistry

Royal Society of Chemistry (RSC) Educational resources. The main UK chemistry
organization.

http://www.rsc.org/learn-chemistry

American Chemical Society (ACS) Education links on main website.
http://www.acs.org

International Union of Pure and Applied Chemistry (IUPAC). Home page.
http://www.iupac.org/

Compendium of chemical terminology.
http://old.iupac.org/publications/compendium/A.html

PubChem A US National Library of Medicine database of small molecule structures
and links to further information.
https://pubchem.ncbi.nlm.nih.gov/

Chemical Entities of Biological Interest (ChEBI) Small molecule database like
PubChem.
http://www.ebi.ac.uk/chebi/

Protein structures. RCSB Protein Data Bank educational resources.
http://pdb101.rcsb.org/

A.1.2.8. Biotechnology

This heading covers all aspects of biology which are relevant to drug discovery but
with an emphasis on modern cell and molecular biology. The references will help
fill in details about how proteins are made in the cell using specifications laid down
in the genetic code and other technical points that were left out of the main book.

A.1.2.9. Biotechnology: Textbooks

Lewin's Genes XII (2017). Jones & Bartlett Publishers
Watson JD, Molecular Biology of the Gene 7th Edition Paperback (2017). Pearson
Education
Alberts B Molecular Biology of the Cell, 6th Edition (2015). Garland Science,
Taylor and Francis Group

A.1.2.10. Biotechnology: Online Resources

Genome editing.
https://ghr.nlm.nih.gov/primer/genomicresearch/genomeediting

Learn Genetics. SNPs.
http://learn.genetics.utah.edu/content/precision/snips/

Learn Genetics. Human Microbiome.
http://learn.genetics.utah.edu/content/microbiome/

Learn Genetics. Epigenetics.
http://learn.genetics.utah.edu/content/epigenetics/

The Human Genome Project.
https://www.genome.gov/human-genome-project

The Sanger Centre. Educational resources.
http://www.yourgenome.org/

A.1.2.11. Clinical Trials

Applied Clinical Trials.
http://www.appliedclinicaltrialsonline.com/

WHO International Clinical Trials Registry Platform (ICTRP).
http://www.who.int/ictrp/en/

US database of clinical trials.
http://www.clinicaltrials.gov/

EU Clinical Trials Register.
https://www.clinicaltrialsregister.eu/ctr-search/search

Appendix 2: Glossary and External Resources

This appendix is laid out in two sections. The first consists of a small glossary of terms (including some acronyms) derived from all parts of the drug discovery and development process. More terms have been used in the main text and can be located by using the Index.

The second part is designed as a jumping off point for readers who want to explore online glossaries and lists of abbreviations and acronyms. These are inevitably more comprehensive than anything that could be included in a printed book, and they also have the advantage of being searchable on the computer.

Glossary

ADMET (adsorption, distribution, metabolism, excretion, toxicology)
Pharmacokinetic properties of a compound that must be optimized prior to use as a medicine.

Agonist

A molecule that activates a physiological process by interacting with a target and mimicking the natural ligand.

AI

Artificial intelligence. Computer-based system using mathematical algorithms to extract information from datasets. Potential uses in all aspects of drug discovery and development.

Antagonist

A molecule that blocks a physiological process by interacting with a target and preventing the natural ligand from exerting its effect.

ANDA

Accelerated New Drug Application. Regulatory submission for generic drugs.

API

Active pharmaceutical ingredient

Assay

The process of testing a chemical sample for activity against a specific target or cellular response.

Bioavailability

A measurement of the amount of a compound absorbed into the bloodstream.

Bioinformatics

Bioinformatics is the use of computers to analyze nucleic acid and protein sequence information.

Chemoinformatics

The use of computers to analyze small chemical molecules.

Combinatorial Chemistry

The generation of large collections, or "libraries," of compounds by synthesizing all possible combinations of a set of smaller chemical structures or "building blocks".

Cytokines

Specialized proteins that allow cells to communicate with each other.

DNA (Deoxyribonucleic Acid)

DNA is a chemical compound present in the nucleus of cells. It consists of two strands of polynucleotides entwined in a double helix.

Druggable Target

These are proteins that can be targeted using small molecules.
They include cell surface receptors, ion channels, and enzymes.

EMA

European Medicines Agency

Enzyme
Complex proteins that are produced by living cells; they catalyze specific biochemical reactions.

Excipient
Inert product(s) added to APIs to formulate drugs in the correct dosage form.

FDA
American Food and Drug Administration.

Genes
The unit of inheritance. Physically present as DNA packaged into 23 pairs of human chromosomes.

Genomics
The analysis of the full complement of genes in a genome. The human genome contains approximately 20,000 genes.

GPCR
G-protein-coupled receptor

High-Throughput Screening (HTS)
Trial-and-error evaluation of hundreds or more compounds in a target-based assay.

Hit Compound
A compound that is active in a biological assay, normally a screen. Needs further optimization to become a lead compound.

Hormone
Messenger molecules that communicate between tissues secreted by endocrine glands. Can be small molecules or peptides/proteins.

IND
Investigational new drug

In Vitro
Experiments carried out in cellular systems or on cellular components, such as genes or proteins or subcellular fractions.

In Vivo
Experiments carried out in living organisms.

Ion Channels
"Holes" in cell membranes that selectively allow transport of charged atoms/molecules (ions).

Lead Compound
A compound that exhibits pharmacological or biochemical properties which suggest its value as a starting point for drug development.

Ligand
A molecule that binds to a receptor. A natural ligand might be a hormone such as adrenaline.

Microplate
A standardized plastic tray with 96 (or 384 or 1536) "wells" or depressions for holding small quantities of material. The 96 wells are uniformly located in 8 rows of 12 wells each.

Model Organism/Model System
Laboratory-grown organism (microbe, plant, or animal), which has representative features of human biology. Is normally inexpensive to maintain and easy to manipulate for the purpose of understanding a complex biological phenomenon.

NCE
New chemical entity

NDA
New drug application

Neurotransmitter
Small molecules that transmit signals across nerves.

NME
New molecular entity

Optimization
The process of synthesizing chemical variations, or analogs, of a lead compound, with the goal of creating those compounds with improved pharmacological properties.

Orally Active Drugs
Drugs that are effective in treating a disease when administered by mouth and absorbed by the digestive system.

Organic Molecules
Molecules containing the element carbon among the atoms that define their structure.

Pathway
Very few proteins act in isolation; they usually function by interacting with other molecules along a defined "pathway." These other components of the pathway may be more appropriate for drug discovery.

Peptide
A molecule composed of two or more amino acids. Larger peptides are generally referred to as polypeptides or proteins.

Pharmacogenetics
The inherited response to medicines.

Pharmacogenomics
A genomic approach to pharmacogenetics in which DNA is analyzed to determine sequences which are responsible for individual responses to medicines.

Proteomics
The equivalent of genomics but analyzing proteins instead of DNA.

Protein
A molecule composed of a long chain of amino acids. Proteins are the principal constituents of cellular material. Examples of proteins are enzymes, hormones, structural elements, and antibodies.

Protein Therapeutics
Protein therapeutics are proteins used as drugs. They are normally produced artificially using genetic engineering but can be isolated from tissues.

Receptor
A molecule within a cell or on a cell surface to which a substance (such as a hormone or a drug) selectively binds, causing a change in the activity of the cell.

Small Molecule
A chemical entity having a molecular weight of less than the range 500–700 Daltons.

Single Nucleotide Polymorphisms (SNPs)
DNA sequences from different individuals show millions of differences of one nucleotide. These polymorphisms can affect protein coding regions and therefore protein structures.

Specificity
Quality of a compound that describes its lack of interaction with targets that are related to the main target of an assay. Compounds with high specificity should have fewer side effects.

Structure-Activity Relationship (SAR)
An analysis which defines the relationship between the structure of a molecule and its ability to affect a biological system.

Substrate
A molecule on which an enzyme effects a biochemical reaction.

Target
A target is a protein upon which a drug could act to correct a disease state.

Ultrahigh-Throughput Screening (uHTS)
High-throughput screening accelerated to greater than 100,000 tests per day.

Validation
A demonstration that a protein acts specifically on a cellular or physiological process or pathway that is relevant to human disease.

Whole Organism Screens/in Vivo Screens

Compounds are screened on animals for their ability to produce measurable responses when the target of interest is affected.

Online Glossaries and Lists of Abbreviations

Cambridge Healthtech Institute Pharmaceutical Glossaries & taxonomies A-Z Index.
http://www.genomicglossaries.com/content/gloss_cat.asp
Clinical Data Interchange Standards Consortium (CDISC) Clinical Research Glossary. Download options from link on webpage.
https://evs.nci.nih.gov/ftp1/CDISC/Glossary/

Index

Printed in the United States
by Baker & Taylor Publisher Services